Advances in PARASITOLOGY

VOLUME 70

Parasitoids of *Drosophila*

Advances in PARASITOLOGY

VOLUME 70

Parasitoids of *Drosophila*

Edited by

GENEVIÈVE PRÉVOST
Laboratoire de Biologie des Entomophages
Université de Picardie–Jules Verne
Amiens cedex, France

ELSEVIER

AMSTERDAM • BOSTON • HEIDELBERG • LONDON
NEW YORK • OXFORD • PARIS • SAN DIEGO
SAN FRANCISCO • SINGAPORE • SYDNEY • TOKYO
Academic Press is an imprint of Elsevier

Academic Press is an imprint of Elsevier

32 Jamestown Road, London, NW1 7BY, UK
525 B Street, Suite 1900, San Diego, CA 92101-4495, USA
30 Corporate Drive, Suite 400, Burlington, MA 01803, USA
Radarweg 29, PO Box 211, 1000 AE Amsterdam, The Netherlands

First edition 2009

ISBN: 978-0-12-374792-1
ISSN: 0065-308X

For information on all Academic Press publications
visit our website at www.elsevierdirect.com

Printed and bound in UK
09 10 11 12 10 9 8 7 6 5 4 3 2 1

CONTENTS

CONTRIBUTORS

Roland Allemand
Université Lyon 1; CNRS, UMR5558, Laboratoire de Biométrie et Biologie Evolutive, F-69622 Villeurbanne, France

C. Anselme
Institut National de la Recherche Agronomique, INRA Sophia Antipolis, UMR 1301; Centre National de la Recherche Scientifique, CNRS, UMR 6243; Université Nice Sophia Antipolis, UFR Sciences, France

Y. Carton
IRD, UR072 Laboratoire Evolution, Génomes et Spéciation/UPR9034, CNRS 91198 Gif-sur-Yvette Cedex, France/Université Paris-Sud 11, 91405 Orsay Cedex, France

Anas Cherqui
Laboratoire de Biologie des Entomophages, EA 3900 BioPI, Université de Picardie-Jules Verne, 33 rue Saint Leu, 80039 Amiens cedex, France

D. Colinet
Institut National de la Recherche Agronomique, INRA Sophia Antipolis, UMR 1301; Centre National de la Recherche Scientifique, CNRS, UMR 6243; Université Nice Sophia Antipolis, UFR Sciences, France

Aude Couty
Unité de Recherche EA3900 BioPI-UPJV; Biologie des plantes et contrôle des insectes ravageurs; Laboratoire de Biologie des Entomophages; Faculté des Sciences, Ilot des Poulies, 33 rue Saint Leu, 80039 Amiens Cedex, France

Géraldine Doury
Laboratoire de Biologie des Entomophages, EA 3900 BioPI, Université de Picardie-Jules Verne, 33 rue Saint Leu, 80039 Amiens cedex, France

A. Dubuffet
Institute of Integrative and Comparative Biology, Faculty of Biological Sciences, University of Leeds, United Kingdom

S. Dupas
IRD, UR072 Laboratoire Evolution, Génomes et Spéciation/UPR9034, CNRS 91198 Gif-sur-Yvette cedex, France/Université Paris-Sud 11, 91405 Orsay cedex, France; and Pontificia Universidad Católica del Ecuador, Facultad de Ciencias Exactas y Naturales, Quito, Ecuador

Patrice Eslin
Laboratoire de Biologie des Entomophages, EA 3900 BioPI, Université de Picardie-Jules Verne, 33 rue Saint Leu, 80039 Amiens cedex, France

Frédéric Fleury
Université Lyon 1; CNRS, UMR5558, Laboratoire de Biométrie et Biologie Evolutive, F-69622 Villeurbanne, France

Sylvain Gandon
Centre d'Ecologie Fonctionnelles et Evolutives (CEFE) – UMR 5175, 1919 route de Mende, F-34293 Montpellier cedex 5, France

Patricia Gibert
Université Lyon 1; CNRS, UMR5558, Laboratoire de Biométrie et Biologie Evolutive, F-69622 Villeurbanne, France

H. Charles J. Godfray
Department of Zoology, University of Oxford, Oxford OX1 3PS, United Kingdom

Shubha Govind
Department of Biology, The City College of New York, New York, NY 10031, USA; and The Graduate Center of the City University of New York, New York, NY 10016, USA

Sébastien Havard
Laboratoire de Biologie des Entomophages, EA 3900 BioPI, Université de Picardie-Jules Verne, 33 rue Saint Leu, 80039 Amiens cedex, France

Thomas S. Hoffmeister
Institute of Ecology, University of Bremen, FB 02, D-28359 Bremen, Germany

Laure Kaiser
INRA Centre de Versailles-Grignon; UMR 1272 Physiologie de l'Insecte Signalisation et Communication, Route de St Cyr, 78026 Versailles Cedex, France; and IRD, UR 072, c/o Laboratoire Evolution, Génomes et Spéciation, UPR 9034, CNRS, 91198 Gif-sur-Yvette Cedex, France

Marta E. Kalamarz
Department of Biology, The City College of New York, New York, NY 10031, USA; and The Graduate Center of the City University of New York, New York, NY 10016, USA

Alex R. Kraaijeveld
University of Southampton, School of Biological Sciences, Southampton SO16 7PX, United Kingdom

Mark J. Lee
Department of Biology, The City College of New York, New York, NY 10031, USA

Sébastien J.M. Moreau
UMR 6035 CNRS, Institut de Recherche sur la Biologie de l'Insecte, Faculté des Sciences et Techniques, Université François-Rabelais, Parc Grandmont, 37200 Tours, France

Alix D.N. Mabiala-Moundoungou
Laboratoire de Biologie des Entomophages, EA 3900 BioPI, Université de Picardie-Jules Verne, 33 rue Saint Leu, 80039 Amiens cedex, France

Laurence Mouton
Université Lyon 1; CNRS, UMR5558, Laboratoire de Biométrie et Biologie Evolutive, F-69622 Villeurbanne, France

A. Nappi
Department of Biology, Loyola University of Chicago, Chicago, IL 60525, USA

Maxime Nardin
Laboratoire de Biométrie et Biologie Evolutive, Université Lyon 1; CNRS; UMR 5558, 43 boulevard du 11 novembre 1918, F-69622 Villeurbanne, France

Indira Paddibhatla
Department of Biology, The City College of New York, New York, NY 10031, USA; and The Graduate Center of the City University of New York, New York, NY 10016, USA

Bart A. Pannebakker
Evolutionary Genetics, Centre for Ecological and Evolutionary Studies, University of Groningen, P.O. Box 14, NL-9750 AA Haren, The Netherlands

Sabine Patot
Laboratoire de Biométrie et Biologie Evolutive, Université Lyon 1; CNRS; UMR 5558, 43 boulevard du 11 Novembre 1918, F-69622 Villeurbanne, France

Raquel Perez-Maluf
Laboratório de Zoologia, DNC/UESB, Estrada do Bem Querer KM 04, Vitoria da Conquista, 45000 BA, Brazil

M. Poirié
UMR Interactions Biotiques et Santé Végétale, Institut Agrobiotech, 06 903 Sophia Antipolis, France. INRA, UMR 1301/CNRS UMR 6243/Université Nice Sophia Antipolis, 28, avenue de Valrose, 06103 Nice Cedex 2, France

Geneviève Prévost
Laboratoire de Biologie des Entomophages, EA 3900 BioPI, Université de Picardie-Jules Verne, 33 rue Saint Leu, 80039 Amiens cedex, France

Roma Rajwani
Department of Biology, The City College of New York, New York, NY 10031, USA

Nicolas Ris
Centre INRA de Sophia Antipolis, Unité Expérimentale 1254 "Lutte Biologique", BP 167, F-06903 Sophia-Antipolis, France

Chiyedza Small
Department of Biology, The City College of New York, New York, NY 10031, USA; and The Graduate Center of the City University of New York, New York, NY 10016, USA

Andra Thiel
Institute of Ecology, University of Bremen, FB 02, D-28359 Bremen, Germany

Julien Varaldi
Laboratoire de Biométrie et Biologie Evolutive, Université Lyon 1; CNRS; UMR 5558, 43 boulevard du 11 novembre 1918, F-69622 Villeurbanne, France

Fabrice Vavre
Université Lyon 1; CNRS, UMR5558, Laboratoire de Biométrie et Biologie Evolutive, F-69622 Villeurbanne, France

Sophie Vinchon
Laboratoire de Biologie des Entomophages, EA 3900 BioPI, Université de Picardie-Jules Verne, 33 rue Saint Leu, 80039 Amiens cedex, France

PREFACE

Extensive studies have been conducted on various parasitoid species, and almost all different approaches of the biology of parasitoids have been, at least once, studied on parasitoids of *Drosophila*. Originally, this particular interest for species parasitizing *Drosophila* hosts has been motivated by the exceptional knowledge we have on *Drosophila* species themselves, and particularly on *Drosophila melanogaster*. Benefiting from this knowledge, the research conducted on *Drosophila* parasitoids has covered very diversified topics, such as physiological and immunity relationships with hosts, reproduction strategies, the role of symbiotic microorganisms, behavioral and chemical ecology, genetics, population dynamics, evolutionary biology, the different approaches of which include field surveys, and laboratory experiments. As a result, we now benefit from a particularly vast and extended knowledge of the biology of these parasitoid species and of their relationships with their *Drosophila* hosts.

Thanks to the authors who kindly contributed to the writing of this special volume, the book highlights the diversity of the fields of research that have been explored on *Drosophila* parasitoids. It also points out that this biological model is of particular interest for an evolutionary approach to parasitism.

Due to circumstances beyond our control, Professor Jacques J.M. Van Alphen (University of Leiden, the Netherlands) could not contribute to this volume in the section 1 on the ecology of *Drosophila* parasitoids. Because of the particularly important work of Pr. J.J.M. Van Alphen on the evolution of life histories and foraging behaviour in larval parasitoids of *Drosophila*, I strongly recommend to study his amazing bibliography.

Geneviève Prévost

INTRODUCTION

Although for a long time parasitoid insects have received much attention, most former studies were focused on their interest as biocontrol agents. Now, however, academic research on parasitoid insects has developed and reached a full and very promising maturity.

As studies led to a better balance between the classic applied and academic sides, researchers turned to other insect species, correlatively, and as well as species of agricultural importance, new biological models progressively obtained increasing favor due to their suitability for laboratory and field studies. Obviously, Hymenopteran *Drosophila* parasitoids offer exceptional advantages in all respects and despite their lack of interest for agriculture, they have been adopted as biological models by more and more, mainly European, teams.

In 1986, they were included in Ashburner's monumental treatise on *Drosophila* spp., and the chapter written by Y. Carton, M. Boulétreau, J. van Lenteren and J. van Alphen in vol. 3e synthesized previous data, together with current work. More than 20 years ago, this paper highlighted a number of questions that, hopefully, *Drosophila* parasitoids could help to answer. It is time now to restate the field, and owing to her enthusiasm and energy, Professor Geneviève Prévost has succeeded in convincing a number of outstanding authors to contribute to the challenge.

Section I considers the "preovipositional" relationships of parasitoids with their *Drosophila* hosts; this encompasses all the steps (mainly ecological and behavioral) that precede the actual physical contact between the parasitoid and its host individual, and which thus correspond to the classic ecological and encounter filters of parasitologists. Populations and communities submit to a number of particular constraints from their common environment that are not evenly distributed over time and space. Chapter 1 considers the temporal and spatial organization of communities on different scales. The seasonal and the daily organization of activities reveal a fine interspecific tuning which could enhance the probability of parasitoids encountering suitable hosts, and also mediate competition among parasitoid species, thus promoting their coexistence and enhancing the stability of the whole community. Comparative studies bring to light significant variations of populations and communities on a narrow geographic scale, suggesting fine local adaptation to ecological conditions that are mainly climatic. Genetic analysis establishes the role of natural selective pressures in shaping life-history traits, suggesting a dynamic equilibrium of populations and communities with their local environmental constraints, which is likely to be open to adaptive variation as a response to predictable long-term climatic changes.

At the individual level, behavioral relationships of parasitoids with their hosts have reached an extreme degree of sophistication, probably unmatched among the whole invertebrate phylum. Two chapters are devoted to this field, each of them taking a particular approach. Understanding of the individual relationships between a parasitoid and its hosts necessarily takes into account the immediate external context where they are living. Behavior leading to host infestation is not a fixed pattern, but will depend first of all on several external conditions that need to be considered on different scales, both spatial and temporal. Those directly arising from insects themselves are of special interest. Among others, host diversity and suitability, host abundance and distribution, the presence of potential competitors, whether conspecific or not, will strongly influence the foraging female and orientate her behavioral decisions. Second, the inner state of the female herself, both physiological and informational, will also change her decision when faced with a given external context. Field observations, experimental analysis and mathematical modeling generally confirm the theoretical prediction that the female behaves so as to optimize her own fitness. That leads us to credit females with a surprising analysis/integration capacity, which is logically taken as resulting from natural selection; however, the origin and acquisition of such a capacity remain doubtful and clearly call for further studies.

Both chapters deal with behavioral phenomenology: which cues, which receptors are involved, and how parasitoids use information to successfully reach and discriminate their hosts. We are not surprised by the specificity of responses to cues emanating from the host habitat and from the host itself, but we discover the complexity of the integrative process that results in a precise decision made by females under given circumstances. Memory; experience; assortative learning; estimates of space, of time and perhaps of numbers; and decision-making have long been considered as a prerogative of vertebrates, but it is now clear that insects are also capable of such achievements. In this field, *Drosophila* parasitoids demonstrate incredible performances, ending in highly efficient host infestation strategies. Adaptiveness of such behaviors is quite obvious, and their variation, whether epigenetic or genetic, allows females to adapt to the precise situation they are faced with perfectly, on the immediate individual scale and on the ecological and evolutionary ones. The consistency of these chapters is mostly based on their common evolutionary approach, which is the actual thread of the whole book. Obviously, the authors have made the most of using *Drosophila* parasitoids in their studies, whereas scattering their efforts over various parasitoids would not have made for such consistency.

Section II is devoted to "postovipositional" host–parasitoid relationships, that is the compatibility filter. In the absence of any study of the physiologic components of the filter (e.g., biochemical, hormonal or developmental), this section focuses on the immunological host–parasitoid relationships. We can regret that only endoparasitoids of host larvae are considered: no reference can be found to the immunological situation of endopupal parasitoids, or to any possible conflict between the host pupa and its ectoparasitoids. That means that the studies only deal with "koinobiotic parasitoids," those species which spare their host's survival as long as they need it for their own development, and thus behave like "true parasites" for a while. We can, however, wonder whether "idiobiotic parasitoids," which generally kill their host at the earliest stages of their own development, suffer any kind of defense from their host. To be honest, the boundary between "host killers" and "host savers" is somewhat fuzzy, and furthermore we perfectly understand and share the special attention that koinobionts have received, since only their biology allows immunological conflicts to take place for a significant duration.

In this immunological field, *Drosophila* parasitoids offer the unique opportunity for researchers to put a symmetrical effort into studying both hosts and parasites. As far as we know, no other insect system allows such a comprehensive analysis of the whole immunological race. Three main parasitoid species are studied: *Leptopilina boulardi*, *Leptopilina heterotoma* and *Asobara tabida*, which belong to different Hymenoptera families: Figitidae and Braconidae, respectively, and which exploit a few *Drosophila* spp., mainly frugivorous, often in sympatry. They have rather similar within-host developmental strategies: infestation at the very beginning of the larval host life, then solitary endoparasitic development until host pupation, and finally saprophytic consumption of the host pupa tissues before nymphosis. Free pharate imagos emerge from the mummified host puparium, the envelope that the host larva had built for its own protection before metamorphosis. The immunological reaction of the host and the reciprocal counter-reaction of the parasitoid occur at an early stage of infestation, and mostly address parasitoid eggs. It is worth keeping in mind that compared to most other host–parasite systems, survival of both the host and the parasite is quite impossible: either the host succeeds in killing its parasite and then survives, perhaps at some cost to itself but with the benefit of life, or the parasitoid evades or overcomes the host's defenses, then carries on with its own development. The alternative is terrible and suffers no exception, at least in *Drosophila* parasitoids. We thus understand that such a vital challenge results in drastic selection pressure on both partners, and has elicited powerful defense and counter-defense mechanisms. Chapters 4–8 explore the biochemical and physiological mechanisms involved in host "resistance" (its ability to survive infestation) and parasitoid counter-resistance or "virulence" (its ability to survive the host's defenses). In hosts, authors give a thorough analysis of the sequence of events, be they molecular, biochemical or cellular, that are elicited by the infestation and will end in parasite elimination. More than one path exists, and the outcome can result from the synergic interplay of several of them. Moreover, different parasitoid species can trigger different defense processes in the same host species. For parasitoids, evading or overcoming the host's defense is an absolute prerequisite, and we understand that they have evolved efficient protective strategies that the females themselves put into operation to protect their descent. At the time of infestation, they inject into the host, together with their eggs, various immunodepressive or immunosuppressive factors that they have developed, and which can consist either of venom (mostly proteinic) and/or of structural entities (virus-like particles). Acting independently or in a coordinated manner, sometimes reinforced by specific surface properties of the parasitoid egg, these "virulence factors" inhibit or annihilate the host's defense either by changing the expression of some specific host genes, or by direct effect on their effectors, or through other mechanisms, which can be quantitative or qualitative. It is worth noting that in all cases, immunoprotection, which is so vital for the parasitic stages, is mostly the duty of their mothers, which thus pay all costs and do not benefit directly from their own investment.

Given the drastic consequences of these processes on the host's and parasitoid's fitness, the question of their specificity and their variability immediately arises, together with their possible genetic determinisms. That is the condition for selective and evolutionary processes to take place. Here again, *Drosophila* parasitoids provide unequalled facilities. Comparative studies reveal that the intensity and specificity of the *Drosophila* defense against parasitoids (their "resistance") can vary within species and are controlled by a few genes, some of which could be specifically oriented against a given parasitoid species. Reciprocally, the "virulence" of parasitoids can vary within species, and appears to be under the control of a few genes that could be specific toward the defense reactions of different host species. Obviously, the hypothesis of resistance- and virulence-genes being species specific, based on the absence of clear cross-resistance and cross-virulence among the only few species so far studied, is strengthened when more species are analyzed, especially among the very rich and diverse tropical *Drosophila*–parasitoid communities. In any case, we have the demonstration that in all partners, several genetic systems are in operation, each specialized in a certain immunological path or combination of paths. Thus, the *Drosophila*–parasitoid system offers exceptional conditions for the study of evolution and coevolution of host–pathogen systems.

Section III deals with the evolution of host–parasitoid relationships. This classical chapter of evolutionary parasitology receives special attention here, thanks to the unique evolutionary and genetic status of *Drosophila* spp. Predictably, the authors focus on the immunological processes involved in host resistance and parasitoid counter-resistance (or ''virulence''). Indeed, these traits do decide which partner will survive the infestation, resulting in drastic – but nonsymmetrical – selective pressures. Their crucial involvement in individual fitness, together with the rather clear genetic basis of their within- and between-population variations, make them ideal targets for evolutionary and coevolutionary processes. Comparative analysis, field data and experimental results fuel exciting discussions that keep within – and illustrate – current theories on host–parasite coevolution. However, we must admit that the diversity of the mechanisms involved in host resistance and the reciprocal parasitoid counter-resistance, the variety of specificity levels among the species studied here, the complexity of population structures and of ecological situations as yet poorly documented, do not lead to clear and definitive conclusions. The cost of being resistant (for hosts) and of being counter-resistant (for parasitoids), tradeoffs with other unrelated traits that are either physiological, or behavioral, etc., are likely to play a major role in the evolution of immunological traits and need many more studies. In my opinion, forthcoming analysis of the evolution of a given trait – here immunological relationships – benefits from keeping within inclusive comprehension of the whole selective and evolutionary process including life styles of species, their reproductive strategies, ecological and behavioral features, population structures, local histories and many other factors that act on a variety of scales.

Clearly, this fascinating topic is far from being exhausted. We do not expect a unique conclusion, and only the patient accumulation of concrete data and of conflicting hypotheses will bring more and more light in the field.

Section IV approaches a new topic: the association of *Drosophila* parasitoids with symbiotic prokaryotes. We now know that a number of insect species, perhaps most of them, harbor *Wolbachia*, those astonishing intracellular bacteria which interfere with their host's reproduction in different ways, all of which enhance their own transmission rate. Thus, it is not surprising that most *Drosophila* parasitoids are infected. However, the study of their interaction with *Wolbachia* is of special interest in several respects. First, Hymenopteran parasitoids reproduce parthenogenetically. Interference of *Wolbachia* with their haplo-diploid sex determination leads to a variety of consequences on offspring sex-ratio, ranging from all female to all male progeny, with or without differential mortality of sexes. Such diversity goes far beyond what has been described in any other insect group. Second, we benefit here from the fact that *Drosophila* is among the best-documented *Wolbachia* partners. Predictably, host–parasitoid relationships are an ideal ground for horizontal transmission to take place. Indeed, the *Drosophila*–parasitoid communities provide a well-documented demonstration that in evolutionary times, *Wolbachia* could transfer from hosts to parasitoids, and furthermore that parasitoids did "accumulate" the different *Wolbachia* strains they acquired from different host species. That accounts for the high frequency of multi-infection among parasitoids, which creates fascinating situations of intra-cellular microbial ecology, with numerous questions about the evolution of virulence. Readers also learn about the consequences of *Wolbachia* infection on the dynamics of parasitoid populations, on the individual host–parasitoid relationships, and on the physiology of the infected female, with this astonishing result: *Asobara* is totally dependent on *Wolbachia* infection for egg production. Again, the unique status of *Drosophila* allows the authors to develop molecular and genomic programs which, hopefully, will soon bring about a better understanding of insect–*Wolbachia* relationships, from both the mechanistic and the evolutionary points of view.

Viruses also infect a number of insect species, causing a wide range of pathogenic effects ranging from near triviality to harsh lethality. Strains of *L. boulardi* have proved to be infected with a strange filamentous virus, vertically transmitted, which causes no severe pathology physiologically-speaking, but which induces a deep change in a critical behavioral step of parasitoid females. While uninfected females discriminate already parasitized host larvae and reject them, which is taken as an optimal behavior that avoids the death of supernumerary parasites and egg wastage, infected ones do accept superparasitism. In doing so, they doom most of their progeny to death and bring down their own offspring production and fitness. Obviously, such behavioral dysfunction is prejudicial to parasitoids, but as far as we understand it benefits the virus which proved able to transfer from one parasitoid larva to another within the same superparasitized *Drosophila* host, and thus to colonize new parasitoid lineages that otherwise would not be reached. Theoretical studies show that adding horizontal transmission to vertical transmission, even casually, allows the virus to invade parasitoid populations and can be considered as an efficient reproductive strategy. Until now, the generality, the mechanisms and the evolution of this new virus–parasitoid association have been poorly understood, but we can wonder whether such manipulation of the female's behavior by viruses questions current theories about the possible adaptiveness of behavioral decisions of wasps that deviate from the expected optimization.

Clearly, microorganisms associated with parasitoids are a very exciting and promising research area, to which *Drosophila* parasites are bringing an outstanding contribution.

It is a pleasure for me to advise entomologists and parasitologists to read this book carefully. Predictably, *Drosophila* parasitoids have allowed the authors to bring new data and/or new hypotheses to a number of fields, most of which extend to other parasitoids. While not exhaustive, the contributions brought together here are essential to the understanding of the functioning and evolution of host-parasitoid relationships, and thus are very helpful to entomologists involved in academic or agricultural research. They are also important for the whole field of parasitology, since they document a number of current theories dealing with the evolution of symbiotic relationships, with the evolution of virulence–resistance interplay, with the dynamics of host–parasite communities and their possible evolution in response to climatic change. Considering the genetic and molecular mechanisms that underlie host–parasite relationships, the unique status of *Drosophila* makes it likely that, provided a reasonable effort is made to study the genomics of parasitoids (including microbes), our knowledge will soon make huge strides forward.

Finally, we now have a clear demonstration that with or without the help of associated microorganisms, insect parasitoids have established highly sophisticated relationships with their hosts. If it is possible that the obligatory death inflicted by parasitoids on their hosts does not basically differ from the severe and possibly lethal harmful effects "true" parasites inflict on hosts, then the popular discrimination between parasitoids and parasites starts to seem less and less obvious.

Michel Boulétreau
Université Lyon 1
Villeurbanne, France

SECTION 1

ECOLOGY OF *DROSOPHILA* PARASITOIDS

CHAPTER 1

Ecology and Life History Evolution of Frugivorous *Drosophila* Parasitoids

Frédéric Fleury,* Patricia Gibert,* Nicolas Ris,† and **Roland Allemand***

Contents

* Université Lyon 1; CNRS, UMR5558, Laboratoire de Biométrie et Biologie Evolutive, F-69622, Villeurbanne, France
† Centre INRA de Sophia Antipolis, Unité Expérimentale 1254 "Lutte Biologique", BP 167, F-06903 Sophia-Antipolis, France

Advances in Parasitology, Volume 70
ISSN 0065-308X, DOI: 10.1016/S0065-308X(09)70001-6

Abstract Parasitoids and their hosts are linked by intimate and harmful interactions that make them well suited to analyze fundamental ecological and evolutionary processes with regard to life histories evolution of parasitic association. *Drosophila* aspects of what parasitoid Hymenoptera have become model organisms to study aspects that cannot be investigated with other associations. These include the genetic bases of fitness traits variations, physiology and genetics of resistance/virulence, and coevolutionary dynamics leading to local adaptation. Recent research on evolutionary ecology of *Drosophila* parasitoids were performed mainly on species that thrive in fermenting fruits (genera *Leptopilina* and *Asobara*). Here, we review information and add original data regarding community ecology of these parasitoids, including species distribution, pattern of abundance and diversity, host range and the nature and intensity of species interactions. Biology and the evolution of life histories in response to habitat heterogeneity and possible local adaptations leading to specialization of these wasps are reported with special emphasis on species living in southern Europe. We expose the diversity and intensity of selective constraints acting on parasitoid life history traits, which vary geographically and highlight the importance of considering both biotic and abiotic factors with their interactions to understand ecological and evolutionary dynamics of host-parasitoid associations.

Within the huge diversity of host–parasite interactions, parasitoidism occupies the higher level of virulence because successful development of parasites classically leads to the host death. Reuter (1913) first used the term "parasitoid" to name this particular way of life observed in about 10–20% of insect species, mainly Hymenoptera, Diptera and Coleoptera (Godfray, 1994). Parasitoidism is however not restricted to the Arthropod taxa and encompasses a large variety of organisms such as parasitic nematodes, bacteria or certain viruses (bacteriophages) that also kill their host as a consequence of parasite multiplication (Forde et al., 2004; Gomez-Gutierrez et al., 2006; Kaya and Gaugler, 1993). The fact that parasitoids and their hosts are linked by tightly and harmful interactions makes the system particularly suited to analyze fundamental ecological and evolutionary processes with regard to life histories evolution of both partners, local adaptation, coevolutionary arms race, maintenance of species and genetic diversity, as well as demographic mechanisms of host population regulation (Boulétreau, 1986; Godfray and Shimada, 1999; Hassel and Waage, 1984; Jessup and Forde, 2008; Rosenheim, 1998). Parasitoids insects overwhelmingly attracted most attention in these fields with studies performed on a wide diversity of associations (Godfray, 1994). Classically, parasitoid insects are free-living as adults and only their larvae develop as parasites of a single insect host that is killed. A number of species were studied with applied goals evaluating their potential to control phytophagous insect populations and their possible use in pest management programs.

With more fundamental objectives, *Drosophila* parasitoid Hymenoptera arose as model organisms to study all aspects of host–parasitoid relationships, particularly those that are hard to investigate in others systems, mainly the genetic bases and variations of fitness traits or processes underlying coevolutionary dynamic and local adaptation. Indeed, as their *Drosophila* host, *Drosophila* parasitoids cumulate many advantages such as rearing facilities and ease of experimental and field works that make them particularly suited to disentangle physiological, genetic and ecological interactions occurring among partners of host–parasitoid association. Moreover, research can benefit from data available on *Drosophila* biology and genetics. In return, parasitoid knowledge may also contribute to understand *Drosophila* ecology and evolution that cannot miss out the possible impact of parasites. Since the synthesis of Carton et al. (1986), our understanding of *Drosophila* parasitoid biology progressed in all fields of parasitic interactions but clearly most attention was paid to host immune resistance, factors of parasitoid virulence and their possible pleiotropic effects (see other chapters in this volume). Field ecology of *Drosophila* parasitoids including species distribution, pattern of abundance and diversity, host range, nature and intensity of interactions in natural communities, evolution of life histories with regard to habitat heterogeneity and possible local adaptations leading to specialization have been investigated less. According to Carton et al. (1986), approximately 50 hymenopterous parasite species of *Drosophila* belonging to four families and at least 16 genera are recognized: the larval parasites Braconidae (*Asobara*, *Aphaereta*, *Phaenocarpa*, *Tanycarpa*, *Aspilota*, *Opius*) and Eucoilidae (*Leptopilina*, *Ganaspis*, *Kleidotoma*, *Dicerataspis*), the pupal parasites Diapriidae (*Trichopria*, *Spilomicrus*) and Pteromalidae (*Pachycrepoideus*, *Spalangia*, *Trichomalopsis*, *Toxomorpha*). Research over these last 20 years on *Drosophila* parasitoid ecology has been performed mainly on species attacking *Drosophila* larvae living in fermenting substrates (rotting fruits, sap fluxes, decaying plants and mushrooms) and by far most knowledge was obtained from frugivorous *Drosophila* parasitoids of the genera *Leptopilina* and *Asobara*, and among them the wasp species *Leptopilina heterotoma*, *Leptopilina boulardi* and *Asobara tabida* that share common hosts (*Drosophila melanogaster*, *D. simulans*, *D. subobscura*). Of these associations, complex patterns of geographical and genetic variation of resistance and virulence with their associated costs and benefits were described (Dubuffet et al., 2007; Fellowes et al., 1999a; Fellowes and Godfray, 2000; Kraaijeveld and Godfray, 1999; Kraaijeveld et al., 1998), but clearly present results cannot be fully understood without a good knowledge of the ecology of interacting species.

In this article, we review information and add original data about community ecology, biology and evolution of these wasp species with special emphasis on the complex *Leptopilina / Asobara / Drosophila* living in rotting fruits in southern Europe. We show evidence of the intensity of selective constraints acting on parasitoid life history traits that show geographical pattern of variations and highlight the importance to consider both biotic and abiotic factors with their interactions to better understand ecological and evolutionary dynamics of the community.

1.1. DISTRIBUTION, COMMUNITY STRUCTURE AND ECOLOGICAL INTERACTIONS

1.1.1. Diversity, biogeography and phylogeny

Geographical distribution, host range and taxonomy are insufficiently documented for the four major groups of *Drosophila* parasitoids (Braconidae, Figitidae, Diapriidae and Chacidoidea) with a few exceptions (Carton et al., 1986). Samplings in tropical regions of America or Africa suggest that the fauna is poorly known and that many other new species remain to be described (Carton and Lachaise, personal observations). Parasitoids that attack frugivorous *Drosophila* are diverse but the most important species are the larval parasites of the genus *Leptopilina* and *Asobara* and the pupal parasites *Spalangia, Pachycrepoideus* and *Trichopria* (Allemand et al., 1999; Carton et al., 1991; Rohlfs and Hoffmeister, 2004; Wertheim et al., 2006). *Tanycarpa punctata* and *Aphaereta scaptomyzae* can be locally abundant in some associations with *L. heterotoma* and *A. tabida* in fruit or sap fluxes (Hardy and Godfray, 1990; Janssen, 1989).

Leptopilina is the best-known genus (Allemand et al., 2002; Nordlander, 1980; Schilthuizen et al., 1998) ahead of *Asobara* (Belokobylskij, 1998) or *Spalangia* (Boucek, 1963) and very few data are available for other genera of *Drosophila* parasitoids. The taxonomic difficulties can be partly solved using molecular tools that also give the possibility to establish phylogenies. In the *Leptopilina* genus, molecular phylogenies agreed with morphological traits (Allemand et al., 2002; Schilthuizen et al., 1998), and separate three groups of species (*longipes, boulardi* and *heterotoma* groups), which correlate with their distribution, ecology or associations with symbiotic microorganisms (Chapter 12 by Vavre et al.). The *longipes* group consists of six described species (*L. longipes, L. clavipes, L. fimbriata, L. australis, L. cupulifera* and *L. mahensis*) specialized on fungi and decaying plant materials (Driessen and Hemerik, 1991; Janssen et al., 1988; van Alphen et al., 1991; van Dijken and van Alphen, 1998; Vet and van Alphen, 1985; Wertheim et al., 2000). The other two groups are species living mainly in fermenting fruits. Taxonomy, recently revisited by Allemand et al. (2002), separates a *L. boulardi* group including two newly described species originating from tropical Africa (*L. orientalis* and *L. freyae*) and extends the *L. heteroma* group to five species (*L. heteroma*, *L. victoriae*, *L. rufipes*, *L. atraticeps* and the newly described *L. guineaensis*). To date, precise information on the ecology and geographical distribution of these species is available for *L. heterotoma* and *L. boulardi* only.

L. heterotoma is clearly the most generalist parasitoid among all other *Leptopilina* species. It colonizes fermenting fruits but also sap fluxes and decaying plant materials in a wide holarctic distribution but seems absent in the afro-tropical region (Carton et al., 1986; Hardy and Godfray, 1990; Janssen et al., 1988; Mitsui et al., 2007; Norlander, 1980; van Alphen et al., 1991). To date, there is no record of this species in the austral hemisphere. In contrast, *L. boulardi*, a specialist of frugivorous *Drosophila*, is recorded worldwide in Mediterranean and intertropical climates including southern Europe, Africa, North and South America, and the West Indies (Allemand et al., 2002; Barbotin et al., 1979; Carton et al., 1991; Chabora et al., 1979; Hertlein, 1986; Nordlander, 1980). The northern limit of the *L. boulardi* geographical range in Europe is precisely localized around 45 °N latitude, separating the continental and Mediterranean climates. Its southern limit remains largely unknown but records have been obtained from South Africa and Brazil. The niches of *L. heterotoma* and *L. boulardi* overlap in a number of countries of the Mediterranean areas where they compete for *D. melanogaster* and *D. simulans* during part of the season (Allemand et al., 1999; Carton et al., 1991; Fleury et al., 2004). Both species were also sympatric in the New World both in the east and west of North America (Schlenke et al., 2007).

L. heterotoma and *L. boulardi* also interact with parasitoids of the genus *Asobara*, which draw together at least nine species among which *A. tabida*, *A. citri*, *A persimilis*, *A. japonica* and *A. gahanii* thrive in rotting fruits (Janssen et al., 1988; Mitsui et al., 2007; Vet and van Alphen, 1985; Vet et al., 1984). By far, *A. tabida* is the most thoroughly studied species and has become a model for life history evolution, chemical and behavioral ecology, coevolutionary interactions of host resistance and parasitoid virulence, and infection by endosymbiotic microorganisms (Dedeine et al., 2001; Ellers et al., 2001; Green et al., 2000; Kraaijeveld and Godfray, 1997, 1999; Pannebakker et al., 2007; Prévost et al., 2005; Vet and van Alphen, 1985). *A. tabida* shows a wide holarctic distribution with reports from the northwest of America (Hoang, 2002), Japan (Mitsui et al., 2007) and all Europe from the Mediterranean cost to northern Scandinavia (Carton et al., 1986). Niche separation among *A.tabida*, *L. heterotoma* and *L. boulardi* is not clearly established and probably varies throughout the season, geographical sites and local conditions.

If distribution of these three main frugivorous *Drosophila* parasitoid species is approximately known, there are very few data available on their relative abundance on different fermenting substrates and intensity of competitive interactions among them. By collecting fermenting fruits in Tunisia over several months, Carton et al. (1991) observed the presence of *L. heterotoma* and *L. boulardi* only with moderate levels of abundance but species are not present at the same time of the season probably because *L. boulardi* excludes *L. heterotoma* from April to December as a consequence of its higher competitive ability. In contrast, fieldwork performed by Hardy and Godfray (1990) in England showed that the northern Europe parasitoid community is dominated by *L. heterotoma*, *A. tabida* and *Tanycarpa punctata* with similar level of abundance and a synchronization of their activity from May to September. In southern France where distribution of the three main *Drosophila* parasitoids overlaps, we sampled during all the warm season (from April to September) several sites distributed along a transect of 300 km by collecting standardized rotting fruits, mainly bananas or apples, submitted to natural colonization by insects over 2 weeks (Allemand et al., 1999). Collected materials were brought back to the laboratory and incubated at 21 °C and all emerging *Drosophila* and parasitoids were identified to the species level. Overall, species composition and abundance reveal a complex community structure in rotting fruits that involved at least seven *Drosophila* spp. dominated by *D. melanogaster*, *D. simulans* and *D. immigrans* with approximately the same level of abundance (Fig. 1.1). Two pupal parasitoids coexist with three larval parasitoids among which a very weak representation of *A. tabida* compared to *L. heterotoma* and *L. boulardi* (Fig. 1.1). Clearly in this area, rotting fruits are colonized by a very high density of *Drosophila* and parasitoids, thus suggesting intense interactions within and among trophic levels that raise the question of mechanisms responsible for the maintenance of this diversity. These high population densities also suggest that *Drosophila* probably experience crowded conditions in the field that can make possible the expression of the cost of resistance genes demonstrated under laboratory condition in both *A. tabida* et *L. heterotoma* spp. (Fellowes et al., 1998; Kraaijeveld and Godfray, 1997). High densities of hosts and parasites, however, did not reach the same level according to localities and ecological conditions that vary among orchards or woodland, as shown by Wertheim et al. (2006), thus resulting in a geographical mosaic of selective pressures that can lead to local adaptation.

1.1.2. Spatial and seasonal variations of communities

In natural conditions, frugivorous *Drosophila* and their parasitoids are present throughout the year except during the colder season. Species are all polyvoltine, their diversity and abundance vary according to the habitat, the available resources and their thermal tolerance. For instance, the patchy distribution of mushrooms result in lower insects

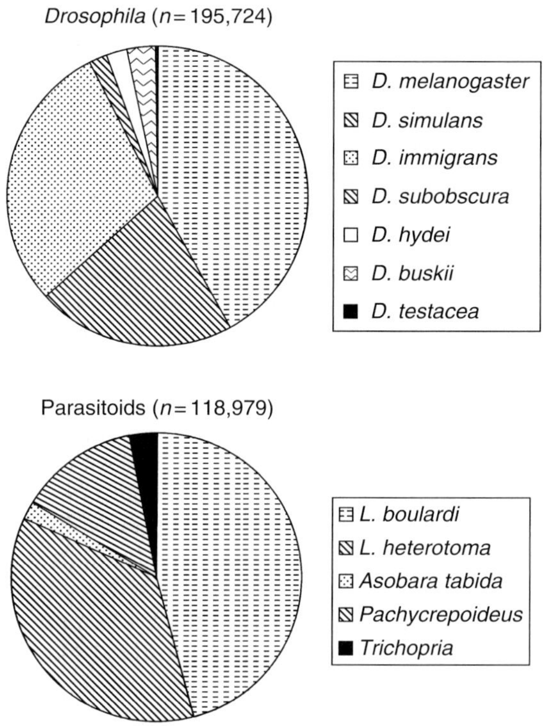

FIGURE 1.1 Relative abundance of *Drosophila* and parasitoids emerging from standardized fruits (bananas) deposited in orchards in the Saône-Rhône valley (southern France). Data are pooled from about 480 fruits (six sites, five periods between April and September, 16 replicates by site and period).

density (Driessen and Hemerik, 1991) that contrasts with the continuous distribution of fruits (peaches, pears or apples in orchards) where important populations of *Drosophila* breed and develop (Fleury et al., 2004; Wertheim et al., 2006). Among studies focusing on *Drosophila* and their parasitoid wasps that develop in fermenting fruits, very few reported how natural assemblage of species varies in space and during the season. Janssen et al. (1988) recorded presence/absence of *Drosophila* species and their parasitoids in various fermenting substrates sampled in several woods in The Netherlands but unfortunately without the composition of local community. However, they noticed high levels of parasitism that reach 50%. A more quantitative study performed on stinkhorn fungi in the same woodlands showed the evidence of short temporal refuge for hosts throughout the season that can stabilize the community experiencing high rates of parasitism rising up 100% in July (Driessen *et al.*, 1990). Hertlein (1986) followed overwintering populations in citrus orchards of California from fall to spring and observed rapid growth of populations for all members of the community, here restricted to *D. melanogaster*, *D. simulans* and *L. boulardi* only. However a lack of synchrony among hosts and parasites was observed as a consequence of longer periods of inactivity of parasites (5 months) that enter hibernal diapause whereas hosts overwinter in a state of quiescence. Reproductive activity of hosts that occurs 2–3 months before parasites allows *Drosophila* populations to build up high levels of densities and cope with high mortality rates induced by parasitoids. Similar pattern of host–parasitoids asynchrony also appears in Tunisia where *Drosophila* hosts thrive in prickly pears of *Opuntia* during the winter, when parasitoids populations fall to very low level abundance (Carton et al., 1991).

In southeastern France, we performed a more detailed survey of *Drosophila*-parasitoid communities from April to the end of October in 4 orchards distributed along a north–south axis at the edge of *L. boulardi* geographical range (Fig. 1.2, see Fleury et al., 2004 for more details). In the north above 45 °N latitude (area around Lyon), *D. melanogaster* always overtook other *Drosophila* spp., whereas in the south *D. simulans* was dominant during all the season. At intermediate sites, a progressive replacement of *D. melanogaster* by *D. simulans* was observed that may be interpreted as a consequence of increased competitive interactions among *Drosophila* spp., due to climate conditions becoming favorable for *D. simulans* or to an effect of parasitoid community composition that also varies within the season. Indeed interestingly, parasitoid species show parallel spatial and seasonal variations. *L. heterotoma* predominates in the north where it develops mainly on *D. melanogaster* whereas its abundance diminishes to only a few percent in the south. The coexistence of *L. heterotoma* facing the competitive superiority of *L. boulardi* in the south may partly result from parasitoid seasonal asynchrony and precedence of *L. heterotoma* that is active several weeks before *L. boulardi*. This probably results from their different thermal requirement and overwintering strategy that occurs as prepupal diapause in *L. boulardi* (Claret and Carton, 1980; Hertlein, 1986) but as adult quiescence in *L. heterotoma* (Eijs, 1999). Clearly in this area, sharp variation of community structure is observed at low spatial and temporal scales that results in a geographical mosaic of selective pressures giving potential for local adaptations. However, this phenomenon may be impeded by a high amplitude of seasonal variation.

1.1.3. Intensity of parasitism, competition and coexistence

1.1.3.1. Impact of parasitoids on *Drosophila* populations

Most studies agreed that *Drosophila* parasitoids induce a high rate of mortality on their host populations despite the fact that the level of parasitism varies according to breeding site, the local situation and the

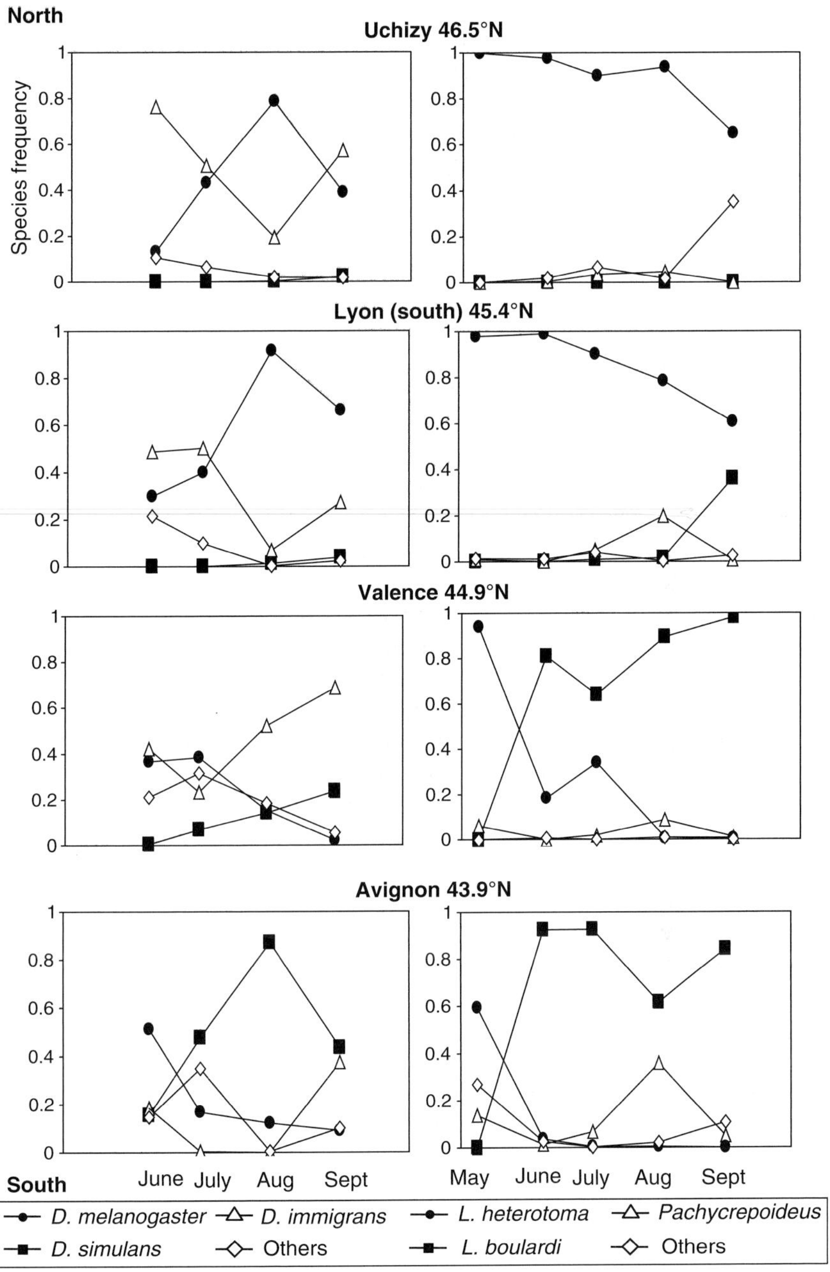

FIGURE 1.2 Seasonal variation of the relative abundance of *Drosophila* species (left) and their parasitoids (right) in four sites distributed along a north–south axis in the Saône-Rhône valley (southern France).

season. At the fruit scale, a high percentage of parasitized *Drosophila* larvae (up to 80%) has been observed in Australia (Parsons, 1977), The Netherlands (Driessen et al., 1990; Janssen et al., 1988), Tunisia (Boulétreau et al., 1991) and in southern France (Fleury et al., 2004). Most often, the natural average rate of parasitism varies between 5% and 40% in midsummer under favorable climate conditions, that is, 30% in temperate woodland of The Netherlands on various fermenting substrates (Janssen et al., 1988), around 15% on plums in northern Germany (Rohlfs and Hoffmeister, 2004), 16% in peaches and 22% in apples sampled in the same orchard in France (Wertheim et al., 2006) and up to 80% in some sites of southern France (Fleury et al., 2004). These high values of parasitism with their local variations indicate that parasitoids may be one of the main factors of mortality in fly populations and thus constitute a selective force acting on their hosts. To date, there is no indication how parasitoids may have shaped *Drosophila* life history traits in the wild, even for the evolution of host resistance, which has been studied mainly under laboratory condition. In experimental cages, we know that selective pressure of parasitoids can modify the polymorphic equilibrium of different genotypes of *D. melanogaster* (Boulétreau, 1986). With some markers, like the *sepia* gene, the evolution of allelic frequency is modified by the presence of parasitoids and the effect is not caused by a discriminatory destruction of one genotype but by indirect effects on population density.

Parasitism is favored by the ability of parasitoid females to use chemical stimuli from the host itself such as larval excrements or feeding traces (Dicke et al., 1985; Galis and van Alphen, 1981; Vet, 1985a; Vet and Bakker, 1985; Vet and van der Hoeven, 1984; Vet et al., 1993) or those from the host habitat such as odors of decaying plant materials (Dicke et al., 1984; van Alphen et al., 1991; Vet, 1985b; Vet et al., 1984). Moreover, responses to cues could be learnt by parasitoids thus enhancing their efficiency in microhabitat and host location. Associative learning was demonstrated for several species of *Asobara* (Vet and van Opzeeland, 1984), *L. heterotoma* (Vet and van Opzeeland, 1985, Vet et al., 1998) and *L. boulardi* (De Jong and Kaiser, 1991; Poolman et al., 1992; Vet, 1985a). *Drosophila* parasitoids also spy on the communication systems of their hosts to locate favorable sites since aggregation pheromones of flies are used by both *L. heterotoma* and *L. boulardi* (Couty et al., 1999; Hedlund et al., 1996; Wertheim et al., 2003; Wiskerke et al., 1993). A positive dose-dependant response to aggregation pheromones may explain higher aggregation of wasps in patches of higher larval density, and thus the positive density-dependent parasitism observed in some situations (Driessen and Hemerik, 1991; Hertlein and Thorarinsson, 1987; Wertheim et al., 2003). Such a clumped distribution of hosts and parasitoids with positive functional response may contribute to the stability of host/parasitoid associations as shown theoretically (Hassel and May, 1973, 1974; Lett et al., 2003; Pacala et al.,

1990). However, this pattern of density-dependent parasitism risk is not the rule and it varies according to the *Drosophila* breeding substrates. For example, Rohlfs and Hoffmeister (2004) showed that both *L. heterotoma* and *A. tabida* express positive density-dependent parasitism on sloes, inverse density-dependent on plums, and a hump-shaped relationship among a proportion of parasitized larvae and host density on apples observed also in other sites (Wertheim et al., 2006). These variations could be explained by the nature of breeding substrates that allow, or not, a proportion of larvae to burrow within the fruits' tissues, thus escaping parasitism as a result of their concealment. The fact that risk of parasitism is decreased with increasing host density may provide an adaptive hypothesis to *Drosophila* aggregation and thus the evolution of aggregation pheromones (Rohlfs and Hoffmeister, 2004). However, spatial aggregation probably also leads to an increased intraspecific competition as suggested by lower fitness traits values observed on individuals emerging from high-density patches (Wertheim et al., 2006). The adaptive significance of *Drosophila* aggregation thus results from a benefit–cost balance between refuge against parasitoids and mortality due to density-dependent competition, thus underlying the key role that parasitoids may have played in the evolutionary outcome of this trait.

At the interspecific level, intraspecific aggregation across fragmented resources may promote local diversity as suggested by several authors (Atkinson and Shorrocks, 1984; Sevenster and van Alphen, 1996; Toda et al., 1999). However, mechanisms of species coexistence are obviously multifactorial and it is likely that parasitoids also participate to the species diversity of their host community as suggested by results from laboratory experiments (Boulétreau et al., 1991; Davis et al., 1998; Fleury et al., 2004). Indeed, in simple experimental ecosystems, when parasitoids are absent, *D. melanogaster* always eliminates *D. simulans* when they compete for limited resources whatever the thermal regime (from 22 °C to 28 °C). The outcome of competition is modified by the presence of parasitoids and the issue varies according to temperature: *D. simulans* remains a poor competitor at 28 °C, but the frequency of this species increases until the extinction of *D. melanogaster* at 22 °C, and both species coexist at 25 °C (Fleury et al., 2004). At 22 °C, parasitoids invert the outcome of competition and allow the coexistence of species that normally exclude each other at 25 °C, thus suggesting their role in the diversity of *Drosophila* community.

1.1.3.2. Competition within parasitoids community

With regard to parasitoids, we also have evidence that competition is severe among species in the wild, thus suggesting that horizontal interactions are probably selective factors that also participate to shape the whole *Drosophila*-parasitoid community. The first line of evidence results from a high parasitization rate that sometimes exceeds 90%, which inevitably leads to strong competitive interactions at least among individuals of the same species. Convincing arguments were provided by field observations showing that a number of *Drosophila* larvae host more than one parasitoid larva (superparasitism), although only one adult can emerge from a single host (Fleury et al., 2004; Wertheim et al., 2003). The variation of relative abundance of species across the season despite favourable conditions also argues in favor of strong interspecific competition among parasitoids. For instance, in southeastern France, *L. heterotoma* is the main species early in the season but its density sharply decreases when *L. boulardi* appears, probably as a result of competitive displacement (Fig. 1.2). Indeed, both biotic and abiotic factors remain suitable for *L. heterotoma* when populations persist, facing *L. boulardi* at very low density over the whole season (Fauvergue et al., 1999), and the geographical range of *L. heterotoma* clearly includes wider habitat conditions than those occurring when populations collapse (e.g., North Africa). A similar pattern of seasonal abundance was observed in Tunisia (Carton et al., 1991). Because intensity of competition is likely to vary geographically, it is expected that competitive selective pressures result in a population differentiation and local adaptation for a number of parasitoid traits and reproductive strategies. How *Drosophila* parasitoids cope with competitive interactions remains underinvestigated. Host range, habitat preference, temporal as well as spatial refuge resulting from parasitoid aggregation are all probably involved in reducing interspecific competition (Vet et al., 1984; Wertheim et al., 2000). However segregation of realized niches in the field remains largely unknown. In some cases, infochemicals can be used by *Drosophila* parasitoids to avoid patches exploited by a superior competitor (Janssen et al., 1995). The presence of a conspecific on a patch also influences oviposition decisions of *L. heterotoma* females less disposed to parasitize hosts (Visser, 1995). Females are also able to recognize parasitized hosts and avoid superparasitism (Bakker et al., 1990; van Alphen and Visser, 1990; van Lenteren, 1976) but to date there is no evidence that parasitoids discriminate hosts attacked by another species and then multiparasitism might occur (Turlings et al., 1985; van Strien-van Liempt and van Alphen, 1981). Parasitoids such as *Leptopilina* and *Asobara* share the same habitat and common hosts in rotting fruits and thus probably compete in the field. However, intensity of interspecific competition may be softened by fine-scale differences in species-specific life histories that can balance competitive ability, thus promoting species coexistence as shown in *Drosophila* (Joshi and Thompson, 1996). For example, despite that the seasonal phenology of frugivorous *Drosophila* parasitoid largely overlaps, all species show different circadian rhythms, leading to fine temporal segregation of activity within a day (Fleury et al., 2000a). Intrinsic inferior competitors then take advantage to be active earlier, a temporal niche that probably cannot exploit the superior competitor because of thermal constraints. Species also differ by host searching behavior and host detection such as antennal or ovipositor searching, host location by vibrotaxis, or the use of infochemicals, which may lead to exploit different microniches or host species, thus contributing to coexistence (van Dijken and van Alphen, 1998; Vet and Bakker, 1985). As the aggregation of parasitism occurs at local scales (Driessen and Hemerik, 1991), microscale aggregation might also probably participate in reducing interspecific interactions. This was clearly suggested in experimental observations where *L. boulardi* and *L. heterotoma* alone or competing by pair (both intra- or interspecific association) were allowed to parasitize four patches of *Drosophila* larvae in a small Petri dish. Results clearly showed that when the female is alone, both species aggregate their attacks mainly on one patch only (high value of aggregation index) and the same result is observed with pair of conspecific females (Fig. 1.3). In contrast, a very low aggregation index was measured in heterospecific pairs (Fig. 1.3). Analysis of parasitoid emergence indicated that observed overdispersion results from microscale spatial segregation of heterospecific females that aggregate their infestation on different patches. This probably might be mediated by infochemical repellents or the consequence of physical contacts among females that drive them on different patches. Thus, a number of factors contribute to microniche differentiation among frugivorous *Drosophila* parasitoids in the wild that might compensate intrinsic competitive differences among species. However, according to the high rate of parasitism in some communities, species differences are probably not broad enough to override competitive interactions. This competition must constitute one of selective forces that drive a large number of life history traits of *Drosophila* parasitoids.

1.2. *DROSOPHILA* PARASITOID LIFE HISTORIES

The adult and larval biology of *Drosophila* parasitoids are known from the initial work of Jenni (1951) and Nöstvik (1954) on *Leptopilina* spp. and are well reviewed by Carton et al. (1986). The biology of *Asobara* genus is less carefully described. More recently, Melk and Govind (1999) described in detail the biology of *Ganaspis xanthopoda*, which appears to be quite similar to the *Leptopilina* genus. These species are all koinobiont and solitary endoparasitoids that attack first and second stages of *Drosophila* larvae. Under favorable conditions, with unlimited number of hosts, *L. heterotoma* adult females deposit only one egg per host which hatch 2 days after oviposition in the host hemocele. The host tissues are progressively consumed by second and third instars. Third-instar parasites become ectoparasites, consuming all host pupa, and metamorphosis

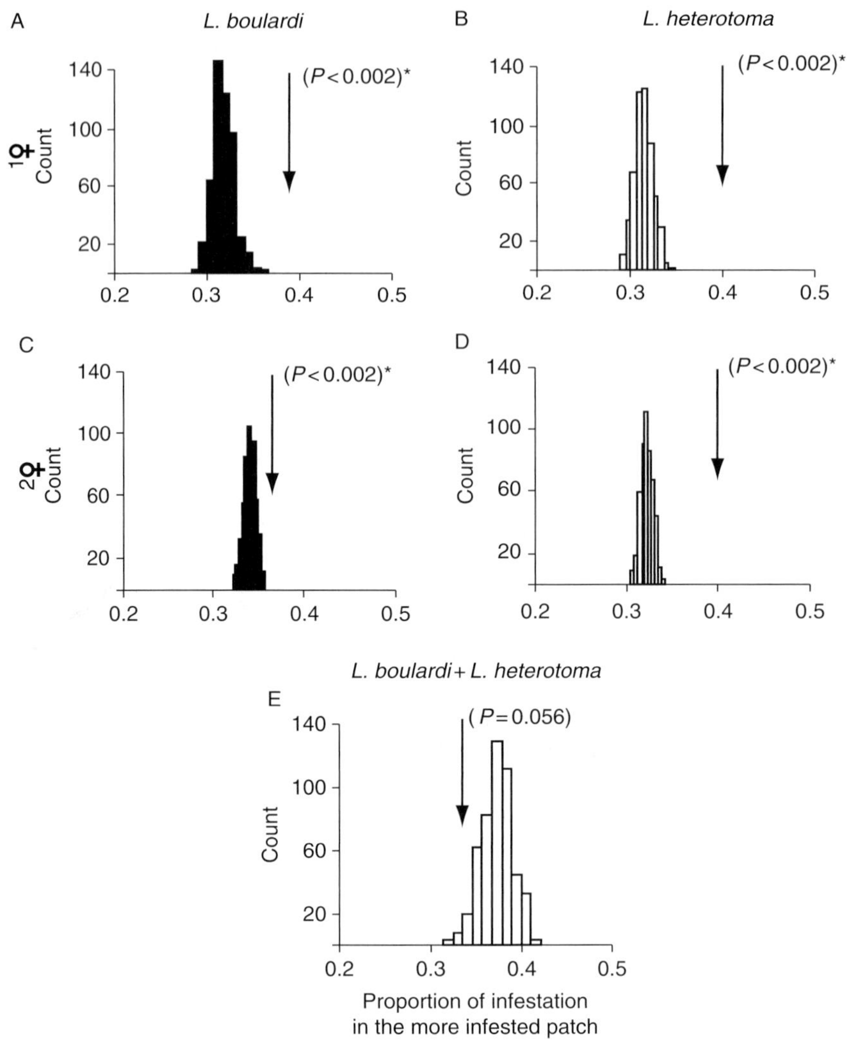

FIGURE 1.3 Aggregation of infestation by isolated or competing *Leptopilina* females. Females were allowed to parasitize 4 patches of 50 *Drosophila* larvae in a small Petri dish (5.5 cm diameter) during 24 h (25 °C, LD 12:12). Each graph indicates the observed value of aggregation index (arrow) compared to the simulated distribution of this index under the hypothesis of random parasitoid attacks on all patches (500 simulations). Results show that alone (A, B) or by conspecific pairs (C, D), both species aggregate their attacks mainly on one patch only. In contrast, interaction between *L. heterotoma* and *L. boulardi* females (E) results in overdispersion of parasitization.

occurs in *Drosophila* puparium approximately 2 weeks after oviposition. Under laboratory conditions, adults emerge after about 3 weeks and emergence spreads over 4–5 days at 25 °C, with males emerging 1–2 days before females. Within the day, protandry is observed with males emerging 1–2 h before females (Fauvergue et al., 1999). In the field, development time can last 35 days in May–June but decreases along the season with rising temperatures (Fauvergue et al., 1999). Development time of the parasitoid is then approximately 1.5 to 2 times longer than those of the unparasitized host. According to the conditions, the preimaginal wasp development does not always succeed and the parasitism may fail in three different ways: (1) a precocious parasite death induced by the immune response of the host called encapsulation, as described in different *Drosophila–Leptopilina* interactions, (2) the death of the parasitoid eggs without any consequence to the adult *Drosophila* (Nappi and Streams, 1970; Streams, 1968) and (3) a failure or "inadequacy", which leads to the death of both host and parasitoid. Both *L. heterotoma* and *Asobara* parasitoids emerge with a number of mature eggs ready to be laid but the degree of proovigeny varies among species. Adult wasp fitness components thus may be experimentally estimated by counting the number of oocytes in ovaries soon after the emergence. Because *Drosophila* parasitoids are all haplodiploid species (haploid males and diploid females produced respectively by unfertilized and fertilized eggs), the offspring sex ratio may vary according to local conditions. Isolated females of *L. heterotoma* produce around 20–30% of males but females are able to modify the sex ratio when wasp densities increase according to sex-allocation theories (Debout et al., 2002). Despite these common characteristics, *Drosophila* parasitoids exhibit differences in several life history traits such as host range, biological rhythms, overwintering strategies, host foraging behavior and thermal sensitivity.

1.2.1. Host range and specialization

Host range of larval *Drosophila* parasitoids differ significantly among species even if they are phylogenetically closely related, such as *L. heterotoma* and *L. boulardi*. The host spectrum is predominantly determined under laboratory conditions on the basis of host acceptation and parasite developmental success. Clearly, the host range actually used in the field by parasitoids needs to be determined. To date, authors agree that *L. boulardi* is a specialist of *D. melanogaster* and *D. simulans* (Barbotin et al., 1979; Carton and Nappi, 1991; Carton et al., 1981, 1987; Fleury et al., 2004). However, particular African strains can develop in *D. yakuba* (Dubuffet et al., 2008), and recent data suggest that *L. boulardi* can also develop in *D. subobscura* and *D. pseudoobscura* at a low rate (Schlenke et al., 2007). *L. heterotoma* is by far the more generalist species, since its successful development has been recorded on numerous *Drosophila* (*D. busckii*, *D. funebris*, *D. kuntzei*, *D. melanogaster*, *D. obscura*, *D. phalerata*, *D. simulans*, *D. subobscura* and *D. willistoni*) or related genera (*Chymomyza* or *Scaptomyza*) (Carton et al., 1986; Janssen, 1989; Jenni, 1951; Ris, personal observations). To date, determinants of this large host range are not precisely known but it should be hypothesized that this relies on the ability of *L. heterotoma* to cope with different host defenses (Schlenke et al., 2007) or its ability to cope with qualitatively and quantitatively different host resources. *A. tabida* is considered rather as a specialist species attacking *D. subobscura*, *D. obscura* and *D. melanogaster* mainly, but can also develop on *D. tristis* (Janssen, 1989; Kraaijeveld et al., 1995). Recently, Eslin and Prévost (1998) reported the successful development of *A. tabida* in other species of the *D. melanogaster* subgroup, such as *D. sechellia* (58%), *D. simulans* and *D. mauritiana* (18%) that could enlarge the potential host spectrum of this species. However, other studies have reported no development in *D. simulans* that expresses a complete resistance to *A. tabida*, whose eggs are encapsulated (Kraaijeveld and van der Wel, 1994), thus explaining the weak abundance of this species in orchards of Mediterranean area where *D. simulans* is dominant. The origin of such discrepancies remains unclear, but it is likely that variability results from the origin of host or parasite strains. Indeed, a number of studies reported genetic variability within or among populations in either host suitability or the ability of a parasite to develop in a particular host species (Boulétreau, 1986). For example, while suitability of *D. subobscura* is fairly constant whatever the geographical origin of *A. tabida*, only southern European wasp populations show a good survival rate in *D. melanogaster* (40–80% compared to few percent). This is interpreted as an adaptation to local conditions where *D. melanogaster* is more abundant (Kraaijeveld and van der Wel, 1994). Genetic variability was also observed from the host side on their ability to enable parasite development. Such variation was reported mainly for *D. melanogaster* that show different suitability for *L. heterotoma* (Boulétreau and Wajnberg, 1986), *L. boulardi* (Boulétreau and Fouillet, 1982; Carton et al., 1989; Wajnberg et al., 1985) and *A. tabida* (Kraaijeveld and van Alphen, 1995a). The origin of these variations relies on both immunological resistance and parasite virulence ability (see for a review Kraaijeveld and Godfray, 1999; Kraaijeveld et al., 1998 and this volume) but can also result from physiological inadequacy of the host with regard to parasite requirement (Boulétreau, 1986). Of course, a number of other factors participate to larval survivorship of parasites in a given host, mainly temperature, giving rise to complex interactions (Fleury et al., 2004; Kraaijeveld and van der Wel, 1994; Ris et al., 2004). Crowding appears also important since success of parasite development could rise from 40% to 90% with increased larval density (Boulétreau and Wajnberg, 1986; Wajnberg et al., 1990).

1.2.2. Effect of developmental host on parasite life histories

Besides the developmental success, host species can influence a number of adult parasitoid traits either on emerging wasps as a result of phenotypic plasticity, or on offspring of these wasps by maternal effects. Hosts also constitute a selective force that can genetically modify parasitoid populations in a number of traits including parasite success. Surprisingly, the extent to which host species influence parasitoid phenotype by developmental plasticity was studied for a limited number of host–parasite combinations. Moreover, complex interactions with other biotic or abiotic factors such as temperature or crowding can mix up conclusions, as well as the origin of the strains (Boulétreau and Wajnberg, 1986; Fleury et al., 2004; Kraaijeveld and van der Wel, 1994). For example, *L. boulardi* females produce many more offspring at 25 °C after development in *D. melanogaster* (mean total progeny 293) than on *D. simulans* (mean total progeny 184). The lesser quality of *D. simulans* as a host was confirmed for both *L. boulardi* and *L. heterotoma* at 25 °C but the difference is sharply reduced at 22 °C, where the quality of the two host species is quite similar (Fleury et al., 2004; Ris et al., 2004). Moreover, parasitoid genotype is also involved, since the southern genotype of *L. heterotoma* performs much better in *D. simulans* than in northern ones originating from the area where this host species is less abundant. This indicates a possible local adaptation of the parasite to cope with the most abundant host in local sites (Fleury et al., 2004). Traits other than egg load or total offspring count such as preimaginal development time, growth rate body size, adult longevity or fat reserve are also influenced by developmental host species (Eijs and van Alphen, 1999). Using a wider host range, we compared the influence of the five sympatric *Drosophila* species potentially used as a host by *L. heterotoma* in the southeast of France (21 °C, photoperiod LD 16:8). Under experimental condition, survival of all host species is high but they differ in size with the following order using *D. simulans* as reference: *D. melanogaster* (120% dry weight), *D. subobscura* (150%), *D. immigrans* (230%) and *D. hydei* (310%) with the same trend observed for development time. These five *Drosophila* species clearly do not offer the same developmental conditions to *L. heterotoma* since *D. melanogaster* and *D. simulans* appear to be as expected the two most suitable hosts, with almost 80% of parasite survival, while *D. subobscura* and *D. hydei* show intermediate quality (parasite survival ranging from 40% to 60%), and *D. immigrans* (20%) clearly appears as a rather unsuitable hosts (Fig. 1.4). Parasitism failure on *D. immigrans* is associated with the death of the parasitized host and no encapsulation was observed. However, less suitable hosts for parasite development give rise to females with higher egg loads, probably as a consequence of a direct effect of host size on parasite size, which

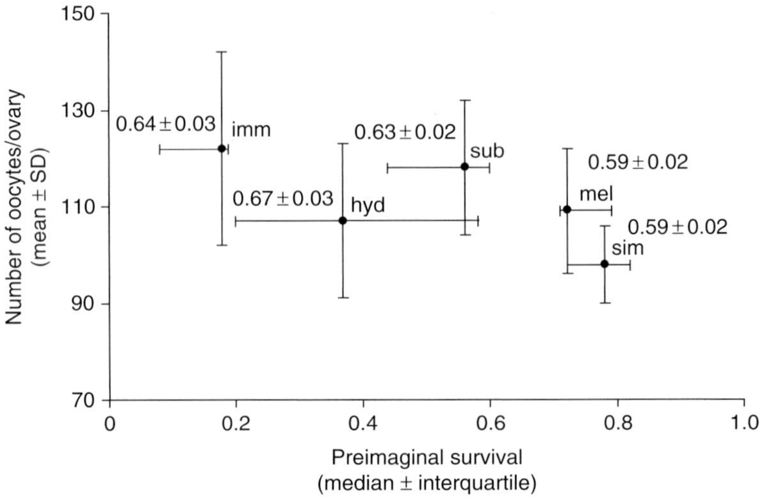

FIGURE 1.4 Influence of five *Drosophila* hosts on preimaginal survival, egg load of 5-day-old females and the tibia length (values on the graph in mm, mean±SD) of *Leptopilina heterotoma* (sympatric strains). imm *D. immigrans*, hyd *D. hydei*, sub *D. subobscura*, mel *D. melanogaster*, sim *D. simulans*. SD, standard deviation.

appears closely correlated. This negative correlation among parasite development success and egg load calls for careful classification of host quality. Interestingly, development time of parasitoid from egg to adult follows more closely the ability of the host to give rise to an adult parasite than host development time or host size themselves. This suggests that despite *L. heterotoma* is a "conformer", with parasite traits directly influenced by those of their host, host species cannot only be viewed as a "temporal scheduler" and/or a "food resource", but more complex physiological interactions are involved in the host–parasite relationship. Future progress should better understand the physiological determinants of this host suitability and, for instance, the puzzling inadequacy of *L. heterotoma* to develop in *D. immigrans*.

Drosophila parasitoids probably switch from one host species to another, as consequences of temporal or spatial variation of community composition. In addition to the direct influence of the host on the emerging parasitoid, the host species experienced by the parents could also influence the phenotype of their progeny by maternal effects. Surprisingly, such phenomenon remains poorly documented for koinobiont parasitoids including *Drosophila* parasitoids. Influence of the parental host has been investigated for *L. heterotoma* using *D. melanogaster* and *D. immigrans*, producing four combinations of developmental and parental hosts. Maternal effects were detected on two of three life history parameters under study (Fig. 1.5). The egg load of 5-day-old females hatching from *D. melanogaster* is lower when their mother had developed in *D. immigrans* whereas such an effect is not observed when *D. immigrans* is used as the developmental host. This demonstrates that host shift from *D. immigrans* to *D. melanogaster* is clearly detrimental for the wasp with possible consequence on host range evolution. More complex maternal effects were also observed in wasp size (estimated by tibia length), with a gain of size when the mother had developed on a different host species than the daughters but no evident explanation arises about underlying physiological processes involved. Only preimaginal parasitoid survival is not influenced by maternal effect, but drastically drops when *D. immigrans* is used as the developmental host. These results clearly demonstrate that the host of the parents can influence the *L. heterotoma* offspring as already reported in other animals (Bernardo, 1996; Rossiter, 1996). The physiological basis of this cross-generational phenotypical plasticity remains however unknown insofar as both size and fecundity are not affected in the same ways. These traits vary in opposite ways (in term of fitness) when development occurs in *D. melanogaster*, which could be a consequence of different allocation of the resources between size and fecundity in response to environmental conditions.

1.2.3. Adult parasitic strategies and life history covariation

Faced with the multitude of factors that can modify wasp phenotype including temperature, crowding, host species and the associated maternal effect and given the fact that traits also vary according to both host and parasite genotype, it is not realistic to review in detail life histories that show wide phenotypic or genotypic variation such as fecundity, longevity or size of *Drosophila* parasitoids. More interesting is our knowledge about how parasitoids behave to locate, select and exploit an optimal host (optimal foraging) and how they trade off the advantage to gain on one trait by the cost sent back on another trait with regard to life histories theory (Roff, 1992; Stearns, 1992).

1.2.3.1. Hosts finding

Because parasitoid searching behavior correlates directly with offspring production, it is expected that host foraging by parasitoids is optimized by natural selection solving the tremendous difficulty of finding hosts. A number of studies reported adaptive foraging decisions observed in *Drosophila* parasitic wasp. As previously discussed, parasitoids of frugivorous *Drosophila* locate host habitat and select suitable hosts using a number of cues, mainly infochemicals. Olfactory microhabitat selection is mediated by the emanation of fermenting fruits, decaying plants or

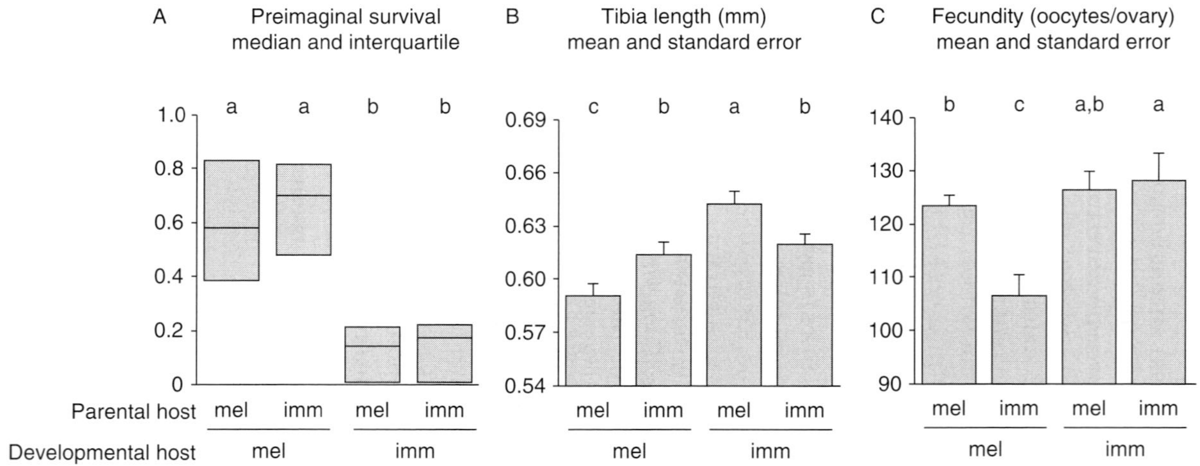

FIGURE 1.5 Influence of *Drosophila immigrans* (imm) and *D. melanogaster* (mel) as developmental host or parental host on the *Leptopilina heteroma* phenotype. Sample sizes for the four combinations of parental and developmental hosts (from left to right on the figures) are (A) for preimaginal survival: 7, 9, 6 and 11; (B) for tibia length: 54, 50, 42 and 27; (C) for fecundity: 56, 42, 46 and 26.

fungi. Laboratory experiments often showed that live baker's yeast odor is attractive (Vet, 1985a). *L. heterotoma* (Dicke et al., 1984; Papaj and Vet, 1990; Vet and van Opzeeland, 1985; Vet et al., 1998), *L. boulardi* (Carton, 1978; Couty et al., 1999; Vet, 1985a) and *A. tabida* (Kraaijeveld and van der Wel, 1994; Vet et al., 1984) all perform olfactory microhabitat selection as well as other *Drosophila* parasitoids (van Alphen et al., 1991). Odor habitat alone is not spontaneously attractive and parasitoids must learn to memorize the fruit odor by associative learning with host cues (Couty et al., 1999; De Jong and Kaiser, 1992; Kaiser et al., 2003; Vet et al., 1998). Odors of the host itself appear much more attractive than substance emanating from the microhabitat. Several studies reported that most wasp species show a very strong innate attraction to *Drosophila* aggregation pheromone (Wertheim et al., 2003; Wiskerke et al., 1993). Molecule was identified as (Z)-11-octadecenyl acetate produced by *Drosophila* male and transferred to female during mating. In contrast to the innate response of *Leptopilina*, *A. tabida* needs to learn this host's odor source (Hedlund et al., 1996). As we could expect, the generalist parasitoid *L. heterotoma* innately responds to odor cues of all *Drosophila* species within its host range, while the specialist *L. boulardi* is attracted by odors of a limited number of *Drosophila* species, mainly those inhabiting fermenting fruit, among which are its natural hosts, *D. melanogaster* and *D. simulans* (Hedlund et al., 1996; Vet et al., 1993). The fact that *L. boulardi* responds also to odor of nonhost species suggests common compounds in odor blend emanating from host and nonhost species. Specialists and generalists also might differ in their ability to learn, with a stronger learning capacity in generalist species (Poolman et al., 1992).

1.2.3.2. Host patch exploitation

Once potential host batches are successfully localized, *Drosophila* parasitoids need to select suitable preys, that is, species of its host range and healthy larvae. *Drosophila* parasitoids exhibit a species-specific searching strategy to localize host larvae that classically involves the use of host-induced vibrations transmitted by the substrate (vibrotaxis), ovipositor searching (walking while probing), antennal searching and local arrestment to perceive high concentration of kairomones (van Dijken and van Alphen, 1998; Vet and Bakker, 1985; Vet and van Alphen, 1985). *A. tabida* uses mainly vibrotaxis (Sokolowski and Turling, 1987; van Alphen and Drijver, 1982) to detect its host that is thought to be the most efficient strategy when hosts are scarce or buried deeply in the substrate. However, this becomes confusing at high host density. *L. heterotoma* use ovipositor searches (Vet and van Alphen, 1985) that can be more efficient to find *Drosophila* larvae at high host density, a situation that is actively sought by this species as suggested by its rapid departure from lower density patches (van Lenteren and Bakker, 1978). *L. boulardi* shows an intermediate strategy with mainly the use of ovipositor but also the movement of larva as host detection cues (Vet and Bakker, 1985). The difference in such searching behavior among species with overlapping niches is thought to favor spatial partitioning of competing species and thus their coexistence (van Dijken and van Alphen, 1998). This could also participate in the choice of the most suitable host species for parasite development in nature where several *Drosophila* species thrive in the same fruits.

1.2.3.3. Hosts acceptation and superparasitism

The following step in the process of host selection that focused extensive, empirical and theoretical studies is the distinction by the parasite among parasitized and unparasitized hosts (Gandon et al., 2006; Godfray, 1992; van Alphen and Visser, 1990; van Lenteren, 1981; Visser et al., 1992a). *Drosophila* parasitoids were largely used to analyze the process of discrimination between healthy and parasitized hosts, and superparasitism, when a female accepts a previously parasitized host. Most results were obtained on *L. heterotoma*, which clearly distinguishes and avoids parasitized hosts with the capacity to determine the number of parasite eggs in a particular host (Bakker et al., 1972, 1990; Hemerik and van der Hoeven, 2003; van Lenteren, 1976). *A. tabida* also discriminates parasitized hosts but lacks the ability to count the number of parasite eggs they contain (Hemerik and van der Hoeven, 2003; van Alphen and Nell, 1982). Host discrimination may prevent the waste of eggs, and it may be involved in the time budget conservation and gives cues about patch quality used by females to leave low profitability patches. However, the fact that superparasitism is often observed, probably arising from the female's decision, initiated theoretical works demonstrating the adaptive value of superparasitism under competition, if there is a fitness pay-off from an egg laid in a parasitized host (Gandon et al., 2006; Hemerik et al., 2002; van Alphen and Visser, 1990; Visser et al., 1992b). Theoretical predictions were supported by results from experiments on *L. heterotoma*, which often behaves as an optimal model of superparasitism (Visser, 1995; Visser et al., 1990, 1992b,c). However, this adaptive view of superparasitism was recently called into question by results obtained on *L. boulardi*, which can show very high level of clearly nonadaptive superparasitism. In this species, excessive superparasitism is surprisingly triggered by a virus (LbFV), that is vertically and horizontally (contagiously) transmitted (Varaldi et al., 2003, 2006; Chapter 13 by Varaldi et al.). The differential adaptive interests of the virus and the wasp can create a conflict that can deeply modify the outcome of superparastism level (Gandon et al., 2006). To date, this viral origin of superparasitism does not challenge adaptive interpretation of this behavior made in other *Drosophila* parasitoid species since the LbFV virus shows a strict specificity to *L. boulardi* (Patot et al., 2009). According to all information gathered within a patch (host kairomone, oviposition experience, patch depletion, time spent) and possible knowledge about quality of the surrounding habitat, parasitoids are supposed to allocate optimally their patch resident time, that can be reached by different procedures (review in van Alphen et al., 2003). In *A. tabida*, ovipositions increase patch resident time (incremental mechanism) and females use several cues to optimize time allocation on a patch (Galis and van Alphen, 1981; van Alphen and Galis, 1983). *L. heterotoma* also shows an incremental mechanism of oviposition whereas rejection of parasitized hosts increases the patch leaving tendency, which is consistent with aggregated distribution of *Drosophila* larvae (Haccou et al., 1991; Varaldi et al., 2005). In contrast, oviposition and rejection have no effect on patch resident time and leaving tendency in *L. boulardi*, thus suggesting a different issue of selection, probably in response to the high level of superparasitism induced by LbFV virus infection (Varaldi et al., 2005).

1.2.3.4. Virulence and resistance

The covariation of life histories and the underlying trade-off that can balance overall parasitoid fitness with different combination of traits according to life histories theory (Roff, 1992; Stearns, 1992) were far less investigated and probably remains a promising issue for the future. More results are available on *Drosophila*, particularly the cost to develop genetic resistance against parasitoids (encapsulation) that can decrease fitness related trait, among which the competitive ability of *Drosophila*, their mating success and resistance to stresses (Fellowes et al., 1998; Hoang, 2001; Kraaijeveld and Godfray, 1997; Kraaijeveld et al., 2001, 2002; Rolff and Kraaijeveld, 2003). With regard to parasitoids, enhanced virulence obtained by artificial selection has only a very slight effect on egg stage duration with no consequence on other fitness traits, but the difference seems sufficient to potentially reduce survival probability in case of superparasitism and then can be considered as an evolutionary trade-off (Kraaijeveld et al., 2001). Extensive progress has been made since the late 1990s on immune response of *Drosophila* against parasitoids, and strategies of wasps used to overcome host resistance, including factor of virulence (Carton and Nappi, 1997, 2001). This aspect of host–parasitoid relationship falls beyond the scope of this review (see section III this volume), but it is, however, noticeable that mechanisms by which parasitoids defeat the *Drosophila* immune response vary hugely among species. *L. heterotoma* suppresses host encapsulation of its egg by injecting virus-like particles (VLPs) that cause cellular immune depression. *L. boulardi* harbors morphologically different VLPs and uses virulence factors present in the venom to weaken the immune competency of their host. Virulence factors of these two *Leptopilina* species have quite different effects on the up- and downregulation of transcriptional activity of the *Drosophila* immune gene (Schlenke et al., 2007). In contrast, *A. tabida* does not provoke any host immune depression but uses sticky eggs that bind to host tissues allowing parasitoids to avoid encapsulation (Eslin et al., 1996; Prévost et al., 2005). How and why these different strategies have evolved remains an open question that could be solved only by integrating precise knowledge about host range of parasitoids and community structure and functioning.

1.2.3.5. Adult life history traits

Whatever the mechanism of virulence used, it seems that the ability of *Drosophila* parasitoids to overcome host immune responses is pervasive in natural populations, giving the opportunity for natural selection to shape other life history traits whose covariation remains poorly understood. Using five *Leptopilina* species, Eijs and van Alphen (1999) studied the correlation among several life history traits (development time, adult life span, body size, egg load, growth rate and fat reserve) when parasitoids are reared on two host species. A classic correlation among body size and egg load was observed in *Leptopilina* species that was previously demonstrated for a number of other parasitoids whose fitness increased with female size (Visser, 1994). The authors however failed to detect any relationship between development time and adult life span, a classic trade-off predicted by life history theory (Roff, 1992). This confirms earlier findings in Hymenoptera parasitoids (Blackburn, 1991) and the authors put forward the hypothesis that the particular way of life of parasitoids, especially the growth rate of their immature stage that shows a host-related plasticity, can explain this result. *A. tabida* also exhibits a positive correlation between size and egg load (Ellers et al., 1998; Kraaijeveld and van der Wel, 1994). Nevertheless, size–fitness relationships could change during the season moderating even impeding selection for large individuals (Ellers et al., 2001).

1.2.4. Effects of temperature and overwintering

In all ectotherms, temperature plays a major role on phenotypic expression of traits and life histories (Cossins and Bowler, 1987; Leather et al., 1993; Precht et al., 1973). Insect hosts and their parasitoids may show different thermal requirements with complex and unexpected effects on the nature of interaction and stability of the associations. Several studies reported temperature-induced variation on phenotypic traits of *Drosophila* parasitoids (Boulétreau et al., 1994; Kraaijeveld and van der Wel, 1994), but very few analyzed precise reaction norms looking for difference in thermal specialization of host and parasite. This is, however, an important issue to understand the geographical distribution of species, the community structure and the impact of climate change on the stability of parasitic associations. Using a range of temperature from 14 °C to 26 °C and the three main host species, Ris et al. (2004) showed that *L. heterotoma* has a narrow thermal niche compared to all their host species (*D. melanogaster*, *D. simulans* and *D. subobscura*). This probably explains the variation of species abundance throughout the season and probably participates to the stability of host–parasitoid association. Interestingly, different genotypes of *L. heterotoma* do not show the same ability to cope with temperature variation. Complex genotype-by-temperature interactions thus appear, revealing local adaptation with southern strains more adapted to warmer temperatures when the most common local host species is used (*D. simulans*). These results illustrate how parasitoids genotypes could be locally specialized to the host species, the temperature and their interaction, thus suggesting that weak environmental variation may destabilize the association. Another issue of parasitoid thermal biology lies in the way species cope with cold temperatures during winter. *L. boulardi* and *A. tabida* have been described as developing a larval diapause. Claret and Carton (1980) demonstrated the occurrence of facultative diapause at the prepupa stage of *L. boulardi* induced by a relatively low temperature, which was also observed in other populations (Hertlein, 1986). Almost 98% of the larvae enter diapause at 17.5 °C, 14% at 22.5 °C and no diapause is observed at 25 °C. Diapause induction is independent of photoperiod and thermoperiod. Diapause termination is hastened by transferring larvae to 25 °C. No indication of any genetic differentiation between tropical and temperate populations has been found on diapause induction in *L. boulardi* (Carton and Claret, 1982). In *A. tabida*, diapause occurs as a prepupa inside the host's puparium and the photoperiod clearly influences diapause induction (Baker, 1979; Jenni, 1951). The termination of diapause requires exposure to 4 °C for at least 6 weeks. Percentage diapause in *A. tabida* is influenced by host species with more diapause in *D. melanogaster* than in *D. subobscura* (Kraaijeveld and van Alphen, 1995b) and more in *D. pseudoobscura* than in *D. subobscura*, *D. ambigua* and *D. athabasca* (Kraaijeveld and van Alphen, 1993). Diapause in *A. tabida* is clearly influenced by both low (15 °C) and high temperatures (25 °C) that both increase its occurence (Kraaijeveld and van Alphen, 1995b). In this species, diapause is associated with energetic costs such as a substantial reduction in egg load, fat reserves and dry weight of the emerging adult females (Ellers and van Alphen, 2002). As with *L. boulardi*, no latitudinal geographical cline on percentage diapause has been observed despite the variation that occurs among populations (Kraaijeveld and van Alphen, 1995b). In contrast, *L. heterotoma* does not show larval or prepupal diapause (Carton et al., 1991) but this species overwinters as an adult (Eijs, 1999). Overwintering females are readily active early in the season as soon as favorable climatic conditions reappear. In spite of a high winter adult mortality, by breeding early in the season, *L. heterotoma* is able to produce up to four generations in temperate areas (Eijs, 1999).

1.3. GEOGRAPHICAL DIFFERENTIATION AND LOCAL ADAPTATION

1.3.1. Geographical variations and host–parasitoids relationship

Abiotic factors and the structure of the *Drosophila* parasitoid community and their interactions vary geographically even at a lower scale (e.g., the situation in the southeast of France). Under the hypothesis that genetic variations exist within populations, geographical differentiation and possible local adaptation are expected and important issues include identifying the selective forces involved in the adaptive processes and what traits actually respond to selection. Significant heritability of host and parasitoid features involved in the interaction, mainly host suitability and resistance, has been revealed by the isofemale lines technique for *L. boulardi* (Boulétreau and Fouillet, 1982; Boulétreau and Wajnberg, 1986; Boulétreau et al., 1987; Carton and Nappi, 1991; Carton et al., 1989; Wajnberg et al., 1985), *L. heterotoma* (Boulétreau and Wajnberg, 1986; Boulétreau et al., 1987; Carton and Boulétreau, 1985; Delpuech et al., 1994) and *A. tabida* (Mollema, 1991; Orr and Irving, 1997). Additive genetic variance within populations was confirmed by artificial selection experiments on traits such as host resistance (encapsulation) that can show rapid response to selection (Fellowes et al., 1998; Hughes and Sokolovski, 1996; Kraaijeveld and Godfray, 1997; see Fellowes and Godfray, 2000 and Kraaijeveld and Godfray, 1999 for reviews). Few investigations focused on genetic variability of parasitoid virulence despite our knowledge that these variations do occur (Carton and Nappi, 1991; Carton et al., 1989; Walker, 1962). These natural variations were then used to identify the genetic basis of the immune response of *Drosophila* and factors of parasitoid virulence which put forward candidate genes involved in the interaction (Benassi et al., 1998; Colinet et al., 2007, 2009; Dubuffet et al., 2008; Dupas and Carton, 1999; Dupas et al., 1998; Fellowes and Godfray, 2000; Labrosse et al., 2003; Nappi et al., 1991, 1992; Poirié et al., 2000; Vass et al., 1993). Unfortunately, in return, these findings on resistance and virulence genes have not yet been used to analyze how they may explain additive genetic variation observed within a population or if these genes are involved in variation among natural populations. A number of studies demonstrated that geographical populations can show marked differences, thus suggesting that local selective forces acted to shape host and parasitoid traits. Virulence of *L. boulardi*, that is, ability to evade *D. melanogaster* encapsulation, differs among populations originating from Europe or central Africa (Carton and Nappi, 1991), but subsequent studies failed to extend this variation to worldwide populations that are clearly nearly all immunosuppressive with the exception of those originating from Central Africa (Dupas and Boscaro, 1999). In *A. tabida*, virulence on *D. melanogaster* shows a north–south clinal variation across Europe (Kraaijeveld and van Alphen, 1994) with a much higher virulence of southern Mediterranean populations. The fact that these variations parallel geographical variation of the relative abundance of encapsulating (*D. melanogaster*) and nonencapsulating (*D. subobscura*) hosts suggests a local adaptation of the parasitoid to *D. melanogaster*, its main host species in the south. On the host side, *D. melanogaster* also shows geographical variation of resistance against both *L. boulardi* (Boulétreau, 1986; Boulétreau and Fouillet, 1982) and *A. tabida* (Kraaijeveld and van Alphen, 1995a), without any correlation between resistances against the two parasitoid species (Kraaijeveld and Godfray, 1999). However, the significance of the geographical pattern of resistance variation is unclear on an adaptive point of view.

Because the host community composition varies locally and the suitability for parasitoid development also varies within one host species, differential selective pressures may lead to local variation in host selection behavior. By comparing choices made by females with those predicted by optimal foraging models, it has been shown that parasitoids indeed behave optimally with regard to host selection (Janssen, 1989; Kraaijeveld et al., 1995). Comparative studies revealed that this could lead to genetic differentiation in the choice of the most suitable local host. Due to geographical variation in the capacity of *A. tabida* to survive in *D. melanogaster*, Kraaijeveld et al. (1995) demonstrated that the northern population rejects this host and prefers to oviposit in *D. subobscura*, whereas the southern populations accept both host species indistinctly as expected in accordance with the to local abundance of host species. A similar conclusion was drawn from a quite a different situation bringing into play two *L. boulardi* strains that can develop either on *D. melanogaster* (Mediterranean strain encapsulated by *D. yakuba*) or on *D. yakuba* (Central Africa strain encapsulated by *D. melanogaster*). In agreement with optimal foraging models, the Mediterranean strain prefers *D. melanogaster* in choice experiments, whereas the other prefers *D. yakuba* (Dubuffet et al., 2006). In this case, more precise information on the local ecological situation is needed to interpret these results in terms of local adaptations. Pannebakker et al. (2008) also showed genetic differentiation in *L. clavipes* between northwestern and southern European strains. Parasitoids from southern Europe accepted all tested *Drosophila* species while northwestern parasitoids appeared to be more specialist and preferred hosts that thrive in fungi, rejecting *D. melanogaster*.

This difference is interpreted by the authors as an adaptation to conditions most often encountered by parasitoids in the field.

1.3.2. Small scale geographical variations and competitive interaction

1.3.2.1. Geographical variation in southeastern France

The ecological situation followed in southeastern France (see section 1.1) shows that selective forces other than single host–parasitoid interactions act in natural communities – mainly competition. As community structures vary locally at low geographical scales (less than 4° of latitude), mainly because this area encompasses the northern limit of *L. boulardi* geographical range, other parasitoid species experience quite different competitive interactions with higher selective pressure in the south, as suggested by the rate of parasitized hosts (80% in some sites). We performed a comparative analysis of life history traits of *L. heterotoma* by sampling populations distributed along a north–south axis that included different levels of competition. Strains were bred in the laboratory and traits were measured under the same conditions (25 °C, photoperiod LD 12:12) to reveal the expression of genetic variation. Huge variations were observed for a number of traits that correlate latitudinal variation of *Drosophila*–parasitoid communities. Activity and its circadian rhythm show genetic differentiation with southern populations active both at the beginning and the end of the day with higher rate of activity, whereas northern populations are less active only during the afternoon (Fleury et al., 1995). These results were confirmed on a number of other *L. heterotoma* populations (Fleury, unpublished observation) and could be interpreted as a better capacity of southern population to disperse and then to explore a more fragmented habitat also exploited by *L. boulardi*. Populations of *L. heterotoma* also differ for fecundity with a very high, significant effect of latitude. Populations from the north show a low fecundity compared to populations from the south that produce almost 100 eggs more despite populations being separated by a distance of less than 500 km. The fact that the variation of fecundity follows a cline that is inverted compared to what is generally observed in insect responses to temperature (Boulétreau-Merle, 1992; Capy et al., 1993; Karan et al., 1998; Mitrovski and Hoffmann, 2001) suggested that other selective factors than a direct response to temperature are involved. We put forward the hypothesis that these variations result from high competitive interactions induced by the presence of the superior competitor *L. boulardi* in the south but absent in the north. Higher fecundity may increase the competitive ability of *L. heteroma* and then balance the competitive ability of interacting species, which could participates with other mechanisms of resource partitioning such as host range and/or temporal segregation of activity to the coexistence of competing species.

Since a classical expectation in evolutionary biology is that organisms are not able to maximize all fitness-related characters (Roff, 1992; Stearns, 1992), negative correlation (trade-off) between life history traits were explored. To test whether the higher fecundity of the southern genotype of *L. heterotoma* is counter-balanced by a lower performance in other traits, several characters were measured in two extreme lines of *L. heterotoma*, one originating from the south (A7 line from Antibes 43.57 °N) and one from the north (SF4 line from Lyon 45.74 °N). These two lines were obtained after regular sib-mating for more than 20 generations, which eliminates the within-line genetic variability (Fleury et al., 2004; Ris et al., 2004). The comparison failed to detect any trade-off (at least for the traits studied), and the A7 southern line always exhibited a significantly higher performance than the northern genotype (Lyon). Fecundity was much higher in the southern genotype (difference of 116 eggs), adult survival was longer (difference of 1.4 days), females were bigger with a slower development time and had a higher lipid content. This study demonstrates that *L. heterotoma* do not follow classic correlation between life history traits. We cannot rule out any possible trade-offs with unmeasured traits, or that the genetic constraints that determine trade-offs do not apply at this scale of geographical differentiation.

1.3.2.2. Local adaptation in competitive ability

In order to test the hypothesis that competition induced by *L. boulardi* may shape *L. heterotoma* life history traits and that southern populations can be locally adapted to the presence of a stronger competitor, the competitive ability of the northern and southern genotypes of *L. heterotoma* against *L. boulardi* was compared. Four populations were used, two from the north that had never experienced competition (Saint-Germain-au-Mont-d'Or, 45.88 °N and Saint-Maurice-de-Beynost, 45.83 °N) and two from the south that live in sympatry with *L. boulardi* (Antibes 43.57 °N and Spain, Tarragona, 41.15 °N). The homozygous genotypes A7 from the south and SF4 from the north were also included in the experiment. For each *L. heterotoma* strain, four *L. heterotoma* and four *L. boulardi* females (from Antibes) were allowed to compete for 24 h on 150 *Drosophila* larvae at 25 °C. Under these conditions, a strong competition exists between females. The same experiment was performed on the two hosts, *D. melanogaster* or *D. simulans*, and the issue of competition was assessed by percentage of each species in the emerging offspring. Results clearly showed that the competitive ability of *L. heterotoma* varies geographically with southern genotypes performing better than northern genotypes

(Fig. 1.6). *L. boulardi* overwhelmingly outcompeted *L. heterotoma* in all conditions with an abundance of offspring that varied from 70% to 95% in some conditions. This is in agreement with the previous study that demonstrated that *L. boulardi* is a stronger competitor (Carton et al., 1991). This may also explain why *L. heterotoma* is almost excluded during the season when *L. boulardi* populations are growing (Fig. 1.2). However, while less than 20% of *L. heterotoma* is observed in emerging offspring, this rate rose to almost 30% for southern strains more able to compete with *L. boulardi*. The same tendency was observed whatever the *Drosophila* host species. Complementary experiments suggested than higher egg load and better life history traits of southern *L. heterotoma* genotypes were not the only factors explaining the issue of competition but that the competitive ability of parasite larva inside the same host during multiparasitism is also involved. These experiments demonstrate that geographical variation simultaneously concerns a number of traits that covary in the same direction increasing parasitoid fitness as a result of

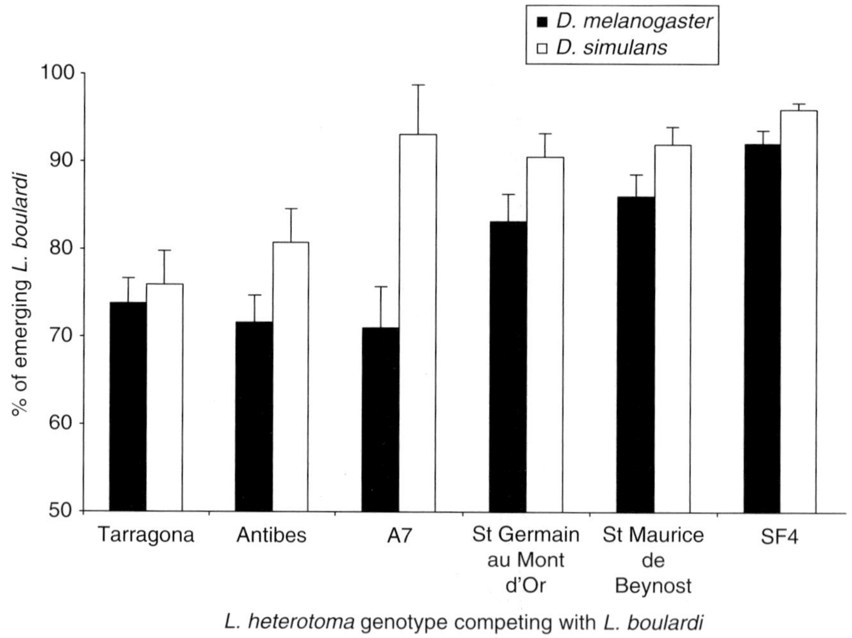

FIGURE 1.6 Competition between *Leptopilina boulardi* and different strains of *L. heterotoma* under laboratory conditions on either *D. melanogaster* or *D. simulans* as hosts. Results are expressed by percentage of emerging *L. boulardi* in the offspring (± standard error); 6 geographical strains of *L. heterotoma* were used: three southern strains (Tarragona, Antibes, A7) and three northern strains (St Germain au Mont d'Or, St Maurice de Beynost, SF4).

strong local selective pressure. Obviously, coexistence is always the result of several mechanisms acting jointly, but it is likely that this higher performance of southern *L. heterotoma* genotypes contributes to the coexistence of *L. heteroma* in areas where *L. boulardi* dominates. Competitive interactions thus also act as selective pressure in *Drosophila* parasitoid communities and this could be the main selective force in some conditions.

1.4. CONCLUDING REMARKS

This update on *Drosophila* parasitoid ecology and their life history evolution underlines progresses achieved since the synthesis of Carton et al. (1986). Since the late 1990s, most advances were obtained in the field of host resistance and parasitoid counter-defenses, on genetic basis, physiological mechanisms and evolutionary constrains that can impede evolution of these traits. However, few data are available on variation within and among populations of candidate genes recently discovered. Evidence of cost and evolutionary trade-off on both resistance and virulence suggests possible local coevolutionary dynamics in relation with spatial variation of community structure. This could also contribute to the maintenance in a same locality of genetic variability by frequency-dependent selection. The issue for the future is to extend laboratory studies to field work in order to understand the evolution of immunological facets of host–parasitoid interactions in the wild better. This should rely on the valuable models that are *Drosophila* parasitoids, which allow both easy sampling of natural populations in different ecological conditions and experimental works.

Progress also concerns the effect of temperature on phenotypic expression of *Drosophila* parasitoid traits and its possible consequences on host–parasitoid interactions and community structure. Temperature probably interacts with other selective forces in the evolutionary dynamics of the association with probably direct and indirect effects via the host responses. However, studies remain scarce and more knowledge is needed. An important issue is to integrate these investigations in the framework of climate change, particularly the observed and expected temperature increase. *Drosophila* parasitoid models, particularly Mediterranean distribution of *L. boulardi*, could be valuable tools, not only to determine how species range vary with increasing temperature, but also to analyze the respective role of phenotypic plasticity and genetic adaptation in thermal response of a population. This requires development of both ecophysiological and genetic studies with the benefit to work on host–parasite models where both partners could exhibit differential responses that might at least temporally destabilize the association.

Promising issues in *Drosophila* parasitoids' ecology and evolution is to understand the role of microparasites, such as virus or endosymbiotic bacteria, which are pervasive in these communities. These endosymbiotic microorganisms that infect both *Drosophila* host and their parasitoids were deliberately excluded from the scope of this analysis. They now need to be integrated as influential partners to better understand variations of host and parasitoid phenotype, evolution of their life histories and the functioning of the whole community (Chapter 12 by Vavre et al.). Results of this last decade revealed a very high rate of *Wolbachia* infection in some *Drosophila* parasitoids species (Vavre et al., 1999, 2002), whereas others such as *L. boulardi* are uninfected by this bacteria but can harbor symbiotic viruses that manipulate the behavior of females (Varaldi et al., 2003). According to the cost of infection demonstrated on *L. heterotoma–Wolbachia* association (Fleury et al., 2000b), this can play a role on relative competitive ability of infected and uninfected species. Interestingly these endosymbionts may also interfere with expression of *Drosophila* resistance and parasitoid virulence and thus on the evolutionary dynamics of the association (Fytrou et al., 2006). Another outstanding effect was demonstrated in *A. tabida* infected by three *Wolbachia* variants among which one is obligatory to complete oogenesis (Dedeine et al., 2001). Clearly, studies on *Drosophila* parasitoid biology and ecology cannot leave out endosymbionts which can question some current knowledge.

Presence of endosymbiotic microorganisms, sometimes as an obligatory partner of a species, indicates that the structure of the *Drosophila* parasitoid community is more complex than expected. More field research is needed to determine how these communities work, what their structure is, which selective pressure dominates and how they vary geographically. The high rate of parasitoid attacks inducing host limitation and occurrence of parasitoid species in the same site at the same time demonstrated that both host–parasitoids (vertical) and competition (horizontal) interactions among species are selective forces that probably act jointly. They vary from small to large scale with other selective forces (abiotic factors) thus resulting in a mosaic pattern of selection. We need more knowledge about how parasitoids respond to selection, and require more studies on geographical variations of parasitoid traits, since among the few investigations available, such differentiations were often observed. This allows the identification of traits that are actually under selection and how these traits evolve according to local conditions. This is an important issue to determine how local adaptation may explain the stability and diversity of the community. In addition, these studies are of interest to understand how life history traits of parasitoids evolve in comparison with those of other insects, and to determine if some special features are associated to the parasitic way of life. The valuable tools that constitute *Drosophila* parasitoids allow us to investigate the genetic determinism and constraints of these traits and we should not miss the possible progress that allows genomic and postgenomic data. Production of genomic data (EST) will increase on all these wasp species with the opportunity to decipher their full genome, which could help to better understand the evolution of *Drosophila* parasitoids.

REFERENCES

Allemand, R., Fleury, F., Lemaître, C., Boulétreau, M., 1999. Dynamique des populations et interactions compétitives chez deux espèces de *Leptopilina*, parasitoïdes de drosophiles, dans la vallée du Rhône (Hymenoptera: Figitidae). Ann. Soc. Entomol. Fr. 35, 97–103.

Allemand, R., Lemaître, C., Frey, F., Boulétreau, M., Vavre, F., Nordlander, G., et al., 2002. Phylogeny of six African *Leptopilina* species (Hymenoptera : Cynipoidea Figitidae) parasitoids of *Drosophila* spp. with descriptions of three new species. Ann. Soc. Entomol. Fr. 38, 319–332.

Atkinson, W.D., Shorrocks, B., 1984. Aggregation of larval Diptera over discrete and ephemeral breeding sites: the implication for coexistence. Am. Nat. 124, 336–351.

Baker, R.H.A., 1979. "Studies on the interactions between *Drosophila* parasites." Oxford University, Oxford.

Bakker, K., Peulet, P., Visser, M.E., 1990. The ability to distinguish between hosts containing different numbers of parasitoid eggs by the solitary parasitoid *Leptopilina heterotoma* (Thomson) (Hym, Cynip). Neth. J. Zool. 40, 514–520.

Bakker, K., van Lente, J.C., Meelis, E., Eijsacke, H.J., 1972. Some models describing distribution of eggs of parasite *Pseudeucoila bochei* (Hym Cynip) over its hosts, larvae of *Drosophila melanogaster*. Oecologia 10, 29.

Barbotin, F., Carton, Y., Kelner-Pillault, S., 1979. Morphologie et biologie de *Cothonaspis boulardi* n. sp. parasite de drosophiles. Bull. Soc. Entomol. Fr. 84, 20–26.

Belokobylskij, S.A., 1998. Tribe Alysiini. In: Lehr, (Ed.), Key to the Insects of Russian Far East Vol. 4. Dal'nauka, Vladivostok, pp. 163–298.

Benassi, V., Frey, F., Carton, Y., 1998. A new specific gene for wasp cellular immune resistance in *Drosophila*. Heredity 80, 347–352.

Bernardo, J., 1996. Maternal effect in animal ecology. Am. Zool. 36, 83–105.

Blackburn, T.M., 1991. Evidence for a fast slow continuum of life-history traits among parasitoid Hymenoptera. Funct. Ecol. 5, 65–74.

Boucek, Z., 1963. A taxonomic study in *Spalangia* Latr. (Hymenoptera, Chalcidoidea). Acta Entomol. Musei. Natl. Prag. 35, 429–512.

Boulétreau, M., 1986. The genetic and coevolutionary interactions between parasitoids and their hosts. In: Waage, Greathead, (Eds.), Insect parasitoids. Academic Press, London, pp. 169–200.

Boulétreau, M., Fouillet, P., 1982. Genetic variability of suitability for a Hymenopteran parasite in a natural population of *Drosophila melanogaster*. Comp. R. Acad. Sci. 295, 775–778.

Boulétreau, M., Wajnberg, E., 1986. Comparative responses of two sympatric parasitoid cynipids to the genetic and epigenetic variations of the larvae of their host, *Drosophila melanogaster*. Entomol. Exp. Appl. 41, 107–114.

Boulétreau, M., Carton, Y., Fouillet, P., Prévost, G., Wajnberg, E., 1987. Interactions génétiques entre les drosophiles (*D. melanogaster*) et leurs parasites Cynipides *Leptopilina boulardi* et *L. heterotoma*. In: Actes du Colloque Biologie des Populations. IASBSE (Eds.) Université Claude-Bernard-Lyon 1, Lyon, pp. 181–186.

Boulétreau, M., Fleury, F., Fouillet, P., 1994. Temperature affects competitive differentially the suitabilities of the two sibling species, *Drosophila melanogaster* and *D. simulans*, to their common larval parasitoid, *Leptopilina boulardi* (Hym.: Cynipidae). Norw. J. Agr. Sci. (Supp. 1), 313–319.

Boulétreau, M., Fouillet, P., Allemand, R., 1991. Parasitoids affect competitive interactions between the sibling species, *Drosophila melanogaster* and *D. simulans*. Redia 74, 171–177.

Boulétreau-Merle, J., Fouillet, P., Terrier, O., 1992. Clinal and seasonal-variations in initial retention capacity of virgin *Drosophila melanogaster* females as a strategy for fitness. Evol. Ecol. 6, 223–242.

Capy, P., Pla, E., David, J.R., 1993. Phenotypic and genetic variability of morphometrical traits in natural populations of *Drosophila melanogaster* and *D. simulans*. I. Geographic variations. Genet. Sel. Evol. 25, 517–536.

Carton, Y., 1978. Search for host habitat by entomophagous insect. Ann. Zool. Ecol. Anim. 10, 353–355.

Carton, Y., Boulétreau, M., 1985. Encapsulation ability of *Drosophila melanogaster*. A genetic analysis. Dev. Comp. Immunol. 9, 211–219.

Carton, Y., Claret, J., 1982. Adaptative significance of a temperature induced diapause in a cosmopolitan parasitoid of *Drosophila*. Ecol. Entomol. 7, 239–247.

Carton, Y., Nappi, A., 1991. The *Drosophila* immune reaction and the parasitoid capacity to evade it: Genetic and coevolutionary aspects. Acta Oecol. 12, 89–104.

Carton, Y., Nappi, A., 2001. Immunogenetic aspects of the cellular immune response of *Drosophila* against parasitoids. Immunogenetics 52, 157–164.

Carton, Y., Nappi, A.J., 1997. *Drosophila* cellular immunity against parasitoids. Parasitol. Today 13, 218–227.

Carton, Y., Boulétreau, M., van Alphen, J.J., van Lenteren, J.C., 1986. The *Drosophila* parasitic wasps. In: Ashburner, Carson, Thompson, (Eds.), The Genetics and Biology of *Drosophila*. Academic Press, London, pp. 347–393.

Carton, Y., Capy, P., Nappi, A.J., 1989. Genetic variability of host parasite relationship traits. Utilization of isofemale lines in a *Drosophila simulans* parasitic wasp. Genet. Sel. Evol. 21, 437–446.

Carton, Y., Chibani, F., Haouas, S., Marrakchi, M., 1987. Eggs-laying strategy under natural conditions of *Leptopilina boulardi*, a Hymenopteran parasitoid of *Drosophila* spp. Entomol. Exp. Appl. 43, 193–201.

Carton, Y., Haouas, S., Marrakchi, M., Hochberg, M., 1991. Two competing parasitoid species coexist in sympatry. Oikos 60, 222–230.

Carton, Y., Rouault, J., Kitano, H., 1981. Susceptibility of the seven sibling species of subgroup *melanogaster* infected with a cynipide parasite. Drosophila Inf. Serv. 56, 24–25.

Chabora, P.C., Smolin, J.S., Kopelman, A.H., 1979. Life-history of *Pseudeucoila* sp (Hymenoptera, Cynipidae) a protelean parasite of *Drosophila* (Diptera, Drosophilidae). Ann. Entomol. Soc. Am. 72, 495–499.

Claret, J., Carton, Y., 1980. Diapause in a tropical species, *Cothonaspis boulardi* (Parasitic Hymenoptera). Oecologia 45, 32–34.

Colinet, D., Dubuffet, A., Cazes, D., Moreau, S.J.M., Drezen, J.M., Poirié, M., 2009. A serpin from the parasitoid wasp *Leptopilina boulardi* targets the *Drosophila* phenoloxidase cascade. Dev. Comp. Immunol. 33, 681–689.

Colinet, D., Schmitz, A., Depoix, D., Crochard, D., Poirié, M., 2007. Convergent use of RhoGAP toxins by eukaryotic parasites and bacterial pathogens. PLoS Path. 3, 2029–2037.

Cossins, A.R., Bowler, K., 1987. Temperature Biology of Animals. Chapman & Hall, London.

Couty, A., Kaiser, L., Huet, D., Pham-Delègue, M.H., 1999. The attractiveness of different odour sources from the fruit–host complex on *Leptopilina boulardi*, a larval parasitoid of frugivorous *Drosophila* spp. Physiol. Entomol. 24, 76–82.

Davis, A.J., Lawton, J.H., Shorrocks, B., Jenkinson, L.S., 1998. Individualistic species responses invalidate simple physiological models of community dynamics under global environmental change. J. Anim. Ecol. 67, 600–612.

De Jong, R., Kaiser, L., 1991. Odor learning by *Leptopilina boulardi*, a specialist parasitoid (Hymenoptera: Eucoilidae). J. Insect Behav. 4, 743–750.

De Jong, R., Kaiser, L., 1992. Odour preference of a parasitic wasp depends on order of learning. Experientia 48, 902–904.

Debout, G., Fauvergue, X., Fleury, F., 2002. The effect of foundress number on sex ratio under partial local mate competition. Ecol. Entomol. 27, 242–246.

Dedeine, F., Vavre, F., Fleury, F., Loppin, B., Hochberg, M.E., Boulétreau, M., 2001. Removing symbiotic *Wolbachia* bacteria specifically inhibits oogenesis in a parasitic wasp. Proc. Natl. Acad. Sci. USA 98, 6247–6252.

Delpuech, J.M., Frey, F., Carton, Y., 1994. Genetic and epigenetic variation in suitability of a *Drosophila* host to three parasitoid species. Can. J. Zool. 72, 1940–1944.

Dicke, M., van Lanteren, J.C., Boskamp, G.J.F., 1984. Chemical stimuli in host habitat location by *Leptopilina heterotoma* (Thomson) (Hymenoptera, Eucoilidae), a parasite of *Drosophila*. J. Chem. Ecol. 10, 695–712.

Dicke, M., van Lanteren, J.C., Boskamp, G.J.F., 1985. Intensification and prolongation of host searching in *Leptopilina heterotoma* (Thomson) (Hymenoptera, Eucoilidae) through a kairomone produced by *Drosophila melanogaster*. J. Chem. Ecol. 11, 125–136.

Driessen, G., Hemerik, L., 1991. Aggregative responses of parasitoids and parasitism in populations of *Drosophila* breeding in fungi. Oikos 61, 96–107.

Driessen, G., Hemerik, L., van Alphen, J.J.M., 1990. *Drosophila* species, breeding in the stinkhorn (*Phallus impidicus* Pers) and their larval parasitoids. Neth. J. Zool. 40, 409–427.

Dubuffet, A., Doury, G., Labrousse, C., Drezen, J., Carton, Y., Poirié, M., 2008. Variation of success of *Leptopilina boulardi* in *Drosophila yakuba*: The mechanisms explored. Dev. Comp. Immunol. 32, 597–602.

Dubuffet, A., Dupas, S., Frey, F., Drezen, J.M., Poirié, M., Carton, Y., 2007. Genetic interactions between the parasitoid wasp *Leptopilina boulardi* and its *Drosophila* hosts. Heredity 98, 21–27.

Dubuffet, A., Rodriguez Alvarez, C.I., Drezen, J.M., van Alphen, J.J.M., Poirié, M., 2006. Do parasitoid preferences for different host species match virulence? Physiol. Entomol. 31, 170–177.

Dupas, S., Boscaro, R., 1999. Geographic variation and evolution of immunosuppressive genes in a *Drosophila* parasitoid. Ecography 22, 284–291.

Dupas, S., Carton, Y., 1999. Two non-linked genes for specific virulence of *Leptopilina boulardi* against *Drosophila melanogaster* and *D. yakuba*. Evol. Ecol. 13, 211–220.

Dupas, S., Frey, F., Carton, Y., 1998. A single parasitoid segregating factor controls immune suppression in *Drosophila*. J. Heredity 89, 306–311.

Eijs, I.E.M., 1999. The Evolution of Resource Utilisation in *Leptopilina* Species. Universiteit Leiden, Leiden [PhD thesis].

Eijs, I.E.M., van Alphen, J.J.M., 1999. Life history correlations: why are hymenopteran parasitoids an exception? Ecol. Lett. 2, 27–35.

Ellers, J., van Alphen, J.J., 2002. A trade-off between diapause duration and fitness in female parasitoids. Ecol. Entomol. 27, 279–284.

Ellers, J., Bax, M., van Alphen, J.J.M., 2001. Seasonal changes in female size and its relation to reproduction in the parasitoid *Asobara tabida*. Oikos 92, 309–314.

Ellers, J., van Alphen, J.J.M., Sevenster, J.G., 1998. A field study of size–fitness relationships in the parasitoid *Asobara tabida*. J. Animal Ecol. 67, 318–324.

Eslin, P., Prévost, G., 1998. Hemocyte load and immune resistance to *Asobara tabida* are correlated in species of the *Drosophila melanogaster* subgroup. J. Insect Physiol. 44, 807–816.

Eslin, P., Giordanengo, P., Fourdrain, Y., Prévost, G., 1996. Avoidance of encapsulation in the absence of VLP by a braconid parasitoid of *Drosophila* larvae: an ultrastructural study. Can. J. Zool. 74, 2193–2198.

Fauvergue, X., Fleury, F., Lemaître, C., Allemand, R., 1999. Parasitoid mating structures when hosts are patchily distributed: field and laboratory experiments with *Leptopilina boulardi* and *L. heterotoma*. Oikos 86, 344–356.

Fellowes, M.D.E., Godfray, H., 2000. The evolutionary ecology of resistance to parasitoids by *Drosophila*. Heredity 84, 1–8.

Fellowes, M.D.E., Kraaijeveld, A.R., Godfray, H., 1999b. Cross-resistance following artificial selection for increased defense against parasitoids in *Drosophila melanogaster*. Evolution 53, 966–972.

Fellowes, M.D.E., Kraaijeveld, A.R., Godfray, H.C.J., 1998. Trade-off associated with selection for increased ability to resist parasitoid attack in *Drosophila melanogaster*. Proc. R. Soc. Lond. B 265, 1553–1558.

Fellowes, M.D.E., Kraaijeveld, A.R., Godfray, H.C.J., 1999a. The relative fitness of *Drosophila melanogaster* (Diptera: Drosophilidae) that have successfully defended themselves against the parasitoid *Asobara tabida* (Hymenoptera: Braconidae). J. Evol. Biol. 12, 123–128.

Fleury, F., Allemand, R., Fouillet, P., Boulétreau, M., 1995. Genetic variation in locomotor activity rhythm among populations of *Leptopilina heterotoma* (Hymenoptera: Eucoilidae), a larval parasitoid of *Drosophila* species. Behav. Genet. 25, 81–89.

Fleury, F., Allemand, R., Vavre, F., Fouillet, P., Boulétreau, M., 2000a. Adaptive significance of a circadian clock: temporal segregation of activities reduces intrinsic competitive inferiority in *Drosophila parasitoids*. Proc. R. Soc. Lond. B 267, 1005–1010.

Fleury, F., Ris, N., Allemand, R., Fouillet, P., Carton, Y., Boulétreau, M., 2004. Ecological and genetic interactions in *Drosophila*–parasitoids communities: a case study with *D. melanogaster*, *D. simulans* and their common *Leptopilina* parasitoids in south-eastern France. Genetica 120, 181–194.

Fleury, F., Vavre, F., Ris, N., Fouillet, P., Boulétreau, M., 2000b. Physiological cost induced by the maternally-transmitted endosymbiont *Wolbachia* in the *Drosophila* parasitoid *Leptopilina heterotoma*. Parasitology 121, 493–500.

Forde, S.E., Thompson, J.N., Bohannan, B.J.M., 2004. Adaptation varies through space and time in a coevolving host–parasitoid interaction. Nature 431, 841–844.

Fytrou, A., Schofield, P.G., Kraaijfeld, A.R., Hubbard, S.F., 2006. *Wolbachia* infection suppresses both host defence and parasitoid counter-defence. Proc. R. Soc. Lond. B 273, 791–796.

Galis, F., van Alphen, J.J.M., 1981. Patch time allocation and search intensity of *Asobara tabida* Nees (Braconidae), a larval parasitoid of *Drosophila*. Neth. J. Zool. 31, 596–611.

Gandon, S., Rivero, A., Varaldi, J., 2006. Superparasitism evolution: Adaptation or manipulation? Am. Nat. 167, E1–E22.

Godfray, H.C.J., 1992. Evolutionary biology: strife among siblings. Nature 360, 213–214.

Godfray, H.C.J., 1994. "Parasitoids: Behavioral and Evolutionary Ecology". Princeton University Press, Princeton, NJ.

Godfray, H.C.J., Shimada, M., 1999. Parasitoids as model organisms for ecologists. Res. Popul. Ecol. 41, 3–10.

Gomez-Gutierrez, J., Peterson, W.T., Morado, J.F., 2006. Discovery of a ciliate parasitoid of euphausiids of Oregon, USA: *Collinia oregonensis* n. sp. (Apostomatida: Colliniidae). Dis. Aquat. Organ. 71, 33–49.

Green, D.M., Kraaijeveld, A.R., Godfray, H.C.J., 2000. Evolutionary interactions between *Drosophila melanogaster* and its parasitoid *Asobara tabida*. Heredity 85, 450–458.

Haccou, P., Devlas, S.J., van Alphen, J.J.M., 1991. Information processing by foragers – effects of intra patch experience on the leaving tendency of *Leptopilina heterotoma*. J. Animal Ecol. 60, 93–106.

Hardy, I.C.W., Godfray, H.C.J., 1990. Estimating the frequency of constrained sex allocation in field populations of Hymenoptera. Behaviour 114, 1–4.
Hassell, M.P., May, R.M., 1973. Stability in insect host–parasite models. J. Animal Ecol. 42, 693–726.
Hassell, M.P., May, R., 1974. Aggregation of predators and insect parasites and its effect on stability. J. Animal Ecol. 43, 567–594.
Hassell, M.P., Waage, J., 1984. Host parasitoid population interactions. Ann. Rev. Entomol. 29, 89–114.
Hedlund, K., Vet, L.E.M., Dicke, M., 1996. Generalist and specialist parasitoid strategies of using odours of adult drosophilid flies when searching for larval hosts. Oikos 77, 390–398.
Hemerik, L., van der Hoeven, N., 2003. Egg distributions of solitary parasitoids revisited *Entomol*. Exp. Appl. 107, 81–86.
Hemerik, L., van der Hoeven, N., van Alphen, J.J.M., 2002. Egg distribution and the information a solitary parasitoid has and uses for its oviposition decisions. Acta Biotheor. 50, 167–188.
Hertlein, M.B., 1986. Seasonal development of *Leptopilina boulardi* (Hymenoptera, Eucoilidae) and its hosts, *Drosophila melanogaster* and *Drosophila simulans* (Diptera, Drosophilidae), in California. Environ. Entomol. 15, 859–866.
Hertlein, M.B., Thorarinsson, K., 1987. Variable patch times and the functional response of *Leptopilina boulardi* (Hymenoptera, Eucoilidae). Environ. Entomol. 16, 593–598.
Hoang, A., 2001. Immune response to parasitism reduces resistance of *Drosophila melanogaster* to desiccation and starvation. Evolution 55, 2353–2358.
Hoang, A., 2002. Physiological consequences of immune response by *Drosophila melanogaster* (Diptera : Drosophilidae) against the parasitoid *Asobara tabida* (Hymenoptera : Braconidae). J. Evol. Biol. 15, 537–543.
Hughes, K., Sokolowski, M.B., 1996. Natural selection in the laboratory for a change in resistance by *Drosophila melanogaster* to the parasitoid wasp *Asobara tabida*. J. Insect Behav. 9, 477–491.
Janssen, A., 1989. Optimal host selection by *Drosophila* parasitoids in the field. Funct. Ecol. 3, 469–479.
Janssen, A., Driessen, G., Haan, M.D., Roodbol, N., 1988. The impact of parasitoids on natural populations of temperate woodland *Drosophila*. Neth. J. Zool. 38, 61–73.
Janssen, A., van Alphen, J.J.M., Sabelis, M.W., Bakker, K., 1995. Odor mediated avoidance of competition in *Drosophila* parasitoids: the ghost of competition. Oikos 73, 356–366.
Jenni, W., 1951. Beitrag zur Morphologie und Biologie der Cynipide *Pseudeucoila bochei* Weld, eines Larvenparasiten von *Drosophila melanogaster* Meig. Acta. Zool. 32, 177–254.
Jessup, C.A., Forde, S.E., 2008. Ecology and evolution in microbial systems: the generation and maintenance of diversity in phage–host interactions. Res. Microbiol. 159, 382–389.
Joshi, A., Thompson, J., 1996. Evolution of broad and specific competitive ability in novel versus familiar environments in *Drosophila* species. Evolution 50, 188–194.
Kaiser, L., Perez-Maluf, R., Sandoz, J.C., Pham-Delègue, M.H., 2003. Dynamics of odour learning in *Leptopilina boulardi*, a hymenopterous parasitoid. Animal Behav. 66, 1077–1084.
Karan, D., Dahiya, N., Munjal, A.K., Gibert, P., Moreteau, B., Parkash, R., et al., 1998. Desiccation and starvation tolerance of adult *Drosophila*: opposite latitudinal clines in natural populations of three different species. Evolution 52, 825–831.
Kaya, H.K., Gaugler, R., 1993. Entomopathohenic nematodes. Ann. Rev. Entomol. 38, 181–206.
Kraaijeveld, A.R., Godfray, H.C.J., 1997. Trade-off between parasitoid resistance and larval competitive ability in *Drosophila melanogaster*. Nature 389, 278–280.
Kraaijeveld, A.R., Godfray, H.C.J., 1999. Geographic patterns in the evolution of resistance and virulence in *Drosophila* and its parasitoids. Am. Nat. 153, S61–S74.

Kraaijeveld, A.R., van Alphen, J.J.M., 1993. Successful invasion of North America by two palearctic *Drosophila* species (Diptera: Drosophilidae): a matter of immunity against local parasitoids? Neth. J. Zool. 43, 235–241.

Kraaijeveld, A.R., van Alphen, J.J.M., 1994. Geographical variation in resistance of the parasitoid *Asobara tabida* against encapsulation by *Drosophila melanogaster* larvae. The mechanism explored. Physiol. Entomol. 19, 9–14.

Kraaijeveld, A.R., van Alphen, J.J.M., 1995a. Geographical variation in encapsulation ability of *Drosophila melanogaster* larvae and evidence for parasitoid specific components. Evol. Ecol. 9, 10–17.

Kraaijeveld, A.R., van Alphen, J.J.M., 1995b. Variation in diapause and sex ratio in the parasitoid *Asobara tabida*. Entomol. Exp. Appl. 74, 259–265.

Kraaijeveld, A.R., van der Wel, N., 1994. Geographic variation in reproductive success of the parasitoid *Asobara tabida* in larvae of several *Drosophila* species. Ecol. Entomol. 19, 221–229.

Kraaijeveld, A.R., Adriaanse, I.C.T., van den Bergh, B., 1999. Parasitoid size as a function of host sex: potential for different sex allocation strategies. Entomol. Exp. Appl. 92, 289–294.

Kraaijeveld, A.R., Ferrari, J., Godfray, H.C.J., 2002. Costs of resistance in insect–parasite and insect–parasitoid interactions. Parasitology 125, S71–S82.

Kraaijeveld, A.R., Limentani, E.C., Godfray, H.C.J., 2001. Basis of the trade-off between parasitoid resistance and larval competitive ability in *Drosophila melanogaster*. Proc. R. Soc. Lond. B 268, 259–261.

Kraaijeveld, A.R., Nowee, B., Najem, R.W., 1995. Adaptive variation in host-selection behavior of *Asobara tabida*, a parasitoid of *Drosophila* larvae. Funct. Ecol. 9, 113–118.

Kraaijeveld, A.R., van Alphen, J.J.M., Godfray, H.C.J., 1998. The coevolution of host resistance and parasitoid virulence. Parasitology 116, S29–S45.

Labrosse, C., Carton, Y., Dubuffet, A., Drezen, J., Poirié, M., 2003. Active suppression of *D. melanogaster* immune response by long gland products of the parasitic wasp *Leptopilina boulardi*. J. Insect Physiol. 49, 513–522.

Leather, S.R., Walters, K.F.A., Bale, J.S., 1993. The Ecology of Insect Overwintering,. Cambridge University Press, Cambridge.

Lett, C., Auger, P., Bravo De La Parra, R., 2003. Migration frequency and the persistence of host–parasitoid interactions. J. Theoret. Biol. 221, 639–654.

Melk, J.P., Govind, S., 1999. Developmental analysis of *Ganaspis xanthopoda*, a larval parasitoid of *Drosophila melanogaster*. J. Exp. Biol. 202, 1885–1896.

Mitrovski, P., Hoffmann, A.A., 2001. Postponed reproduction as an adaptation to winter conditions in *Drosophila melanogaster*: evidence for clinal variation under semi-natural conditions. Proc. R. Soc. Lond. B 268, 2163–2168.

Mitsui, H., van Achterberg, K., Nordlander, G., Kimura, M.T., 2007. Geographical distributions and host associations of larval parasitoids of frugivorous Drosophilidae in Japan. J. Nat. History 41, 1731–1738.

Mollema, C., 1991. Heritability of host selection behavior by the *Drosophila* parasitoid *Asobara tabida*. Neth. J. Zool. 41, 174–183.

Nappi, A., Streams, F., 1970. Abortive development of cynipid parasite *Pseudeucoila bochei* (Hymenoptera) in species of *Drosophila melanica* group. Ann. Entomol. Soc. Am. 63, 321–327.

Nappi, A.J., Carton, Y., Frey, F., 1991. Parasite induced enhancement of hemolymph tyrosinase activity in a selected immune reactive strain of *Drosophila melanogaster*. Arch. Insect Biochem. Physiol. 18, 159–168.

Nappi, A.J., Vass, E., Carton, Y., 1992. Identification of 3,4-dihydroxyphenylalanine, 5-6-dihydroxyinole and *N*-acetylarterenone during eumelanin formation in immune reactive larvae of *Drosophila melanogaster*. Arch. Insect Biochem. Physiol. 20, 181–191.

Nordlander, G., 1980. Revision of the genus *Leptopilina* Förster, 1869, with notes on the status of some other genera (Hymenoptera, Cynipoidea, Eucoilidae). Entomol. Scand. 11, 428–453.

Nöstvik, E., 1954. A study of *Pseudeucoila bochei* Weld (Hymnoptera: Eucoilidae) and its relationship to *Drosophila melanogaster* Meig. (Diptera: Drosophilidae). Genet. Entomol. 2, 139–160.
Orr, H.A., Irving, S., 1997. The genetics of adaptation: The genetic basis of resistance to wasp parasitism in *Drosophila melanogaster*. Evolution 51, 1877–1885.
Pacala, S.W., Hassell, M.P., May, R.M., 1990. Host parasitoid associations in patchly environments. Nature 344, 150–153.
Pannebakker, B.A., Garrido, N.R.T., Zwaan, B.J., van Alphen, J.J.M., 2008. Geographic variation in host-selection behaviour in the *Drosophila* parasitoid *Leptopilina clavipes*. Entomol. Exp. Appl. 127, 48–54.
Pannebakker, B.A., Loppin, B., Elemans, C., Humblot, L., Vavre, F., 2007. Parasitic inhibition of cell death facilitates symbiosis. Proc. Natl Acad. Sci. USA 104, 213–215.
Papaj, D.R., Vet, L.E.M., 1990. Odor learning and foraging success in the parasitoid, *Leptopilina heterotoma*. J. Chem. Ecol. 16, 3137–3150.
Parsons, P.A., 1977. Cosmopolitan, exotic and endemic *Drosophila:* Their comparative evolutionary biology especially in southern Australia. In: Anderson, (Ed.), Exotic Species in Australia. Their Establishment and Success. Proc. Ecol. Soc. Austr. 10, 62–75.
Patot, S., Lepetit, D., Charif, D., Varaldi, J., Fleury, F., 2009. Molecular detection, penetrance, and transmission of an inherited virus responsible for behavioral manipulation of an insect parasitoid. Appl. Environ. Microbiol. 75, 703–710.
Poirié, M., Frey, F., Hita, M., Huguet, E., Lemeunier, F., Périquet, G., et al., 2000. *Drosophila* resistance genes to parasitoids: chromosomal location and linkage analysis. Proc. R. Soc. Lond. B 267, 1417–1421.
Poolman, S.M.T.T., Suverkropp, B.P., Vet, L.E.M., de Moed, G., 1992. Comparison of learning in related generalist and specialist eucoilid parasitoids. Entomol. Exp. Appl. 64, 117–124.
Precht, H., Christophersen, J., Hensel, H., Larcher, W., 1973. Temperature and Life. Springer Verlag, Berlin.
Prévost, G., Eslin, P., Doury, G., Moreau, S., Guillot, S., 2005. *Asobara*, braconid parasitoids of *Drosophila* larvae: unusual strategies to avoid encapsulation without VLPs. J. Insect Physiol. 51, 171–179.
Reuter, O.M., 1913. Lebens gewohnheiten und Instinkte der Insekten. Friedlander, Berlin.
Ris, N., Allemand, R., Fouillet, P., Fleury, F., 2004. The joint effect of temperature and host species induce complex genotype-by-environment interactions in the larval parasitoid of *Drosophila, Leptopilina heterotoma* (Hymenoptera: Figitidae). Oikos 106, 451–456.
Roff, D.A., 1992. The Evolution of Life Histories. Theory and Analysis. Chapman & Hall, London.
Rohlfs, M., Hoffmeister, T.S., 2004. Spatial aggregation across ephemeral resource patches in insect communities: an adaptive response to natural enemies? Oecologia 140, 654–661.
Rolff, J., Kraaijeveld, A.R., 2003. Selection for parasitoid resistance alters mating success in *Drosophila*. Proc. R. Soc. Lond. B 270, S154–S155.
Rosenheim, J.A., 1998. Higher-order predators and the regulation of insect herbivore populations. Ann. Rev. Entomol. 43, 421–447.
Rossiter, M.C., 1996. Incidence and consequences of inherited environmental effects. Ann. Rev. Ecol. Syst. 27, 451–476.
Schilthuizen, M., Nordlander, G., Stouthamer, R., van Alphen, J.J.M., 1998. Morphological and molecular phylogenetics in the genus *Leptopilina* (Hymenoptera: Cynipoidea: Eucoilidae). Syst. Entomol. 23, 253–264.
Schlenke, T.A., Morales, J., Govind, S., Clark, A.G., 2007. Contrasting infection strategies in generalist and specialist wasp parasitoids of *Drosophila melanogaster*. PLoS Path. 3, 1486–1501.
Sevenster, J.A., van Alphen, J.J.M., 1996. Aggregation and coexistence. II A neotropical *Drosophila* community. J. Animal Ecol. 65, 308–324.

Sokolowski, M.B., Turlings, T.C.J., 1987. *Drosophila* parasitoid–host interactions: vibrotaxis and ovipositor searching from the host's perspective. Can. J. Zool. 65, 461–464.
Stearns, S.C., 1992. The Evolution of Life Histories. Oxford University Press, Oxford.
Streams, F.A., 1968. Factors affecting susceptibility of *Pseudeucoila bochei* eggs to encapsulations by *Drosophila melanogaster*. J. Invert. Path. 12, 379.
Toda, M.J., Kimura, M.T., Tuno, N., 1999. Coexistence mechanisms of mycophagous drosophilids on multispecies fungal hosts: aggregation and resource partitioning. J. Animal Ecol. 68, 794–803.
Turlings, T.C.J., van Batenburg, F.D.H., van Strien-van Liempt, W.T.F.H., 1985. Why is there no interspecific host discrimination in the two coexisting larval parasitoids of *Drosophila* species *Leptopilina heterotoma* (Thomson) and *Asobara tabida* (Nees)? Oecologia 67, 352–359.
van Alphen, J., Nell, H.W., 1982. Superparasitism and host discrimination by *Asobara tabida* Nees (Braconidae: Alyssinae) a larval parasitoid of Drosophilidae. Neth. J. Zool. 32, 232–260.
van Alphen, J.J.M., Drijver, R.A., 1982. Host selection by *Asobara tabida* Nees (Braconidae, Alysiinae), a larval parasitoid of fruit inhabiting *Drosophila* species. Neth. J. Zool. 32, 194–214.
van Alphen, J.J.M., Galis, F., 1983. Patch time allocation and parasitization efficiency of *Asobara tabida*, a larval parasitoid of *Drosophila*. J. Animal Ecol. 52, 937–952.
van Alphen, J., Visser, M.E., 1990. Superparasitism as an adaptive strategy for insect parasitoids. Ann. Rev. Entomol. 35, 59.
van Alphen, J.J.M., Bernstein, C., Driesen, G., 2003. Information acquisition and patch time allocation in insects parasitoids. Trends Ecol. Evol. 18, 81–87.
van Alphen, J.J.M., Nordlander, G., Eijs, I.E.M., 1991. Host habitat finding and host selection of the *Drosophila* parasitoid *Leptopilina australis* (Hymenoptera, Eucoilidae), with a comparison of the niches of European *Leptopilina* species. Oecologia 87, 324–329.
van Dijken, M.J., van Alphen, J.J.M., 1998. The ecological significance of differences in host detection behaviour in coexisting parasitoid species. Ecol. Entomol. 23, 265–270.
van Lenteren, J.C., 1976. The development of host discrimination and the prevention of superparasitism in the parasite *Pseudeucoila bochei* Weld (Hym.: Cynipidae). Neth. J. Zool. 26, 1–83.
van Lenteren, J.C., 1981. Host discrimination by parasitoids. In: Nordlund, Jones, Lewes, (Eds.), Semiochemicals: Their Role in Pest Control. Wiley, New York, pp. 153–179.
van Lenteren, J.C., Bakker, K., 1978. Behavioral aspects of functional responses of a parasite (*Pseudeucoila bochei* Weld) to its host (*Drosophila melanogaster*). Neth. J. Zool. 28, 213–223.
van Strien-van Liempt, W.T.F.H., van Alphen, J.J.M., 1981. The absence of interspecific host discrimination in *Asobara tabida* Nees and *Leptopilina heterotoma* (Thomson), coexisting larval parasitoids of *Drosophila* species. Neth. J. Zool. 31, 701–712.
Varaldi, J., Fouillet, P., Boulétreau, M., Fleury, F., 2005. Superparasitism acceptance and patch-leaving mechanisms in parasitoids: a comparison between two sympatric wasps. Animal Behav. 69, 1227–1234.
Varaldi, J., Fouillet, P., Ravallec, M., Lopez-Ferber, M., Boulétreau, M., Fleury, F., 2003. Infectious behavior in a parasitoid. Science 302, 1930.
Varaldi, J., Petit, S., Boulétreau, M., Fleury, F., 2006. The virus infecting the parasitoid *Leptopilina boulardi* exerts a specific action on superparasitism behaviour. Parasitology 132, 747–756.
Vass, E., Nappi, A.J., Carton, Y., 1993. Alterations in the activities of tyrosinase, N-acetyltransferase and tyrosine aminotransferase in immune reactive larvae of *Drosophila melanogaster*. Dev. Comp. Immunol. 17, 109–118.
Vavre, F., Fleury, F., Lepetit, D., Fouillet, P., Boulétreau, M., 1999. Phylogenetic evidence for horizontal transfers of *Wolbachia* in host–parasitoid associations. Mol. Biol. Evol. 16, 1711–1723.

Vavre, F., Fleury, F., Varaldi, J., Fouillet, P., Boulétreau, M., 2002. Infection polymorphism and cytoplasmic incompatibility in Hymenoptera–*Wolbachia* associations. Heredity 88, 361–365.

Vet, L.E.M., 1985a. Olfactory microhabitat location in some eucoilid and alysiine species (Hymenoptera) larval parasitoids of Diptera. Neth. J. Zool. 35, 720–730.

Vet, L.E.M., 1985b. Response to kairomones by some alysiine and eucoilid parasitoid species (Hymenoptera). Neth. J. Zool. 35, 486–496.

Vet, L.E.M., Bakker, K., 1985. A comparative functional approach to the host detection behavior of parasitic wasps. 2. Quantitative study on eight eucoilid species. Oikos 44, 487–498.

Vet, L.E.M., van Alphen, J.J.M., 1985. A comparative functional approach to test the host detection behavior of parasitic wasps 1. A qualitative study on Eucoilidae and Alysiinae. Oikos 44, 478–486.

Vet, L.E.M., van der Hoeven, R., 1984. Comparison of the behavioral response of two *Leptopilina* species (Hymenoptera, Eucoilidae), living in different microhabitats, to kairomone of their host (Drosophilidae). Neth. J. Zool. 34, 220–227.

Vet, L.E.M., van Opzeeland, K., 1984. The influence of conditioning on olfactory microhabitat and host location in *Asobara tabida* (Nees) and *A. rufescens* (Foerster) (Braconidae, Alysiinae) larval parasitoids of Drosophilidae. Oecologia 63, 171–177.

Vet, L.E.M., van Opzeeland, K., 1985. Olfactory microhabitat selection in *Leptopilina heterotoma* (Thomson) (Hym, Eucoilidae), a parasitoid of Drosophilidae. Neth. J. Zool. 35, 497–504.

Vet, L.E.M., de Jong, A.G., Franchi, E., Papaj, D.R., 1998. The effect of complete versus incomplete information on odour discrimination in a parasitic wasp. Animal Behav. 55, 1271–1279.

Vet, L.E.M., Meyer, M., Bakker, K., van Alphen, J.J.M., 1984. Intraspecific and interspecific host discrimination in *Asobara* (Hymenoptera) larval endo-parasitoids of Drosophilidae. Comparison between closely related and less closely related species. Animal Behav. 32, 871–874.

Vet, L.E.M., Sokolowski, M.B., MacDonald, D.E., 1993. Responses of a generalist and a specialist parasitoid (Hymenoptera, Eucoilidae) to Drosophilid larval kairomones. J. Insect Behav. 6, 615–624.

Visser, M.E., 1994. The importance of being large: the relationship between size and fitness in females of the parasitoid *Aphaereta minuta* (Hymenoptera, Braconidae). J. Animal Ecol. 63, 963–978.

Visser, M.E., 1995. The effect of competition on oviposition decisions of *Leptopilina heterotoma* (Hymenoptera: Eucoilidae). Animal Behav. 49, 1677–1687.

Visser, M.E., Luyckx, B., Nell, H.W., Boskamp, G.J.F., 1992a. Adaptive superparasitism in solitary parasitoids: marking of parasitized hosts in relation to the pay-off from superparasitism. Ecol. Entomol. 17, 76–82.

Visser, M.E., van Alphen, J.J.M., Hemerik, L., 1992b. Adaptive superparasitism and patch time allocation in solitary parasitoids: an ESS model. J. Animal Ecol. 61, 93–101.

Visser, M.E., van Alphen, J.J.M., Nell, H.W., 1990. Adaptive superparasitism and patch time allocation in solitary parasitoids: the influence of the number of parasitoids depleting a patch. Behaviour 114, 1–4.

Visser, M.E., van Alphen, J.J.M., Nell, H.W., 1992c. Adaptive superparasitism and patch time allocation in solitary parasitoids: the influence of prepatch experience. Behav. Ecol. Sociobiol. 31, 163–171.

Wajnberg, E., Boulétreau, M., Prévost, G., Fouillet, P., 1990. Developmental relationships between *Drosophila* larvae and their endoparasitoid *Leptopilina* (Hymenoptera, Cynipidae) as affected by crowding. Arch. Insect Biochem. Physiol. 13, 239–245.

Wajnberg, E., Prévost, G., Boulétreau, M., 1985. Genetic and epigenetic variation in *Drosophila* larvae suitability to a hymenopterous endoparasitoid. Entomophaga 30, 187–191.

Walker, I., 1962. *Drosophila* und *Pseudeucoila*. III. Selektionsversuche zur Steigerung der Resistenz des Parasites gegen die Abwehrreaktion des Wirtes. Rev. Suisse Zool. 69, 209–227.
Wertheim, B., Allemand, R., Vet, L.E.M., Dicke, M., 2006. Effects of aggregation pheromone on individual behaviour and food web interactions: a field study on *Drosophila*. Ecol. Entomol. 31, 216–226.
Wertheim, B., Sevenster, J.G., Eijs, I.E.M., van Alphen, J.J.M., 2000. Species diversity in a mycophagous insect community: the case of spatial aggregation vs. resource partitioning. J. Animal Ecol. 69, 335–351.
Wertheim, B., Vet, L.E.M., Dicke, M., 2003. Increased risk of parasitism as ecological costs of using aggregation pheromones: laboratory and field study of *Drosophila–Leptopilina* interaction. Oikos 100, 269–282.
Wiskerke, J.S.C., Dicke, M., Vet, L.E.M., 1993. Larval parasitoid uses aggregation pheromone of adult hosts in foraging behavior a solution to the reliability detectability problem. Oecologia 93, 145–148.

CHAPTER 2

Decision-Making Dynamics in Parasitoids of *Drosophila*

Andra Thiel and **Thomas S. Hoffmeister**

Contents

Abstract

Drosophilids and their associated parasitoids live in environments that vary in resource availability and quality within and between generations. The use of information to adapt behavior to the current environment is a key feature under such circumstances and *Drosophila* parasitic wasps are excellent model systems to study learning and information use. They are among the few parasitoid model species that have been tested in a wide array of situations. Moreover, several related species have been tested under similar conditions, allowing the analysis of within and between species variability, the effect of natural selection in a typical environment, the current physiological status, and previous experience of the individual. This holds for host habitat and host

Institute of Ecology, University of Bremen, FB 02, D-28359 Bremen, Germany

Advances in Parasitology, Volume 70
ISSN 0065-308X, DOI: 10.1016/S0065-308X(09)70002-8

location as well as for host choice and search time allocation. Here, we review patterns of learning and memory, of information use and updating mechanisms, and we point out that information use itself is under strong selective pressure and thus, optimized by parasitic wasps.

2.1. INTRODUCTION

The intricate relationship between foraging behavior and lifetime reproductive success will select behavioral traits that are well adapted to the environmental conditions a forager experiences in its life (Stephens and Krebs, 1986). Since environments are subject to change, fixed strategies are often suboptimal and animals can be expected to gather information that aids in reducing uncertainty throughout their lives and use this information in the foraging process (Dall et al., 2005). The link between foraging success and fitness payoff is particularly close in parasitoids where the number of hosts a female is able to parasitize during her lifetime is linked directly to the number of offspring she is likely to have. Thus, a strong selection pressure on optimizing foraging strategies can be expected in these organisms (van Alphen and Vet, 1986). Accordingly, parasitoids use several cues from their environment in patch-leaving decisions, host acceptance and the like (van Alphen and Bernstein, 2008; Wajnberg, 2006) and have therefore provided very valuable systems for studying the evolution of learning and information use (e.g., Godfray and Waage, 1988; Liu et al., 2009; Smid et al., 2007; Steidle and van Loon, 2003; Thiel et al., 2006). Those experiments have shown that the type of cue used and its effects on the expressed behavior not only differ between parasitoid species, but may even change within a wasp's lifetime due to her experience and physiological state. *Drosophila* parasitoids have been studied thoroughly in the laboratory and provide an excellent basis to focus on the dynamics of decision-making processes, on the adaptive value of behavioral plasticity and on how flexibility is achieved within the framework of a parasitoid's cognitive abilities.

2.2. LEVELS OF PLASTICITY

Behavioral plasticity can be achieved on different time scales. The long-term scale often acts on (genetically) fixed behavioral responses to certain cues, which might even act in the differentiation and niche segregation of species (Vet et al., 1984). Different environmental conditions may drive behavioral differences. This can be seen, for example, in the host acceptance behavior of *Drosophila* parasitoids. As outlined in Chapter 10 by Kraaijeveld and

Godfray, populations of *D. melanogaster* vary in the ability to encapsulate, and therefore, kill, a parasitoid's egg. Females of the parasitoid *Asobara tabida* (Hymenoptera; Braconidae) from different geographical areas have been shown to vary in host acceptance behavior accordingly (Rolff and Kraaijeveld, 2001). Resources used by the fly larvae differ between, for example, Northern and Southern populations in Europe and substrate preferences of *Leptopilina clavipes* (Hymenoptera; Figitidae) females mirror this pattern (Pannebakker et al., 2008). In a similar vein, two sibling species of *Asobara* occur sympatrically, but each species demonstrates olfactory preferences in choice tests that correlate with their optimal microhabitat (Vet et al., 1984).

Short-term behavioral differences, in contrast, depend on the history of the individual, reflecting its past experiences and its current physiological state. Learning of host-associated odors, as is outlined in detail in Chapter 3 by Kaiser et al., is probably the best-studied phenomenon of this kind. Females of *L. boulardi*, for example, are able to learn several host-associated odors but prefer the last one learned and thus the one that gave them the most recent positive reinforcement (De Jong and Kaiser, 1992; Kaiser and De Jong, 1993). How strongly an *L. boulardi* female responds to the learned odor is also influenced by her current state, for example, her mating status (Perez-Maluf and Kaiser, 1998). Short-term behavioral plasticity, however, is not limited to situations of conditioned responses to odor cues. Information processing in the context of patch time allocation or host acceptance decisions, which involves the integration of several external as well as internal cues into the response pattern, also often varies within a parasitoid's lifetime. This kind of short-term behavioral plasticity will be covered in the following.

Both level, short-term and long-term behavioral plasticity, have certain advantages and disadvantages to their bearers. While an innate response to a certain host-associated odor allows correct responses right after hatching, it might hinder host-switching behavior if the preferred species becomes less abundant. However, short-term behavioral plasticity is expensive, since there are most likely time costs until the right cues are learned (Eliassen et al., 2007; Raine and Chittka, 2008; Vet et al., 1995), and physiological costs for processing the information and maintaining the cognitive system (Kolss and Kawecki, 2008; Mery and Kawecki, 2003, 2004). It is, therefore, assumed that learning ability evolves, and is maintained, for cues that show intermediate levels of variability (Stephens, 1989): learning ability is most advantageous when cues vary between generations but are more or less stable within the lifetime of an individual. If changes occur too fast, that it, the predictive potential of a cue changes soon after it was learned, this cue does not reduce the animal's uncertainty about the future and should, therefore, not be learned (Dall et al., 2005; Dukas, 2008; Stephens, 1993). If changes occur hardly at all, wasps would fare better with hard-wired behavioral mechanisms, that is, inherited preferences.

2.3. RELATIVE VALUE OF HOSTS AND PATCHES

Larvae of *Drosophila* usually hide within the substrate and their parasitoids have evolved various searching modes, like ovipositor probing or vibrotactic search, to detect them on various substrates (Vet and Bakker, 1985; Vet and van Alphen, 1985; Vet and van der Hoeven, 1984). After detection of the host, its quality has to be assessed, which usually takes place by ovipositor probing (e.g., van Lenteren, 1972). Larval parasitoids seem to prefer second instar *Drosophila* larvae; this preference mainly comes from physical constraints: first instars are so small that they are difficult to hit correctly, and third instars usually have a thick cuticle that is difficult to penetrate (van Alphen and Drijver, 1982). There is also a strong correlation between host species preference and survival probability of the parasitoid eggs in field studies (Janssen, 1989) as well as in laboratory settings (van Alphen and Janssen, 1982). The acceptance of different host species might be influenced by, for example, the thickness of the cuticle, however, in addition, there is often a true decision underlying the acceptance or rejection, since the response can be rather plastic. The host species *D. subobscura* Collin is a host of high quality, as it lacks an effective immune response against parasitoid eggs and more than 80% of all wasp eggs complete their development successfully. In contrast, *D. immigrans* Sturtevant is of low value because of its effective immune response. A third species, *D. melanogaster* Meigen, shows some encapsulation ability but is generally quite suitable as a host. While the rank order in which these three species are preferred by the wasps seems to be a fixed response found in several populations, the acceptance threshold for less preferred species often varies with experience (Mollema, 1991). For example, *A. tabida* females reared on *D. subobscura* are less likely to reject *D. melanogaster* larvae when they have experienced only *D. melanogaster* larvae the day before (van Alphen and van Harsel, 1982). This shows that information on host species presence (or absence) is remembered for at least 24 h in *A. tabida*. In a more recent study, we showed that experience with different host species also influences patch time allocation in *A. tabida* (Thiel and Hoffmeister, 2006): females were confined for 30 min to an experimental yeast patch where they experienced either 10 larvae of *D. melanogaster*, *D. subobscura*, or *D. immigrans*. When subsequently searching a patch with 10 *D. melanogaster* larvae, females that had previously parasitized *D. immigrans* stayed significantly longer than females that had experienced *D. melanogaster* or *D. subobscura* before. Apparently, wasps that previously experienced a poor-quality host estimated the value of the current patch higher. If two host species are present in a patch at the same time, *L. boulardi* increasingly rejects the less preferred species with repeated encounters of the preferred host (Dubuffet et al., 2006).

Another important factor influencing host quality is the parasitization status. Most *Drosophila* parasitoids are solitary species, that is, only one adult parasitoid can develop per host larva. Already parasitized hosts are, therefore, of lower quality and as age difference increases between the first and second parasitoid's offspring inside the host the death rate of the newcomer increases (e.g., Bakker et al., 1985; Visser, 1993). If a host is accepted for oviposition, it is marked in most species of *Drosophila* parasitoids with an internal marking substance that enables the parasitoid female to recognize it upon re-encounter as already being parasitized, and to distinguish hosts parasitized by herself from hosts parasitized by competing females (e.g., van Alphen and Nell, 1982; Visser, 1993). Already parasitized hosts can either be rejected when the mark is detected, or be parasitized for a second time (so called superparasitism). While in the early years of optimal foraging theory superparasitism was often attributed to a female's inability to discriminate, subsequent work clearly showed the adaptive nature of superparasitism, for example, when unparasitized hosts are rare (van Alphen and Visser, 1990). It therefore comes as no surprise that the acceptance threshold for already parasitized hosts depends on previous experience of a wasp: while the first host encountered by an *L. heterotoma* female is almost always parasitized independently of its status (van Lenteren and Bakker, 1975), a female wasp is more likely to reject an already parasitized host, if she has parasitized several hosts in that patch before (Henneman et al., 1995). Thus, superparasitism in *L. heterotoma* has been interpreted as a response to the host availability experienced by the wasp. In agreement with this explanation, *L. heterotoma* is also more likely to reject parasitized hosts if during a prepatch experience, she encountered healthy hosts instead of parasitized ones (Visser et al., 1992). Interestingly, this wasp more readily accepts parasitized hosts if she has been kept with conspecifics before visiting patches (Visser et al., 1992). This seems to indicate that she associates the presence of other females with increased competition for hosts and a reduced probability of encountering healthy hosts throughout her later life. Increased superparasitism is also likely to occur if females search simultaneously on the same patch, mainly because the female leaving the patch first would leave the hosts she has already parasitized prone to superparasitism by the other female (e.g., Haccou and van Alphen, 2008). This would result in reduced offspring survival, unless the female compensates for it by also starting to superparasitize. Behavior in line with these theoretical predictions has been shown to occur in *L. heterotoma* (Visser, 1995; Visser et al., 1990) as well as in *A. tabida* (van Alphen et al., 1992). The response seems to be triggered by odors emitted by the females as well as by direct physical contact. Additionally, the tendency to superparasitize depends on the number of patches available in the habitat, since the parasitoids stay and superparasitize only if the number of available patches in the habitat is low, but leave in search for a better patch otherwise (Visser et al., 1992).

2.4. HOST PATCH DETECTION

Host patches are difficult to detect in nature and can also vary in quality. Parasitoids have been found to use various cues to assess the quality of the current patch, from a distance and also upon arrival, such that they direct their search effort only to the most promising areas in their habitat and, additionally, neither stay too long on already depleted patches nor spend too much time traveling between patches (Godfray, 1994). In general, host-associated odors play a central role in guiding the parasitoids to their hosts (Papaj, 1993).

One major problem with using host cues for direction, however, is that the most reliable cues are often the least detectable ones, and vice versa, because hosts are under selection to be as inconspicuous as possible (Gould, 1993; Stephens, 1989; Vet et al., 1991). This is reflected in the searching behavior of parasitoids. For example, *L. heterotoma* females are able to detect (innately) attractive substrate odors from a distance to locate patches. Their behavioral responses toward a natural odor source strongly increased for at least 24 h after they had experienced the cue to be profitable (Vet and Schoonman, 1988), indicating that a learned enhancement of the response to odor took place and not only general sensitization. The learned preference can also be reversed by unrewarding experience, that is, the absence of hosts (Papaj et al., 1994). Increased searching efficiency after previous experience with either mushroom or apple odor was even detectable under field conditions in *L. heterotoma* (Papaj and Vet, 1990). The improved ability to detect host patches resulted from wasps walking faster and straighter, making narrower turns and spending more time walking altogether when encountering an odor plume that had been rewarding before, while naive wasps were more or less unaffected when encountering the same odor (Vet and Papaj, 1992). It even appeared that *L. heterotoma* females are parsimonious learners: when they learn apple odor to be associated with a host-containing patch, they do not discriminate between different apple varieties at first. If, however, one of the varieties turns out not to be associated with host presence, the females can clearly discriminate and prefer the profitable apple variety (Vet et al., 1998).

There is obviously a hierarchy involved in which cues to use and females generally use the cue that provides the best cost–benefit ratio (Vet et al., 1990). As soon as a more reliable cue becomes accessible at low additional cost, the previous cue loses influence immediately. For example, the aggregation pheromone of *Drosophila* females is a cue that quite reliably indicates the (former) presence of *Drosophila* females on a patch. If the aggregation pheromone is available together with the general blend of host patch odors, this is highly preferred by the searching parasitoids (Hedlund et al., 1996; Wiskerke et al., 1993). However, this preference is only visible if enough yeast is available on the patches to allow the growth of *Drosophila* larvae (Wertheim et al., 2003). Upon entering a patch, the presence of aggregation pheromone loses significance, if *Drosophila* females have been walking on the patch, leaving their "footprints" (Wertheim et al., 2003), which are probably an even more reliable cue for host presence in the patch.

2.5. PREPATCH EXPERIENCE AND (INITIAL) LEAVING TENDENCY

The simplest situation of patch-leaving studies is a parasitoid searching a patch that does not actually contain any hosts. Even though wasps usually stay longer in empty patches when these contain more kairomone (host traces, e.g., Dicke et al., 1985; Galis and van Alphen, 1981), the willingness to search an empty patch should decrease with increasing estimated probability to find a better patch somewhere else soon. However, this valuable measure of parasitoid search-motivation has rarely been used in this context (but see Thiel et al. (2006) for studies on the ichneumonid *Venturia canescens*). For *Drosophila* parasitoids, we are aware only of two studies: (1) When a female of *L. heterotoma* had been exposed to a sublethal dose of an insecticide, she is subsequently more reluctant to leave a kairomone-containing patch (Delpuech et al., 2005). (2) *A. tabida* females searching a patch containing one or 12 hosts with the corresponding amount of host odor before searching on an empty yeast patch without any kairomone did not show any effect of previous patch quality on their leaving tendency on the empty patch (Thiel, 2004). This lack in responsiveness to previous host encounters seems to be due to a generally weak response of *A. tabida* to previous host density variation, even when tested on host-containing patches (Thiel and Hoffmeister, 2006), thus supporting the idea that behavior on empty patches can nicely mirror changes in parasitoid search motivation.

The effects of certain events on parasitoid search motivation can also be studied on host-containing patches if the effects of on-patch host encounters can be disentangled from a female's prepatch experience. It is, for example, generally considered adaptive if parasitic wasps adjust patch time allocation to patch availability in the habitat, that is, if they stay longer and try to parasitize even the last host that may be hiding somewhere, when host-containing patches are rare (reviewed by van Alphen et al., 2003; Wajnberg, 2006). The distribution of *Drosophila* larvae across host-containing patches and also the distribution of patches within the habitat varies with year, season and location (Janssen et al., 1988; Rohlfs and Hoffmeister, 2004; Shorrocks, 1982), thus variation in host and patch availability is not only existent between generations of wasps but might also be experienced by individuals. Theoretical models have shown that foragers can adjust quickly to this kind of variance, if information is sampled during the foraging process (Mc Namara and Houston, 1985; 1987). The parasitoids of *Drosophila* can therefore be expected to collect information while searching, in order to adjust their patch-leaving strategy according to the prevailing conditions. Indeed, adjusting patch-leaving behavior with each new experience seems to be exactly how the braconid *A. tabida* responds to patch encounter rate (Thiel and Hoffmeister, 2004): females visited several host-containing patches of equal quality in sequence, but had to wait either 5 min or 24 h between successive visits. While wasps from both treatments stayed for the same long time on the first patch and exploited it fully, those with the 5-min waiting time reduced patch residence time and patch exploitation rate in perfect qualitative agreement with optimal foraging theory (Fig. 2.1), and therefore most likely estimated the overall habitat quality increasingly higher. This response was strongest on the second patch but continued

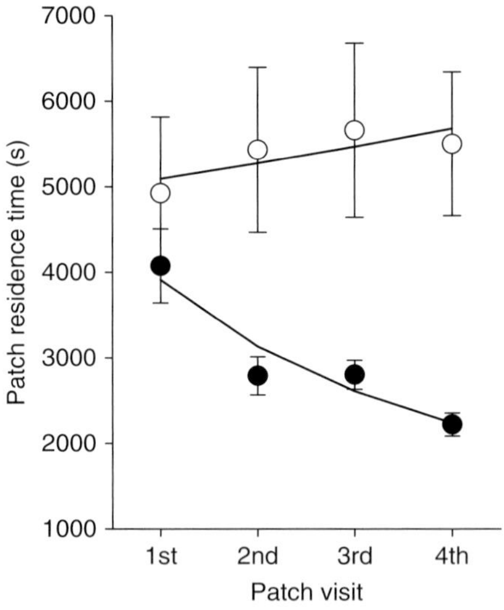

FIGURE 2.1 Patch residence times of *A. tabida* females for four successive patch visits, with females waiting either 5 min (filled circles) or 24 h (open circles) between visits. *Note: Bars indicate the standard error and solid lines show the corresponding predicted values from a generalized linear modeling (GLM) analysis (see* Thiel and Hoffmeister (2004) *for a different approach to analyze the data).*

across the following visits, as the wasps continued to encounter patches at a high rate. Since the patches were all of the same quality and should provide the searching females with similar information, and because we could exclude egg depletion or other confounding factors as possible explanations (Thiel and Hoffmeister, 2004), this response should reflect the wasps' estimate of habitat quality, according to their experience.

In another study, the effect of waiting time interval on the subsequent leaving decision was analyzed when interval durations varied between 5 min and 8 h (Thiel, 2004). From this experiment it became obvious that the increase in leaving tendency on the next visited patch was greatest when the interval between visits was short (Fig. 2.2). However, if waiting times became as long as approximately 2 h, the time on the second patch equaled the time spent on the first (Fig. 2.2). This threshold might again depend on experience, that is, on the number of patches visited, or it might reflect the distribution pattern of patches in the natural situation, where the travel between fruits aggregated under a fruit-bearing tree may take only a few minutes, but the travel time between clusters of patches, or solitary occurring patches, may take much longer, and is connected with a high probability of not even reaching another patch at all (Ellers et al., 1998; Janssen, 1989).

Aging can be regarded as a prepatch experience as well. In particular, older wasps that have spent a long time without finding any hosts at all tend to stay comparatively long on host-containing patches (Thiel and Hoffmeister, 2004). Of course, physical changes may also occur, like increased egg load, reduced energetic reserves, or senescence-related degradations, and disentangling these factors by comparing older and

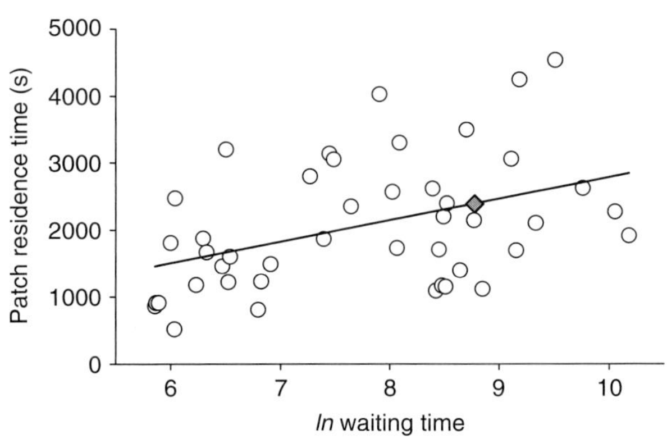

FIGURE 2.2 Patch residence times of *A. tabida* females when females experienced different waiting times between visits (open circles). The average time spent on the first patch is indicated by the filled diamond, the solid line shows the values predicted from a generalized linear modeling (GLM) analysis.

younger wasps is only partly possible (but see Thiel and Hoffmeister, 2004). The same holds for the interpretation of the long residence times observed in *L. heterotoma* females that survived sublethal doses of insecticides (Delpuech et al., 2005). Even more striking are examples where parasitoids responded to the mere possibility of reduced life expectancy. If females of *L. heterotoma* are exposed to circumstances that indicate unfavorable environmental conditions in the future, for example, dropping air pressure heralding a thunderstorm or decreasing day length indicating the onset of winter, they stay longer on a given patch and are more likely to accept already parasitized hosts (Roitberg et al., 1992, 1993). This is in line with predictions from dynamic theory for situations where they cannot expect to find many more patches before dying (e.g., Wajnberg et al., 2006).

2.6. THE EFFECTS OF INTRAPATCH EXPERIENCE

What we have seen so far is that parasitic wasps, upon entering a patch, have an initial leaving tendency set by the amount of host odor present and by previous experience. This initial leaving tendency may represent a wasp's first estimator of relative patch quality. The amount of kairomone, however, is only an indirect cue of absolute patch quality, since it does not provide the wasp with information on actual host availability and suitability, that is, age or parasitism status. Only encounters with host larvae, which require search time investment, can provide this information. This relates to the reliability/detectability problem (Gould, 1993; Stephens, 1989; Vet et al., 1991) mentioned earlier. Several parasitoid species, including the *Drosophila* parasitic wasps, have been shown to prefer more reliable cues when they are available (e.g., Vet and Papaj, 1992; Wertheim et al., 2003). Thus, it is likely that the parasitoids also respond to cues of real host encounters over the cue of kairomone concentration as soon as it becomes available to them. The only indication for this that we are aware of comes from a study where *A. tabida* females on a patch were exposed to kairomone concentrations and host densities that were independent of each other (Thiel, unpublished observation). While *A. tabida* tends to stay significantly longer on empty patches that contain kairomone of 12 hosts compared to one (Galis and van Alphen, 1981), wasps in our experiment adjusted their residence times always according to the density of the larvae present, irrespective of the actual kairomone concentration in a patch (Fig. 2.3).

The introduction of proportional hazard models (= Cox regressions) as a tool for analyzing parasitoid patch-leaving decisions, pioneered by Haccou and Hemerik (Haccou et al., 1991; Hemerik et al., 1993) has given valuable insights into the kind of on-patch cues that influence a

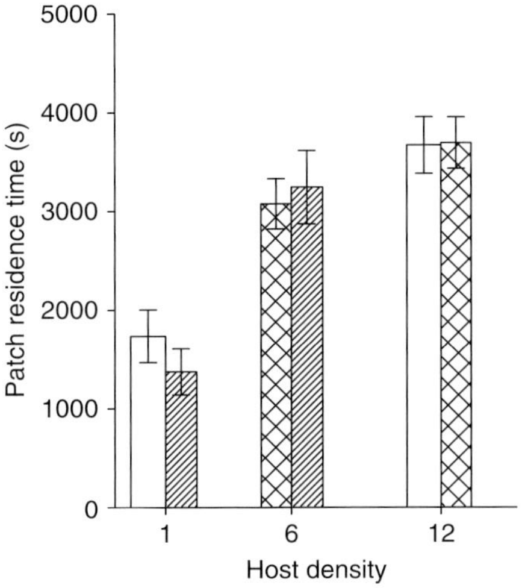

FIGURE 2.3 Patch residence times (with standard errors) of *A. tabida* females on patches with different host densities that varied in kairomone concentration. *Notes: striped bars = kairomone of 1 host larva, white bars = kairomone of 6 host larvae, crossed bars = kairomone of 12 host larvae.*

parasitoid's leaving tendency (van Alphen and Bernstein, 2008; van Alphen et al., 2003; Wajnberg, 2006). However, the best analysis obtained by a Cox regression remains a model description of parasitoid behavior. It provides neither information about how a wasp actually responds to the cue perceived nor about the behavioral change that then results in increased or decreased patch residence time. A way to obtain additional information about the decision-making process of wasps was found by video-tracking females of *L. heterotoma* when they searched a kairomone-containing patch with a single exactly timed host encounter (Schmitz, 2006). In agreement with earlier experiments (Haccou et al., 1991), the parasitization of a healthy host increased the residence time of the female and the video analysis showed that this resulted from a reduction in walking speed and increased turning angles and thus, an enhanced pattern of area-restricted search. After rejecting an already parasitized host, however, no such changes became obvious and patch residence times were not influenced. The effect of encounters with already parasitized hosts on the leaving tendency of parasitoids in general, and *L. heterotoma* in particular, has been controversially discussed in theoretical contributions as well as in the interpretation of experimental data (Gandon et al., 2006; Haccou et al., 1991; Kolss et al., 2006; van Alphen, 1993; van Lenteren, 1991; Varaldi et al., 2005). The reason for the controversial results may be that encountering an already parasitized host can convey different pieces of information: (1) it may indicate that hosts are present, even though they are not of best type possible; (2) it may indicate that another parasitoid has been there before and that the patch might be already depleted; and (3) it may indicate that conspecifics forage in the same habitat and therefore, the whole habitat might be of relatively bad quality. While points 1 and 3 could lead to superparasitism acceptance and longer residence times, point 2 would favor early patch leaving. From everything described so far in this review, it should be obvious that whether increased or decreased or unchanged residence times result with encounters with parasitized hosts will strongly depend on other information the searching parasitoid already has: whilst in young or "optimistic" females an early leaving response would be favored, older or "pessimistic" wasps might value the opportunity to superparasitize as an additional chance for producing offspring. The coexistence of different inherited strategies among wasps of the same population might also be possible (Roitberg, 1990a). However, some strains of *L. boulardi* have been observed to superparasitize much more often than would be adaptive for the parasitoid female under any circumstances (Varaldi et al., 2003). This puzzle was solved by discovering that via superparasitism, the horizontal transmission of an associated virus is enhanced, and this virus seems to have evolved the ability to somehow manipulate the parasitoid's decision making (Varaldi et al., 2006).

2.7. THE PATCH-LEAVING DECISION

While the decision of accepting or rejecting a host of a certain quality is unequivocal usually even to the human observer, the decision-making process of patch leaving is comparatively complicated and also difficult to interpret. Particularly, since many parasitoid species engage in short off-patch excursions while actually searching a patch and these become longer and longer until the patch is finally left. Residence times therefore often depend on the design of the experimental arena and on the observer's definition of "leaving". Despite these problems, thoughtful patch-leaving studies have been proven a powerful tool in the analysis of parasitoid decision making (see van Alphen and Bernstein, 2008; van Alphen et al., 2003 and Wajnberg, 2006 for reviews).

When a parasitoid enters an area containing host-associated odors (kairomones), it usually reduces the speed of walking (orthokinetic response), increases the rate of turning (klinokinetic response) and turns around sharply upon reaching the patch edge, leading to an area-restricted search pattern (Godfray, 1994). In the above-described proportional hazards analysis of patch-leaving decisions, the parasitoid is assumed to have a tendency to leave the patch, which increases over time but is also influenced by certain events, for example, successful ovipositions (Haccou et al., 1991; Hemerik et al., 1993). Which events have an influence on the parasitoids leaving tendency is estimated from behavioral data. Similar a priori models have been proposed by several authors (Green, 1984; Iwasa et al., 1981; Mc Namara, 1982; Pierre and Green, 2008; Waage, 1979). Even though many differences between them exist in detail, they all have the increment-decay mechanism in common with the proportional hazards model. Since many other biological processes are also best described by decay functions (i.e., enzyme kinetics or habituation processes), it is likely that a representation of the decay process assumed in the models is indeed present in parasitoid cognitive processes and thus, may become visible in their searching behavior. For example, as an area-restricted search pattern makes a parasitic wasp stay initially, the cessation of that pattern, that is, turning angles and walking speed returning to their initial values, will probably make her leave. Video tracking and semiautomatic analysis of a wasp's walking path (Ethovision, Noldus) was used to test this idea. In an experiment females of *L. heterotoma* searched on an empty yeast patch, surrounded by plain agar, until they left the patch for more than 30 s (Dieckhoff, 2006; Uhlig, 2008). Total residence times of individual wasps were all divided into 10 identical intervals, to allow for comparison of searching patterns independent of total duration. While great differences were expected to exist between on-patch and off-patch walking speed and turning angles during the first time intervals of a patch visit, these differences should almost disappear during the last interval immediate to the final leave, if the behavior of the females would follow a decay function. However, only the first prediction was met by the data, not the second (Fig. 2.4). Even though wasps did slightly reduce their turning angles and increase their walking speed in the course of a patch visit, obvious differences remained between on-patch and off-patch behavior (Dieckhoff, 2006; Uhlig, 2008). This indicates that the decay function underlying the above-mentioned patch-leaving models might represent a wasp's search motivation, however, wasps most likely took the decision to leave actively while still being able to recognize patch boundaries and did not simply "blunder away" from it, caused by an inability to recognize the patch borders appropriately.

2.8. GENETIC DIFFERENCES IN SEARCHING BEHAVIOR

We can now probably safely assume that parasitoids are generally capable of updating their estimate of habitat quality through previous experience. In a mathematical way, this could be described by Bayesian processes (Mc Namara et al., 2006; Pierre and Green, 2008), even though only a few studies allowed the wasps to gain enough experiences for this

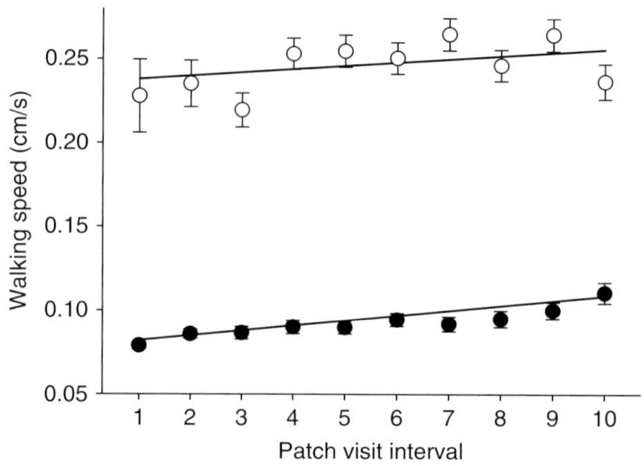

FIGURE 2.4 Walking speed of *L. heterotoma* females when moving either on a yeast patch (filled circles) or, during off-patch excursions, on plain agar (open circles). *Notes: Bars indicate the standard errors and solid lines show the corresponding predicted values from a generalized linear modeling (GLM) analysis. While on-patch walking speed increased significantly in the course of a patch visit, even in the last interval before leaving, the differences between on- and off-patch behavior clearly remained. Figure adapted from* Dieckhoff, 2006.

updating response to become clearly visible (e.g., Pierre et al., 2003; Thiel and Hoffmeister, 2004). If, however, the wasps update their information status, they need a prior experience to start from. This is, most likely, an innate and inherited response to certain cues. Evidence for inherited estimates of habitat quality comes mainly from parasitoid species that do not parasitize *Drosophila*, but should be mentioned in this context: Wajnberg and coworkers showed that several parasitoid species possess genetic variation in walking pattern (Wajnberg and Colazza, 1998) or patch time allocation, when responding to host encounters (Wajnberg et al., 1999) or to the presence of competitors (Wajnberg et al., 2004). Even though we are currently not aware of similar isofemale line studies for parasitoids of *Drosophila*, there is some indication at least that these wasps also differ in their initial search motivation when entering a patch. In *A. tabida*, females that showed long residence times on the first patch were usually also more reluctant to leave the second patch (Thiel, 2004). A similar pattern of individual differences between wasps was apparent in walking speed analyzes of *L. heterotoma* (Dieckhoff, 2006). Observed initial differences between *L. heterotoma* females, however, became smaller with increasing experience, indicating that indeed some updating of the initial response pattern might occur (Vet and Papaj, 1992). Similarly, reduced variance with increased number of patches searched can be seen in *A. tabida* females when patch encounter rate is high (Fig. 2.1). However, these experiments only provide weak evidence for the existence of an innate and inherited search motivation and the corresponding experience-dependent update. Thus, more rigorous studies on this topic are needed to learn more about the impact and interaction pattern of genes and environment on parasitoid behavior and decision-making processes.

2.9. PREDATION AND STARVATION

Parasitoid wasps face a considerable risk of being preyed upon while foraging (e.g., Rosenheim et al., 1995). In comparison to the predictions of optimal foraging models, parasitoids often stay longer than would be optimal, thus, generally overexploiting patches (Nonacs, 2001). It was suggested that long residence times are caused by the significant mortality risks, for example, spider webs, the wasps face when traveling between patches (Völkl and Kraus, 1996). Otherwise, however, the connection between patch residence times and predation risk has only recently made explicit (Roitberg et al., personal communication): in the case where a foraging wasp experiences a cue indicating the presence of a predator, she could either stay on the patch, ignoring the cue in order to produce a few more offspring before leaving (or getting killed), or she could leave immediately, trying to find a safer patch elsewhere. Theory developed by Roitberg et al. suggests that wasps should stay on rich patches but should leave with increasing probability the more the patch is already depleted. When these predictions where tested with searching *A. tabida* females exposed to a puff of formic acid as a proxy for danger, the similarities between the wasp's leaving pattern and the theoretical predictions were striking (Roitberg et al., personal communication).

Being eaten is one problem, finding food is another. Feeding usually greatly enhances parasitoid life expectancy and parasitoid species can be divided into two groups on the basis of their food searching strategies. First, those that find food resources in the same part of the environment as the hosts, second, those that find food resources and hosts in different parts of the environment (Bernstein and Jervis, 2008; Jervis et al., 2008). In particular, the latter is thought to involve considerable costs (Sirot and Bernstein, 1996) and is expected to have important influence on fitness. Individual females might, therefore, decide which option they go for depending on experienced host and patch encounter rate, egg load, starvation level, age and so on (e.g., Bernstein and Jervis, 2008; Jervis et al., 2008). *Drosophila* larvae feed on substrates that might as well be of some nutritional value for parasitoids and Eijs et al. (1998) showed that four species of *Drosophila* parasitic wasps are able to obtain energy from feeding the fruity or leafy substrate.

Even though the tradeoff between host and food searching may be relatively small for *Drosophila* parasitoids, some costs of feeding remain, since feeding and laying eggs are exclusive behaviors, that is, feeding takes time that could otherwise be spent searching for hosts. Indications for the existence of costs are that *L. heterotoma* has adopted a strategy of decreased activity and thus, reduced need for feeding, when not in contact with any host-associated cues (Eijs et al., 1998), whereas other parasitoid species remain active, probably that way enhancing the possibility to encounter new patches. *A. tabida*, when given the choice between either parasitizing the hosts present or feeding from the substrate only starts feeding after approximately 6 h of host searching behavior (Thiel, unpublished observation).

2.10. PROSPECTS AND IMPLICATIONS

From the research reviewed here, it becomes obvious that insect parasitoids use various cues of a variety of information sources in a flexible manner to adjust their foraging decisions. Information use has been called a "key feature to adaptive behavior" (Dall et al., 2005). Now, we can see more and more evidence accumulating that information use itself is under a strong selective pressure and thus, optimized in parasitic wasps. The resulting dynamic effects on behavioral decisions are difficult to analyze and even more difficult to use for predicting parasitoid behavior in, for example, efficiency analyses of potential biocontrol agents or risk assessment for nontarget species (Roitberg, 1990b; van Lenteren et al., 2006). Generally, care must be taken when inferring certain aspects of behavior from laboratory experiments, without assessing the true range under which behavioral responses might occur in nature. Often, predictions from theoretical models can be helpful in guiding experimental approaches, since they allow analysis of the potential fitness consequences of taking into account certain information (e.g., Kolss et al., 2006). However, classic optimization models are often not appropriate, since parasitoids use information to make decisions in a state-dependent or frequency-dependent manner. Stochastic dynamic programming, as it has recently been reviewed by Roitberg and Bernhard (2008), is an important tool when taking state-dependent behavior into account. If the behavior of conspecifics then has an additional impact, genetic algorithms provide a solution (Hoffmeister and Wajnberg, 2008).

When designing experiments and interpreting experimental evidence, it is important to examine critically whether the parasitoids were indeed in the informational state the experimenter assumed they were – it is naïve, somehow, to expect a wasp to be completely naïve at her first patch visit. It is simply impossible to keep wasps in a way such that they would not make any kind of experience prior to their first experimental experience; examples are given in Section 2.5 of this review.

Many parasitoids remember more than just the very last experience made (e.g., Thiel and Hoffmeister, 2004), even though the most recent experience usually has most impact (De Jong and Kaiser, 1992; Kaiser and De Jong, 1993).

We think that studying behavioral dynamics is a very promising route leading to important insights. This review has focused mainly on foraging decisions of female parasitoids. However, patch-defense behavior in *A. citri*, syn. *A. pleuralis*, (van Alphen and Bernstein, 2008) might occur in a state-dependent manner, and could thus be another interesting field of study for information use (e.g., Goubault et al., 2005). Likewise, mate choice might be affected by previous experience, but almost nothing is known about this subject either for parasitoids of *Drosophila*, apart from females and males of *L. clavipes* appearing to be somewhat choosy (Kraaijeveld et al., 2009). The work on parasitoid foraging strategies has almost exclusively dealt with female information use and decision making, and only recently, researchers have started to study male information use and patch time allocation in parasitoid males foraging for mating opportunities (Martel et al., 2008). Sex ratio decisions on a cluster of hosts (Fauvergue et al., 1999) might be promising to further explore the impact of ecological, physiological and informational state on decision-making processes.

ACKNOWLEDGMENTS

We are grateful to J de Boer for valuable suggestions on an earlier version of the manuscript and to K Kraaijeveld and B Roitberg for the opportunity to read their unpublished manuscripts.

REFERENCES

Bakker, K., van Alphen, J.J.M., van Batenburg, F.H.D., van der Hoeven, N., Nell, H.W., van Strien-van Liempt, W.T.F.H., et al., 1985. The function of host discrimination and superparasitization in parasitoids. Oecologia 67, 572–576.

Bernstein, C., Jervis, M.A., 2008. Food searching in parasitoids: The dilemma of choosing between 'immediate' or future fitness gains. In: Wajnberg, Bernstein, van Alphen, (Eds.), Behavioral Ecology of Insect Parasitoids. Blackwell Publishing, Maiden, MA, pp. 129–171.

Dall, S.R.X., Giraldeau, L.-A., Olsson, O., McNamara, J.M., Stephens, D.W., 2005. Information and its use by animals in evolutionary ecology. Trends Ecol. Evol. 20, 187–193.

De Jong, R., Kaiser, L., 1992. Odor preference of a parasitic wasp depends on order of learning. Experientia 48, 902–904.

Delpuech, J.M., Bardon, C., Bouletreau, M., 2005. Increase of the behavioral response to kairomones by the parasitoid wasp *Leptopilina heterotoma* surviving insecticides. Arch. Environ. Contam. Toxicol. 49, 186–191.

Dicke, M., van Lenteren, J.C., Boskamp, G.J.F., van Voorst, R., 1985. Intensification and prolongation of host searching in *Leptopilina heterotoma* (Thomson) (Hymenoptera: Eucoilidae) through a kairomone produced by *Drosophila melanogaster*. J. Chem. Ecol. 11, 125–136.

Dieckhoff, C., 2006. Eine Analyse des Suchverhaltens von Parasitoiden auf Wirtspatches [An analysis of parasitoid searching behavior on host patches] [diploma thesis]. Bremen University.
Dubuffet, A., Alvarez, C.I., Drezen, J.M., van Alphen, J.J.M., Poirie, M., 2006. Do parasitoid preferences for different host species match virulence? Physiol. Entomol. 31, 170–177.
Dukas, R., 2008. Evolutionary biology of insect learning. Annu. Rev. Entomol. 53, 145–160.
Eijs, I.E.M., Ellers, J., van Duinen, G.J., 1998. Feeding strategies in drosophilid parasitoids: the impact of natural food resources on energy reserves in females. Ecol. Entomol. 23, 133–138.
Eliassen, S., Jorgensen, C., Mangel, M., Giske, J., 2007. Exploration or exploitation: life expectancy changes the value of learning in foraging strategies. Oikos 116, 513–523.
Ellers, J., van Alphen, J.J.M., Sevenster, J.G., 1998. A field study of size–fitness relationships in the parasitoid *Asobara tabida*. J. Anim. Ecol. 67, 318–324.
Fauvergue, X., Fleury, F., Lemaitre, C., Allemand, R., 1999. Parasitoid mating structures when hosts are patchily distributed: field and laboratory experiments with *Leptopilina boulardi* and *L.* heterotoma. Oikos 86, 344–356.
Galis, F., van Alphen, J.J.M., 1981. Patch time allocation and search intensity of *Asobara tabida* Nees (Braconidae), a larval parasitoid of *Drosophila*. Neth. J. Zool. 31, 596–611.
Gandon, S., Rivero, A., Varaldi, J., 2006. Superparasitism evolution: Adaptation or manipulation? Am. Nat. 167, E1–E22.
Godfray, H.C.J., 1994. Parasitoids. Behavioral and Evolutionary Ecology. Princeton University Press, Princeton, NY.
Godfray, H.C.J., Waage, J.K., 1988. Learning in parasitic wasps. Nature 331, 211.
Goubault, M., Outreman, Y., Poinsot, D., Cortesero, A.M., 2005. Patch exploitation strategies of parasitic wasps under intraspecific competition. Behav. Ecol. 16, 693–701.
Gould, J.L., 1993. Ethological and comparative perspectives on honey bee learning. In: Papaj, Lewis, (Eds.), Insect Learning: Ecological and Evolutionary Perspectives. Chapman and Hall, New York, pp. 18–50.
Green, R.F., 1984. Stopping rules for optimal foragers. Am. Nat. 123, 30–43.
Haccou, P., van Alphen, J.J.M., 2008. Competition and asymmetric wars of attrition in insect parasitoids. In: Wajnberg, Bernstein, van Alphen, (Eds.), Behavioral Ecology of Insect Parasitoids. Blackwell Publishing, Maiden, MA, pp. 193–211.
Haccou, P., de Vlas, S.J., van Alphen, J.J.M., Visser, M.E., 1991. Information processing by foragers: effects of intra-patch experience on the leaving tendency of *Leptopilina heterotoma*. J. Anim. Ecol. 60, 93–106.
Hedlund, K., Vet, L.E.M., Dicke, M., 1996. Generalist and specialist parasitoid strategies of using odours of adult drosophilid flies when searching for larval hosts. Oikos 77, 390–398.
Hemerik, L., Driessen, G., Haccou, P., 1993. Effects of intra-patch experiences on patch time, search time and searching efficiency of the parasitoid *Leptopilina clavipes*. J. Anim. Ecol. 62, 33–44.
Henneman, M.L., Papaj, D.R., Figueredo, A.J., Vet, L.E.M., 1995. Egg-laying experience and acceptance of parasitized hosts by the parasitoid, *Leptopilina heterotoma* (Hymenoptera: Eucoilidae). J. Insect Behav. 8, 331–342.
Hoffmeister, T.S., Wajnberg, E., 2008. Finding optimal behaviors with genetic algorithms. In: Wajnberg, Bernstein, van Alphen, (Eds.), Behavioral Ecology of Insect Parasitoids. Blackwell Publishing, Maiden, MA, pp. 384–401.
Iwasa, Y., Higashi, M., Yamamura, N., 1981. Prey distribution as a factor determining the choice of optimal foraging strategy. Am. Nat. 117, 710–723.
Janssen, A., 1989. Optimal host selection by *Drosophila* parasitoids in the field. Funct. Ecol. 3, 469–479.
Janssen, A., Driessen, G., de Haan, M., Roodbol, N., 1988. The impact of parasitoids on natural populations of temperate woodland *Drosophila*. Neth. J. Zool. 38, 61–73.

Jervis, M.A., Ellers, J., Harvey, I.F., 2008. Resource acquisition, allocation, and utilization in parasitoid reproductive strategies. Annu. Rev. Entomol. 53, 361–385.
Kaiser, L., De Jong, R., 1993. Multi-odor memory influenced by learning order. Behav. Proc. 30, 175–184.
Kolss, M., Kawecki, T.J., 2008. Reduced learning ability as a consequence of evolutionary adaptation to nutritional stress in *Drosophila melanogaster*. Ecol. Entomol. 33, 583–588.
Kolss, M., Hoffmeister, T.S., Hemerik, L., 2006. The theoretical value of encounters with parasitized hosts for parasitoids. Behav. Ecol. Sociobiol. 61, 291–304.
Kraaijeveld, K., Franco, P., Reumer, B.M.,van Alphen, J.J.M., 2009. Effects of parthenogenesis and geographic isolation on female sexual traits in a parasitoid wasp. Evolution. in press.
Liu, Y.-Q., Bernstein, C., Thiel, A., 2009. Travel duration, energetic expenditure, and patch exploitation in the parasitic wasp *Venturia canescens*. Behav. Ecol. Sociobiol. DOI: 10.1007/s00265-009-0800-z.
Martel, V., Wajnberg, E., Boivin, G., 2008. Patch time allocation in male parasitoids. Ecol. Entomol. 33, 608–613.
Mc Namara, J., 1982. Optimal patch use in a stochastic environment. Theoret. Pop. Biol. 21, 269–288.
Mc Namara, J.M., Houston, A.I., 1985. Optimal foraging and learning. J. Theoret. Biol. 117, 231–249.
Mc Namara, J.M., Houston, A.I., 1987. Memory and the efficient use of information. J. Theoret. Biol. 125, 385–395.
Mc Namara, J.M., Green, R.F., Olsson, O., 2006. Bayes' theorem and its applications in animal behaviour. Oikos 112, 243–251.
Mery, F., Kawecki, T.J., 2003. A fitness cost of learning ability in *Drosophila melanogaster*. Proc. Biol. Sci. 270, 2465–2469.
Mery, F., Kawecki, T.J., 2004. An operating cost of learning in *Drosophila melanogaster*. Anim. Behav. 68, 589–598.
Mollema, C., 1991. Heritability estimates of host selection behavior by the *Drosophila* parasitoid *Asobara tabida*. Neth. J. Zool. 41, 174–183.
Nonacs, P., 2001. State dependent behavior and the marginal value theorem. Behav. Ecol. 12, 71–83.
Pannebakker, B.A., Garrido, N.R.T., Zwaan, B.J., van Alphen, J.J.M., 2008. Geographic variation in host-selection behaviour in the *Drosophila* parasitoid *Leptopilina clavipes*. Entomol. Exp. Appl. 127, 48–54.
Papaj, D.R., 1993. Automatic behavior and the evolution of instinct: lessons from learning in parasitoids. In: Papaj, Lewis, (Eds.), Insect Learning: Ecological and Evolutionary Perspectives.. Chapman and Hall, New York, pp. 243–272.
Papaj, D.R., Vet, L.E.M., 1990. Odor learning and foraging success in the parasitoid, *Leptopilina heterotoma*. J. Chem. Ecol. 16, 3137–3150.
Papaj, D.R., Snellen, H., Swaans, K., Vet, L.E.M., 1994. Unrewarding experiences and their effect on foraging in the parasitic wasp *Leptopilina heterotoma* (Hymenoptera: Eucoilidae). J. Insect Behav. 7, 465–481.
Perez-Maluf, R., Kaiser, L., 1998. Mating and oviposition experience influence odor learning in *Leptopilina boulardi* (Hymenoptera: Eucoilidae), a parasitoid of *Drosophila*. Biol. Control. 11, 154–159.
Pierre, J.S., Green, R.F., 2008. A Bayesian approach to optimal foraging in parasitoids. In: Wajnberg, Bernstein, van Alphen, (Eds.), Behavioral Ecology of Insect Parasitoids.. Blackwell Publishing, Maiden, MA, pp. 357–383.
Pierre, J.S., Van Baaren, J., Boivin, G., 2003. Patch leaving decision rules in parasitoids: do they use sequential decisional sampling? *Behav*. Ecol. Sociobiol. 54, 147–155.
Raine, N.E., Chittka, L., 2008. The correlation of learning speed and natural foraging success in bumble-bees. Proc. Biol. Sci. 275, 803–808.

Rohlfs, M., Hoffmeister, T.S., 2004. Spatial aggregation across ephemeral resource patches in insect communities: an adaptive response to natural enemies? Oecologia 140, 654–661.

Roitberg, B.D., 1990a. Optimistic and pessimistic fruit flies: evaluating fitness consequences of estimation errors. Behaviour 114, 65–82.

Roitberg, B.D., 1990b. Variation in behaviour of individual parasitic insects: bane or boon. In: Mackauer, Ehler, Roland, (Eds.), Critical Issues in Biological Control. Intercept, Hants, United Kingdom, pp. 25–39.

Roitberg, B.D., Bernhard, P., 2008. State-dependent problems for parasitoids: case studies and solutions. In: Wajnberg, Bernstein, van Alphen, (Eds.), Behavioral Ecology of Insect Parasitoids. Blackwell Publishing, Maiden, MA, pp. 337–356.

Roitberg, B.D., Mangel, M., Lalonde, R.G., Roitberg, C.A., van Alphen, J.J.M., Vet, L., 1992. Seasonal dynamic shifts in patch exploitation by parasitic wasps. Behav. Ecol. 3, 156–165.

Roitberg, B.D., Sircom, J., Roitberg, C.A., van Alphen, J.J.M., Mangel, M., 1993. Life expectancy and reproduction. Nature 364, 108.

Rolff, J., Kraaijeveld, A.R., 2001. Host preference and survival in selected lines of a *Drosophila* parasitoid, *Asobara tabida*. J. Evol. Biol. 14, 742–745.

Rosenheim, J.A., Kaya, H.K., Ehler, L.E., Marois, J.J., Jaffee, B.A., 1995. Intraguild predation among biological-control agents – theory and evidence. Biol. Control. 5, 303–335.

Schmitz, H., 2006. Einfluss der ersten Wirtsbegegnung auf die Patchaufenthaltsdauer von Parasitoiden (*Leptopilina heterotoma*). [Influence of the first host encounter on patch residence times in the parasitoid *Leptopilina heterotoma*.] [diploma thesis]. Bremen University.

Shorrocks, B., 1982. The breeding sites of temperate woodland *Drosophila*. In: Ashburner, Carson, Thompson, (Eds.), The Genetics and Biology of *Drosophila*. Volume 3b.. Academic Press, London, pp. 385–428.

Sirot, E., Bernstein, C., 1996. Time sharing between host searching and food searching in parasitoids: State-dependent optimal strategies. Behav. Ecol. 7, 189–194.

Smid, H.M., Wang, G.H., Bukovinszky, T., Steidle, J.L.M., Bleeker, M.A.K., van Loon, J.J.A., Vet, L.E.M., 2007. Species-specific acquisition and consolidation of long-term memory in parasitic wasps. Proc. Biol. Sci. 274, 1539–1546.

Steidle, J.L.M., van Loon, J.J.A., 2003. Dietary specialization and infochemical use in carnivorous arthropods: Testing a concept. Entomol. Exp. Appl. 108, 133–148.

Stephens, D.W., 1989. Variance and the value of information. Am. Nat. 134, 128–140.

Stephens, D.W., 1993. Learning and behavioral ecology: incomplete information and environmental predictability. In: Papaj, Lewis, (Eds.), Insect Learning: Ecological and Evolutionary Perspectives.. Chapman and Hall, New York, pp. 195–218.

Stephens, D.W., Krebs, J.R., 1986. Foraging Theory. Princeton University Press, Princeton.

Thiel, A., 2004. Die Bedeutung von Habitatparametern für das Suchverhalten parasitischer Wespen. [The impact of habitat parameters on the searching behavior of parasitic wasps.] [PhD thesis], CAU Kiel.

Thiel, A., Hoffmeister, T.S., 2004. Knowing your habitat: linking patch-encounter rate and patch exploitation in parasitoids. Behav. Ecol. 15, 419–425.

Thiel, A., Hoffmeister, T.S., 2006. Selective information use in parasitoid wasps. Anim. Biol. 56, 233–245.

Thiel, A., Driessen, G., Hoffmeister, T.S., 2006. Different habitats, different habits? Response to foraging information in the parasitic wasp *Venturia canescens*. Behav. Ecol. Sociobiol. 59, 614–623.

Uhlig, J., 2008. Entscheidung oder Habituation. Verhaltensantworten von Parasitoiden auf Wirte und deren Spuren. [Decision making or habituation. Behavioral responses in parasitoids to hosts and host traces.] [diploma thesis]. Bremen University.

van Alphen, J.J.M., 1993. Patch residence time and encounters with parasitised hosts: A reaction. Neth. J. Zool. 43, 340–349.

van Alphen, J.J.M., Bernstein, C., 2008. Information acquisition, information processing, and patch time allocation in insect parasitoids. In: Wajnberg, Bernstein, van Alphen, (Eds.), Behavioral Ecology of Insect Parasitoids. Blackwell Publishing Maiden, MA, pp. 172–192.

van Alphen, J.J.M., Drijver, R.A.B., 1982. Host selection by *Asobara tabida* Nees (Braconidae; Alysiinae), a larval parasitoid of fruit inhabiting *Drosophila* species. I. Host stage selection with *Drosophila melanogaster* as host species. Neth. J. Zool. 32, 215–231.

van Alphen, J.J.M., Janssen, A.R.M., 1982. Host selection by *Asobara tabida* Nees (Braconidae; Alysiinae), a larval parasitoid of fruit inhabiting *Drosophila* species. II. Host species selection. Neth. J. Zool. 32, 194–214.

van Alphen, J.J.M., Nell, H.W., 1982. Superparasitism and host discrimination by *Asobara tabida* Nees (Braconidae; Alysiinae) a larval parasitoid of Drosophilidae. Neth. J. Zool. 32, 232–260.

van Alphen, J.J.M., van Harsel, H.H., 1982. Host selection by *Asobara tabida* Nees (Braconidae; Alysiinae), a larval parasitoid of fruit inhabiting *Drosophila* species. III. Host species selection and functional response. In: van Alphen, (Ed.), Foraging Behaviour of *Asobara tabida*, a Larval Parasitoid of *Drosophila*. University of Leiden, PhD thesis, pp. 61–93.

van Alphen, J.J.M., Vet, L.E.M., 1986. An evolutionary approach to host finding and selection. In: Waage, Greathead, (Eds.), Insect Parasitoids. Academic Press, London, pp. 23–61.

van Alphen, J.J.M., Visser, M.E., 1990. Superparasitism as an adaptive strategy for insect parasitoids. Annu. Rev. Entomol. 35, 59–79.

van Alphen, J.J.M., Bernstein, C., Driessen, G., 2003. Information acquisition and time allocation in insect parasitoids. Trends Ecol. Evol. 18, 81–87.

van Alphen, J.J.M., Visser, M.E., Nell, H.W., 1992. Adaptive superparasitism and patch time allocation in solitary parasitoids: searching in groups vs sequential patch visits. Funct. Ecol. 6, 528–535.

van Lenteren, J.C., 1972. Contact-chemoreceptors on the ovipositor of *Pseudeucoila bochei* Weld (Cynipidae). Neth. J. Zool. 22, 347–350.

van Lenteren, J.C., 1991. Encounters with parasitized hosts: to leave or not to leave the patch. Neth. J. Zool. 41, 144–157.

van Lenteren, J.C., Bakker, K., 1975. Discrimination between parasitised and unparasitised hosts in the parasitic wasp *Pseudeucoila bochei*: a matter of learning. Nature 254, 417–419.

van Lenteren, J.C., Cock, M.J.W., Hoffmeister, T.S., Sands, D.P.A., 2006. Host specificity in arthropod biological control, methods for testing and interpretation of the data. In: Bigler, Babendreier, Kuhlmann, (Eds.), Environmental Impact of Invertebrates for Biological control of Arthropods. Cabi Bioscience, Delémont, Switzerland, pp. 39–63.

Varaldi, J., Fouillet, P., Bouletreau, M., Fleury, F., 2005. Superparasitism acceptance and patch-leaving mechanisms in parasitoids: a comparison between two sympatric wasps. Anim. Behav. 69, 1227–1234.

Varaldi, J., Fouillet, P., Ravallec, M., López-Ferber, M., Boulétreau, M., Fleury, F., 2003. Infectious behavior in a parasitoid. Science 203, 1930.

Varaldi, J., Petit, S., Boulétreau, M., Fleury, F., 2006. The virus infecting the parasitold *Leptopilina boulardi* exerts a specific action on superparasitism behaviour. Parasitology 132, 747–756.

Vet, L.E.M., Bakker, K., 1985. A comparative functional approach to the host detection behaviour of parasitic wasps. 2. A quantitative study on eight eucoilid species. Oikos 44, 487–498.

Vet, L.E.M., Papaj, D.R., 1992. Effects of experience on parasitoid movement in odour plumes. Physiol. Entomol. 17, 90–96.

Vet, L.E.M., Schoonman, G., 1988. The influence of previous foraging experience on microhabitat acceptance in *Leptopilina heterotoma*. J. Insect Behav. 1, 387–392.

Vet, L.E.M., van Alphen, J.J.M., 1985. A comparative functional approach to the host detection behaviour of parasitic wasps. 1. A qualitative study on Eucoilidae and Alysiinae. Oikos 44, 478–486.

Vet, L.E.M., van der Hoeven, R., 1984. Comparison of the behavioral-response of two *Leptopilina* species (Hymenoptera, Eucoilidae), living in different microhabitats, to kairomone of their host (Drosophilidae). Neth. J. Zool. 34, 220–227.

Vet, L.E.M., Dejong, A.G., Franchi, E., Papaj, D.R., 1998. The effect of complete versus incomplete information on odour discrimination in a parasitic wasp. Anim. Behav. 55, 1271–1279.

Vet, L.E.M., Janse, C., van Achterberg, C., van Alphen, J.J.M., 1984. Microhabitat location and niche segregation in two sibling species of drosophilid parasitoids: *Asobara tabida* (Nees) and *A. rufescens* (Foerster) (Braconidae: Alysiinae). Oecologia 61, 182–188.

Vet, L.E.M., Lewis, W.J., Cardé, R.T., 1995. Parasitoid foraging and learning. In: Cardé, Bell, (Eds.), Chemical Ecology of Insects. Chapman & Hall, New York, pp. 65–101.

Vet, L.E.M., Lewis, W.J., Papaj, D.R., van Lenteren, J.C., 1990. A variable-response model for parasitoid foraging behavior. J. Insect Behav. 3, 471–490.

Vet, L.E.M., Wäckers, F.L., Dicke, M., 1991. How to hunt for hiding hosts: The reliability-detectability problem in foraging parasitoids. Neth. J. Zool. 41, 202–213.

Visser, M.E., 1993. Adaptive self-superparasitism and conspecific superparasitism in the solitary parasitoid *Leptopilina heterotoma* (Hymenoptera, Eucoilidae). Behav. Ecol. Sociobiol. 4, 22–28.

Visser, M.E., 1995. The effect of competition on oviposition decisions of *Leptopilina heterotoma* (Hymenoptera: Eucoilidae). Anim. Behav. 49, 1677–1687.

Visser, M.E., van Alphen, J.J.M., Nell, H.W., 1990. Adaptive superparasitism and patch time allocation in solitary parasitoids: the influence of the number of parasitoids depleting a patch. Behaviour 114, 21–36.

Visser, M.E., van Alphen, J.J.M., Nell, H.W., 1992. Adaptive superparasitism and patch time allocation in solitary parasitoids: The influence of pre-patch experience. Behav. Ecol. Sociobiol. 31, 163–171.

Völkl, W., Kraus, W., 1996. Foraging behaviour and resource utilization of the aphid parasitoid *Pauesia unilachni*: adaptation to host distribution and mortality risks. Entomol. Exp. Appl. 79, 101–109.

Waage, J.K., 1979. Foraging for patchily-distributed hosts by the parasitoid, *Nemeritis canescens*. J. Anim. Ecol. 48, 353–371.

Wajnberg, E., 2006. Time allocation strategies in insect parasitoids: from ultimate predictions to proximate behavioral mechanisms. Behav. Ecol. Sociobiol. 60, 589–611.

Wajnberg, E., Colazza, S., 1998. Genetic variability in the area searched by a parasitic wasp: analysis from automatic video tracking of the walking path. J. Insect Physiol. 44, 437–444.

Wajnberg, E., Bernhard, P., Hamelin, F., Boivin, G., 2006. Optimal patch time allocation for time-limited foragers. Behav. Ecol. Sociobiol. 60, 1–10.

Wajnberg, E., Curty, C., Colazza, S., 2004. Genetic variation in the mechanisms of direct mutual interference in a parasitic wasp: Consequences in terms of patch-time allocation. J. Anim. Ecol. 73, 1179–1189.

Wajnberg, E., Rosi, M.C., Colazza, S., 1999. Genetic variation in patch time allocation in a parasitic wasp. J. Anim. Ecol. 68, 121–133.

Wertheim, B., Vet, L.E.M., Dicke, M., 2003. Increased risk of parasitism as ecological costs of using aggregation pheromones: Laboratory and field study of *Drosophila–Leptopilina* interaction. Oikos 100, 269–282.

Wiskerke, J.S.C., Dicke, M., Vet, L.E.M., 1993. Larval parasitoid uses aggregation pheromone of adult hosts in foraging behaviour: A solution to the reliability-detectability problem. Oecologia 93, 145–148.

CHAPTER 3

Dynamic Use of Fruit Odours to Locate Host Larvae: Individual Learning, Physiological State and Genetic Variability as Adaptive Mechanisms

Laure Kaiser,[*,†] **Aude Couty,**[‡]
and **Raquel Perez-Maluf**[§]

Contents

[*] IRD, UR 072, c/o Laboratoire Evolution, Génomes et Spéciation, UPR 9034, CNRS, 91198 Gif-sur-Yvette Cedex, France
[†] INRA Centre de Versailles-Grignon; UMR 1272 Physiologie de l'Insecte Signalisation et Communication, Route de St Cyr, 78026 Versailles Cedex, France
[‡] Unité de Recherche EA3900 BioPI-UPJV; Biologie des plantes et contrôle des insectes ravageurs; Laboratoire de Biologie des Entomophages; Faculté des Sciences, Ilot des Poulies, 33 rue Saint Leu, 80039 Amiens Cedex, France
[§] Laboratório de Zoologia, DNC/UESB, Estrada do Bem Querer KM 04, Vitoria da Conquista, 45000 BA, Brazil

Advances in Parasitology, Volume 70
ISSN 0065-308X, DOI: 10.1016/S0065-308X(09)70003-X

Abstract

This chapter presents a series of behavioral studies designed to document how *Leptopilina* spp. learn fruit odours in order to find and explore host-infested fruits. Experimental analyses of conditioned responses explored individual learning, physiological changes and genetic variability as adaptive mechanisms of the host searching behavior. Both oriented walking and substrate probing can be easily observed and quantified in laboratory devices. We studied walking in a four-arm olfactometer and probing in an agar substrate in response to olfactory stimulation by fruit odours. We analyzed the odour learning process and the dynamics of the memory. We next investigated how odour memory is influenced by motivation factors such as mating or egg-load, and how much variation is due to inheritance, using isofemale lines. Next, we addressed the adaptive significance of innate and conditioned responses to fruit odour by comparing and crossing populations originating from areas with contrasted levels of host availability.

3.1. INTRODUCTION

Experimental analysis of behavior tells us what an animal can do, to what it responds, and how and when it does. It calls for laboratory devices where animals are exposed to lures or to natural elements representing a simplified environment, and where parameters of both the animal and the environment are controlled. This approach allows analyzing sources of variation of behaviors. In this chapter, we present an application to investigate fruit odour learning by a *Drosophila* parasitic wasp searching for host larvae, and how this behavior can be influenced by individual experience, by physiological state linked to motivation, and by inheritance. Most studies were conducted on *Leptopilina boulardi*

Barbotin et al. (Hymenoptera: Figitidae), except one study addressing the differences of the searching behavior between two genotypes of *L. heterotoma* Thomson.

L. boulardi is a solitary endoparasite, found on two species of frugivorous drosophila, *D. melanogaster* Meigen and *D. simulans* Sturt, in Mediterranean and tropical climates (Carton et al., 1986). In the *Leptopilina* genus, parasitoids are koinobiont, meaning that the larva feeds on and develops in a host that also continues to develop and is only killed when the parasitoid reaches the nymphal stage. The parasitoid nymph then undergoes metamorphosis inside the empty host puparium. Parasitic development lasts for about 3 weeks at 25 °C. Full egg load is mature at emergence (proovigeny), and can reach about 200 eggs (Carton et al., 1986). Oviposition dynamics are not known in natural environments. Van Lenteren (1976) observed complete egg deposition within 8 h, but Fleury et al. (1995) reported a parasitic rhythm limited to a couple of hours a day. We also observed long resting periods following oviposition bouts lasting for about 1 h in *L. boulardi* (unpublished data). Females live for 2–3 weeks in laboratory conditions (22 °C, 60% relative humidity, 16:8 photoperiod). Altogether, these data suggest that a female may deplete her egg load over a period of several days.

When adults emerge, the old fruit is not likely to contain young host larvae so mated females disperse to find new host habitats (Fig. 3.1). They display in-flight and walking orientation to host-infested fruits. Then they probe the fruit with their ovipositor and sting when touching a larva.[1] Both behavioral steps are at least partly chemically mediated. Females are attracted by a host aggregation pheromone (Hedlund et al., 1996, Wertheim et al., 2003; Wiskerke et al., 1993) and the attraction is synergized by fruit odours (Couty et al., 1999); they are also attracted by fermentation odours linked to yeast development in the decayed fruit, due to *Drosophila* infestation (Dicke et al., 1984). The wasps localize host larvae by probing in response to both chemicals and vibrations generated by larvae foraging in the substrate. The aggregation pheromone triggers the probing behavior and can mediate specific recognition of the host, because its composition is host specific (Vet et al., 1993). Past experience of Oviposition attempts in host-infested fruit enables females to use fruit odour in orientation (Couty et al., 1999) and probing behavior (Kaiser et al., 1995). Vet and Dicke (1992) proposed the ecological concept of ''reliability-detectability''. Host-habitat odours are more detectable than host odours because they are more substantial in terms of biomass, and often more volatile (C6 green leaf odours *vs* long hydrocarbon chains of insect pheromone or cuticular

[1] Searching modalities vary between species of larval parasitoids of *Drosophila*, showing for instance probing while walking, antennal search... These variations have been related to foraging variation between host species (Vet and Bakker, 1985; Vet and Van Alphen, 1985).

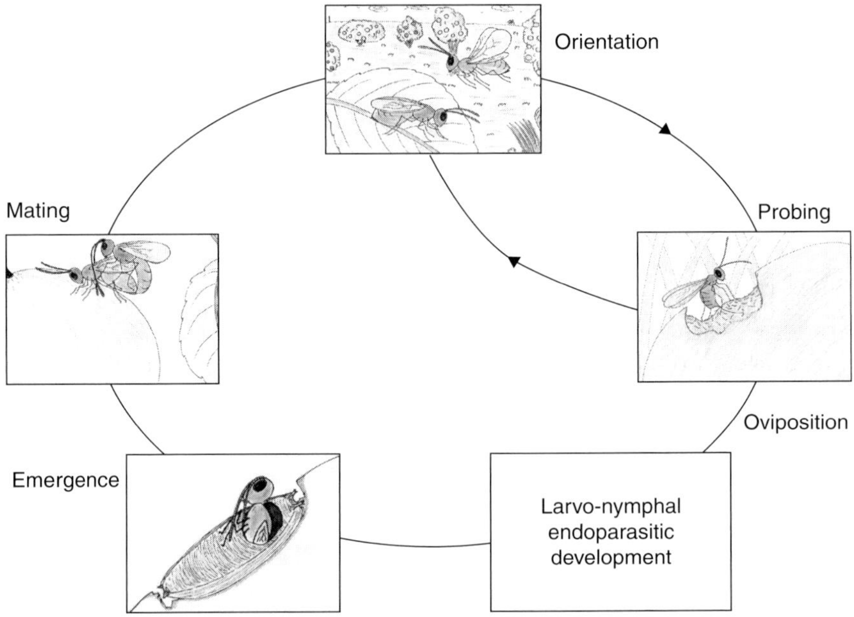

FIGURE 3.1 Life cycle of *Leptopilina* genus. Females emit sex pheromones that are perceived by males through the nymphal case (host puparium) so mating occurs soon after emergence. They appear to disperse to find young host larvae on fresh fruits. Being attracted by odours from *Drosophila*-damaged fruit, they probe the infested area with their ovipositor, and oviposit into the larvae they detect. Probability of host immune reaction and parasite escape mechanism depends mainly on both host and wasp genotype (see Chapter 6 by Dubuffet et al. and Chapter 11 by Dupas et al.). Endoparasitic development is koinobiont, meaning that the larva feeds on and develops into a host that also continues to develop and is only killed when the parasitoid reaches the nymphal stage, after three larval stages. The parasitoid nymph then undergoes metamorphosis inside the empty host puparium. *Note: Drawings from Huet and Kaiser.*

compounds). But they are evidently less reliable. Females would solve the so-called "reliability-detectability" problem by learning to associate host habitat odour with host presence, in the same way that they subsequently learn from ovipositioning attempts.

Both oriented walking and substrate probing can be easily observed and quantified in laboratory devices. We studied walking in a four-arm olfactometer, and probing in an agar substrate, in response to olfactory stimulation by fruit odours (see Section 3.2). We analyzed the odour learning process and the dynamics of the memory. We next investigated how odour memory is influenced by two motivation factors, mating and egg-load, and how much variation is due to inheritance, using isofemale lines. Next, we addressed the adaptive significance of innate and conditioned responses to fruit odour by comparing and crossing strains originating from areas with contrasted levels of host availability.

3.2. GENERAL MATERIAL AND METHODS

3.2.1. Observation apparatus

The odour-guided walking response was observed in a four-arm olfactometer. This apparatus was designed by Petersson (1970) for studies into aphid orientation to pheromones and has since been adapted for many species of parasitic wasps including *Drosophila* parasites (De Jong and Kaiser, 1991; Kaiser and Marion-Poll, in press; Vet et al., 1983). It consists of a star-shaped chamber with four branches. Air is passed through each branch and extracted through a central hole in the bottom of the chamber, thereby creating four distinct fields of equal area. One or two of the fields are usually odourized. Insects are introduced individually into the centre of the chamber where they perceive the four flows, then explore/walk around the chamber and eventually choose one field. It is quite suitable to quantify odour choice, because flows are kept distinct without a physical barrier. It is, however, not suited to study orientation mechanisms, because both odour concentration and speed of the airflows decrease then increase across the fields, due to the shape of the chamber, and the odour is everywhere in the field, so both attraction and arrestment responses[2] are observed. The insect's choice can be quantified between two odours or between one odour versus blank fields, by simply recording the relative amount of time spent in each field (25% of time meaning random presence in a field). The odour source is usually kept in a vial connected upstream of the entrance to the branch.

The device to observe odour-triggered ovipositor search was conceived by Kaiser et al. (1995). It allows females to probe into agarose gel in response to an olfactory stimulus (Fig. 3.2). It was designed for associative conditioning of probing responses to odours (see Section 3.2.4). The probing response is quantified by the percentage of females probing, the latency of the response following onset of odour delivery, and its duration.

3.2.2. Odour sources

In most of our studies, we used commercial extracts to limit uncontrolled variations that are expected with natural odour sources like whole fruits. The fruit aromas® were designed by the producer to be chemically close to the natural fruit odour they mimic.[3] We checked that *Leptopilina* females could not discriminate between natural and commercial banana odour in our experimental conditions. We also decided to use a commercial perfume to investigate odour learning. So-called "novel" odours

[2] See McFarland, 2001a, for definition.
[3] Haarmann and Reimer, Holzminder, Germany.

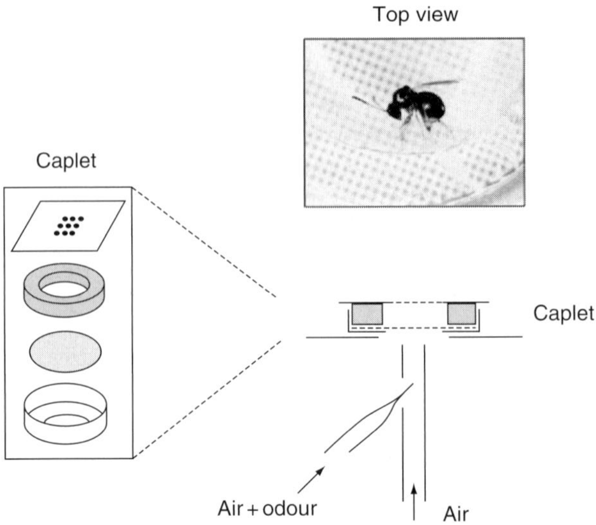

FIGURE 3.2 Caplet device for conditioning and observing odour-triggered searches by ovipositor probing. A female is placed in the middle of a ring of agar enclosed in a transparent plastic cap. To let the air flows through the ring, the bottom of the caplet is pierced in the center and covered with a disc of nylon; the caplet is covered with a piece of plastic sheet perforated with pin holes (left: elements of the caplet). It is positioned over a tube delivering the airflow. The odour can be delivered through a secondary airflow controlled with a manual valve. The female can be easily observed probing the inner part of the agar ring, or ovipositing if larvae have been transferred on the ring for associative conditioning (photo). *Note: Adapted from Kaiser* et al., *1995.*

which are not encountered in natural environment, have been successfully used to investigate wasps' cognitive abilities (Lewis and Takasu, 1990; Vet and Groenewold, 1990).

3.2.3. Insects

Except for the comparison of different natural populations, most studies were conducted on a laboratory strain of *L. boulardi* provided by Y. Carton (LEGS, CNRS, 91190 Gif-sur-Yvette, France). It originated from *D. melanogaster* collected from prickly pears (*Opuntia* sp.) in Tunisia (Nasrallah) and was reared on this fly on a standard *Drosophila* medium. To limit exposure to diet odour during metamorphosis and emergence, a period at which odour learning can occur, pupae of parasitized *Drosophila* were collected from their host fruit, washed in 5% bleach, rinsed in water and dried before storage in tubes containing agar-agar and honey as a food source for emerging adults (De Jong and Kaiser, 1991).

3.2.4. Conditioning methods

Females were conditioned to search in response to fruit odours by previous oviposition experience during the odour delivery. This is an associative conditioning, where the animals learn to respond to a conditioned stimulus (CS), here, an odour, that anticipates an unconditioned stimulus (US), here, the host larvae. We designed two methods, depending on the type and timing of response. The first was used to test the walking response 1 day after the experience. Females were allowed to oviposit for 1 h on a patch containing hundreds of young *Drosophila* larvae, in an odourized airflow. The second method was set to observe both short and long-term effects on the probing activity. This was achieved by placing host larvae on the ring of agar and by delivering the odour only during an oviposition attempt. This association could be repeated to study the effect of the number of oviposition rewards on memory dynamics. The conditioned probing response was subsequently tested by delivering the odour in the absence of host larvae.

3.3. DYNAMICS OF ODOUR MEMORY DISPLAYED IN ODOUR CHOICES

3.3.1. Learning is associative

During oviposition, parasitic wasps can learn cues that they subsequently use in various behavioral activities such as olfactory or visual orientation, host choice and acceptance, host quality perception, host-patch use (reviewed by Turlings et al., 1993). We were interested in analyzing how *Leptopilina* learn host-habitat odours and what were the associated memory dynamics. Associative learning where a stimulus and a reward are associated, for example, an odour and an oviposition experience, is thought to be the main mechanism of memory formation and is supposed to be advantageous to a parasite searching for hosts in a variety of habitats. The importance of the temporal pairing of the conditioned and the unconditioned stimulus is critical to associative learning (McFarland, 2001b) and has been studied in detail in model species such as the honeybee and *D. melanogaster* (Menzel et al., 1993; Tanimoto et al., 2004). At the time we ran the experiments, various studies on parasitoids, especially those on *Microplitis croceipes* (Lewis and Tumlinson, 1988) and *L. heterotoma* (Vet and Groenewold, 1990), strongly suggested that learning in parasitoids was associative. Parasitoid responses such as attraction to a host are usually triggered by unconditioned stimuli released by the host. They become conditioned to other stimuli such as like host-habitat odour or color, if such conditioned stimuli have been rewarded by oviposition or contact with host products. However, in most studies on parasitoid learning, the effects of pseudoconditioning[4] could not be excluded and the temporal relation between the stimulus and the reward had not been proved to be important. Vet and colleagues nicely showed that *L. heterotoma* and *L. boulardi* acquired a strong preference for the odour of the experienced host-habitat (Poolman-Simons et al., 1992; Vet et al., 1983). We controlled the possible effects of pseudoconditioning by comparing effects of single or combined exposure to the conditioned (odour) and unconditioned (host larvae) stimuli.

The behavioral responses of differently experienced females to an artificial odour (perfume *Must de Cartier*, Paris) were analyzed using the four-arm olfactometer. The responses of females with a prior 1-h period of oviposition experience in the presence of the perfume were compared with those of four control groups. As controls we used naïve females, females with an oviposition experience in the absence of odour, females that had been previously exposed for 1 h to perfume but without the oviposition experience, and females with desynchronized oviposition and perfume experience (perfume following oviposition experience). *L. boulardi* females were tested the next day, and while all four control groups spent equal amounts of time in the perfumed field and the non-scented fields, only the group with simultaneous exposure to odour and larvae showed a marked preference for the scented field (Fig. 3.3A). This result demonstrates that *L. boulardi* can learn to respond to an odour by associating this odour with oviposition experience serving as reward.[5] This conditioned response is not a laboratory artifact because Papaj and Vet (1990) checked that laboratory-conditioned *L. heterotoma* females exhibited a preference for the experienced host-habitat odour in outdoor conditions.

3.3.2. Multiodour memory is influenced by learning order

Hosts of *L. boulardi* can be found in a variety of fruits. The wasp might then face a multifruit host resource and experience different fruit species successively. Can they memorize several odours? We investigated this cognitive capacity by exposing females to two successive conditioning periods, each associated with a given fruit odour (banana and strawberry odours) (De Jong and Kaiser, 1992). The next day, conditioned females were tested in the olfactometer for their choice between both odours (Fig. 3.3B). They exhibited a strong preference for the last learned odour, and this was not due to forgetting the first one, which was still as attractive as after a single conditioning, when tested in a no-choice

[4] A response to a conditioned stimulus is called pseudoconditioned when it changes as a result of unpaired experience with unconditioned and/or conditioned stimulus.

[5] In *L. heterotoma* and other parasitoids (*M. croceipes, Cotesia marginiventris*: Turlings et al., 1993), females can memorize an odour when it is associated to contact with host larvae products, but fewer females are efficiently conditioned than when they can oviposit.

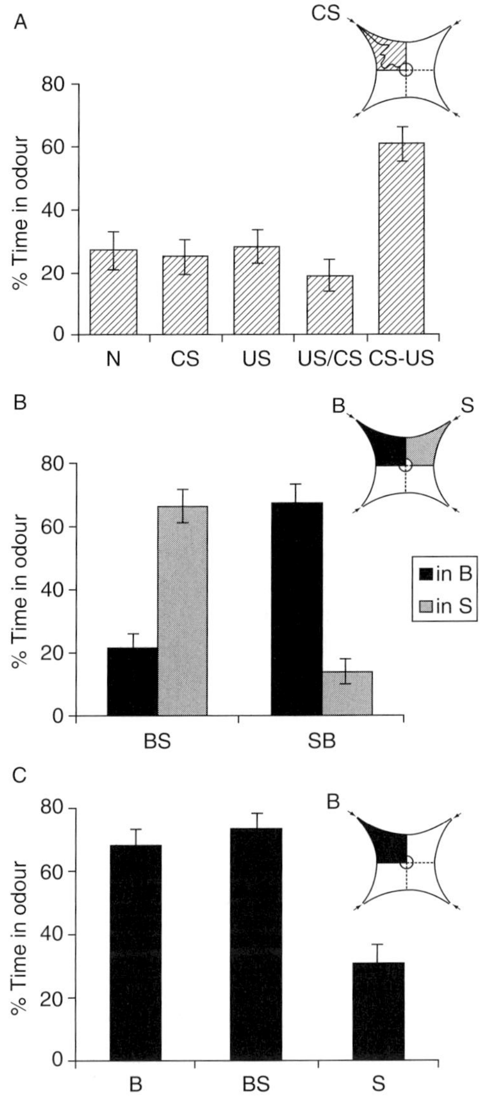

FIGURE 3.3 Dynamics of conditioned attraction studied in a four-arm olfactometer. (A) Synchronized oviposition and odour experience is essential to memorize the odour. The graph represents the percentage of time (mean and standard error of mean) spent by *L. boulardi* females in the field of the olfactometer delivering the perfume odour used as conditioned stimulus (see schematic view of the inner chamber of the olfactometer where one field out of the four carries the odour). Insects are tested individually 1 day after exposure N, group of naïve females; CS, females exposed to the perfume; US, females exposed to the larvae (allowed to oviposit); US/CS, desynchronized exposure of females to the larvae, then to the perfume; CS-US,

situation (Fig. 3.3C). *L. boulardi* shows perfect discriminative ability between two memorized odours. What is the limit of such cognitive abilities? Would this be the case for additional odours? To examine this we conditioned females to three odours in succession (Kaiser and De Jong, 1993). The next day females preferred the last learned odour, remembered the two first ones but without conditioned preference between both.

Ability to memorize several odours associated to the same reward has not been extensively explored in insects. Lewis and Takasu (1990) showed that *M. croceipes*, a caterpillar parasitoid, can learn two novel odours associated either with host or with food, and makes a choice between these odours on the basis of its own relative host and food needs. This memory for a particular odour depended on the physiological state of the insects, as hungry wasps chose the food-associated odour, and well-fed wasps the oviposition-associated odour. We observed that *L. boulardi* can remember different odours associated to the same reward, the host, and make a choice depending on the learning time. Kolterman (1974) described a time-linked memory in bees. Bees relate the time of day to the learning of an odour and can be trained to respond to different odours at different times. Representation of time in insects remains a current question about their cognitive capacities (Gallistel, 1989) and as a mechanism of adaptation to changing resources through their lifetime. Recently, studies on the learning mechanisms involved in the estimate of habitat profitability by parasitoids have begun to increase (e.g., Tentelier et al., 2009; Thiel et al., 2006; and see Chapter 2 by Thiel and Hoffmeister). These experimental studies test the predictions of Bayesian updating, a form of learning by which individuals may estimate the profitability of the patches they encounter to update their estimate of the profitability of the habitat as a whole, and adjust their foraging decisions accordingly. Limited updating capacities were found in the short-lived aphid parasitoid *Lysiphlebus testaceipes* when a series of seven patches were

---

synchronized exposure (associative conditioning) to larvae and perfume. Only this last group shows a significant attraction to the CS. (B) When females are conditioned successively to two odours, they subsequently prefer the last one. Here banana odour is delivered in one field of the olfactometer, and strawberry in an adjacent one. BS, females with a first 20-min oviposition experience in banana odour, and a second 20-min oviposition experience in strawberry odour (with a 30-min resting period in an empty vial in between). SB, females that had the reciprocal experience. Responses are tested the next day. (C) Females conditioned to two odours do not forget the first one. Here, only banana odour is delivered in one field of the olfactometer. BS females show the same attraction to banana odour than B females (females that are only conditioned to banana odour), BS females attraction to banana odour it is not due to conditioning to strawberry because S females (only conditioned to S) are not significantly attracted to banana odour. *Note: Redrawn from data of De Jong and Kaiser (1991, 1992).*

experienced (Tentelier et al., 2009). A relatively long-lived parasitoid with high memory performance such as *L. boulardi* should be a good model to investigate this mechanism of optimal foraging.

In our experiments, we used fruit odours that could be well discriminated between by *L. boulardi*, probably due to their distinct chemical composition. Vet et al. (1998) investigated the performance of discrimination or generalization between chemically similar odours. An Animals' abilities to generalize a response to resembling stimulus is considered as adaptive. For instance, bees that have learned the scent of a nectar-producing flower respond to closely related scents, which enables them to cope with daily variation of flower scent (Smith, 1993). When *L. heterotoma* oviposits on a cultivar of apple, it will generalize its conditioned orientation response to odours of other apple cultivars. However, it has the remarkable ability to discriminate between odours of two different apple cultivars if one has been rewarded by the presence of host larvae, and the other not. The authors proposed that memory is a mechanism to adapt the searching behavior to the state of information wasps collect from their experience, which gives them a more or less complete info about their resources (and see Chapter 2 by Thiel and Hoffmeister).

3.4. DYNAMICS OF ODOUR MEMORY DISPLAYED IN PROBING BEHAVIOR

Learning to prefer the odour of a rewarding host habitat has been particularly well studied (Vet et al., 1995), with the learned response being a marked choice for the experienced host-habitat odour. However, memory for the host-habitat odour has other behavioral expressions that have received little attention. It can influence the amount of time spent searching on plants, and be involved in both plant species recognition (Kester and Barbosa, 1991) and adjustment of patch time to host resources (Tentelier and Fauvergue, 2007). Some studies even reported that attempts to oviposit in a lure and ovipositor probing into a substrate could be released by novel stimuli after this behavior had been associated with the presence of the hosts from previous experiences (?) (chemical stimulus: Vinson et al., 1977; visual stimuli: Arthur, 1966, 1971; Wardle, 1990; textural stimuli: Wardle and Borden, 1985).

We examined whether fruit odour memory could influence *L. boulardi* probing behavior. Odour conditioning and testing of ovipositor probing was studied with a device we designed for this purpose (Fig. 3.2). The system allows experimental control of the number of ovipositions, and the timing of odour delivery. This is crucial for studying dynamics of memory formation and persistence. In animals classically used to investigate neurobiology, neurogenetics and biochemistry of memory, including drosophila

(*D. melanogaster*) and the honeybee (*A. mellifera*), it has been well established that short- or long-term memory phases can form according to the number and the spacing of rewards, due to the activation of different biochemical pathways (Schwärzel and Müller, 2006). Such memory complexity is thought to be a product of natural selection and Menzel et al. (1993) assessed the ecological benefits of the different memory performances of bees. With *L. boulardi*, we compared odour-conditioned probing responses of wasps submitted to nonassociative and associative conditioning, and varied the number of ovipositions and the time elapsed between conditioning and testing to get an insight into memory phases and their characteristics.

3.4.1. Sensitization and associative component of short- and long-lasting memory

Banana flavor was used for the conditioning of odour choice in the olfactometer. Females were naïve before being conditioned for this trial. In a first experiment, we compared females submitted to five odour-associated ovipositions (associative conditioning trial) with four control groups (one of naïve females, one of females exposed to the odour in the absence of larvae, one of females allowed to perform five ovipositions without being exposed to the odour and one of females exposed to the odour and with five oviposition experiences, but not at the same time) (Kaiser et al., 1995). Females were then all kept in clean agar until they were tested for their probing response to banana odour 12 min later. There we observed that both oviposition experience in itself and associative experience induced conditioned probing to the odour, so this response was sensitized by oviposition, at least in the short term. Rare probing responses were observed in naïve females or after odour exposure.

With this setup, conditioned and unconditioned responses are produced with the same behavior (probing), and can thus be compared. We observed that probability and latency of the conditioned response to banana odour were close to the values observed for the unconditioned probing response to host larvae in our experimental conditions. So it appears that conditioned females respond to the learned odour as if it was a larval stimulus, showing that they use the CS fully to anticipate the presence of larvae, as expected from classical associative conditioning (McFarland, 2001b). This has been explained in bees trained to associate an odour to a sugar reward, where the CS activates *de novo* the sugar-sensitive neuron (VUM), and thereby creates the same neuronal activity pattern as that with the sugar reward (Schwärzel and Müller, 2006).

In further experiments (Kaiser et al., 2003), we varied the number of ovipositions and the timing of the test. A single odour–oviposition association (conditioning trial) produced a high subsequent probing response to banana odour when tested shortly afterward (12 min or 2 h).

More associative trials did not always increase the response probability. In these short term studies, the conditioned response to banana was also seen after mere oviposition experience, and the response probability increased with the number of oviposition experiences. Thus, probing responses after short term intervals were the product of two parallel processes with different properties: first an associative memory trace, which developed at the first associative trial, and second, nonassociative sensitization, which increased with increasing trial numbers. The next day, conditioned probing to banana odour was only observed in females submitted to multiple odour–oviposition associations (Fig. 3.4A). Sensitization was thus short lived, and not observed after 24 h, and the formation of long-lasting memory required several reinforcements of the odour by oviposition. Few conditioned responses were observed 2 days after exposure which was also the case for the conditioning of odour choice in the olfactometer, although the conditioning was longer in that case, with an estimate number of 60 oviposition experiences. In both sensitized and associatively conditioned females, if they were re-exposed to banana odour (without a reward) a second and a third time after the first test, their probability to probe decreased drastically, regardless of the spacing between repeated tests (Fig. 3.4B). There was no recovery of the response after a 1-day resting period, showing that repeated presentation of the odour without reward erased the memory.

To relate these data to what we know of the foraging dynamic of *L. boulardi*, we proposed a model for the organization of memory phases (Kaiser et al., 2003), as developed for the honeybee (Menzel, 1999; Menzel et al., 1993). Short-term memory (highly dependent on sensitization) and long-term memory (highly associative) would thus be involved, respectively, in two phases of the parasitic activity. First, throughout the same day, sensitization of the probing activity to fruit odour, lasting for at least 2 h, would be the main process maintaining a high searching activity during a foraging bout on one infested fruit and between foraging bouts.[6] Then, on the next day, associative learning would be involved solely in remembering the fruit odour when the insects resume a period of searching activity. Rapid decline of the conditioned probing after perception of the fruit odour without larvae would help females to leave a fruit with few or no host larvae. Considering that females renew their associative conditioning by daily ovipositions, a memory lasting for more than 24 h would be of little use.

[6] *D. melanogaster* females aggregate on oviposition sites (wound on ripe fruits) where tens to hundreds of larvae can then develop. During the period of host-searching activity, *Leptopilina* females may leave their initial patch to visit several other host patches, even when their host population is far from being fully parasitized (Mangel, 1993).

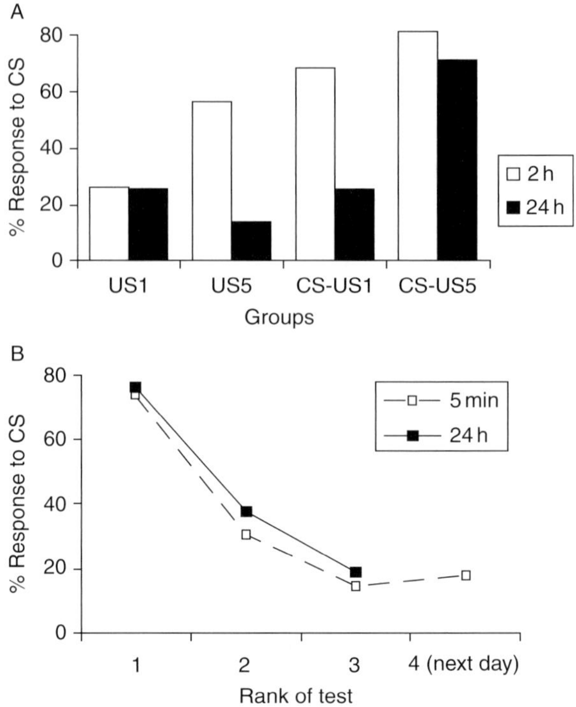

FIGURE 3.4 Dynamics of conditioned probing responses. The graphs represent the probability of probing response to banana odour (conditioned stimulus, CS). Females are tested individually in the caplet device. (A) Role of sensitization and associative memory in short- and long-term responses. In the short term (test performed 2 h after conditioning), the probability of conditioned response is equally increased by sensitization through five oviposition experiences (group US5; US: unconditioned stimulus, here, the host larvae) or one odour–oviposition association (CS-US1: one conditioning trial). But at the long term (test performed 24 h after conditioning), the effect of sensitization or of a single conditioning trial is erased and conditioned responses are only frequent in females that had five conditioning trials (CS-US5). (B) Extinction of conditioned responses due to repeated and nonreinforced exposure to the CS. Probing responses to banana odour are tested three times in succession, in females submitted to five conditioning trials. One group is submitted to an intertest interval of 5 min, compared to 24 h in the other one. The probability of conditioned response decreases rapidly at the second and third test, regardless of the time interval in between. The absence of recovery the next day (test 4) indicates that the odour memory has been erased. *Note: Adapted from Kaiser* et al. *(2003).*

This study revealed that sensitization, associative memory and its extinction produce highly plastic responses to a host-habitat cue, and could be key factors in the optimality of patch exploitation.

3.5. MOTIVATION INFLUENCES THE LEARNED SEARCHING RESPONSES

Intrinsic factors of behavioral variability are classically grouped in three categories: individual history, motivation and genotype (Slater, 1985). In insects, individual history can explain memory for events or stimuli experienced by immature and adult stages, while motivation is linked to a physiological state. In adults, such a state can depend on the maturation of the reproductive system, such as ovocytes maturation, on initial events such as mating, and on daily variation of hunger and egg load for instance (McFarland, 2001b). How learning and memory performance depends on motivation has been well studied in appetitive tasks. In the honeybee (Friedrich et al., 2004) and in *Drosophila* (Chabaud et al., 2006; Colomb et al., 2009), starvation is essential to be able to learn and recall food-associated stimuli. In the wasp *M. croceipes*, hunger makes females recall and prefer a food-associated odour to a host-associated one (Lewis and Takasu, 1990). To our knowledge, the influence of motivation on learning and memory of host-related stimuli has been little investigated. Influence of egg-load has been studied on host-searching activity, for example, in aphid parasitoids (Collins and Dixon, 1986), and on superparasitism strategy, notably in *L. heterotoma* (Van Lenteren, 1976). How egg load and particularly mating influence female behavior is now well understood in *Drosophila*. Drastic changes of receptivity to courting males and egg-laying activity are observed following mating, due to a particular hormonal peptide (the sex peptide) of the seminal fluid, which is cotransferred into the female reproductive tract together with the sperm, and perceived by internal sensory neurons (Chapman & Davies, 2004; Yang et al., 2009).

In the studies presented above to characterize cognitive abilities of *L. boulardi*, females were mated and had their full egg load when used for the experiments. Here, we studied the influence of the mating status and egg load on conditioned probing behavior (Perez-Maluf and Kaiser, 1998).

Females were conditioned and tested as described above (five associative conditioning trials, tests performed over 2 h and then 24 h later). Four groups were compared: mated or virgin, and for both mating statuses, full or decreased egg load. Decreased egg load was obtained by a 1-h-oviposition period on a patch of ad libidum host larvae, 24 h before conditioning. Females are not expected to mature new ovocytes because this species was characterized as pro-ovogenic. They emerge with about 200 mature ovocytes (Carton et al., 1986) and infest about 60 larvae per hour under our experimental conditions, which corresponds to a loss of about one-third of the initial egg load per hour.

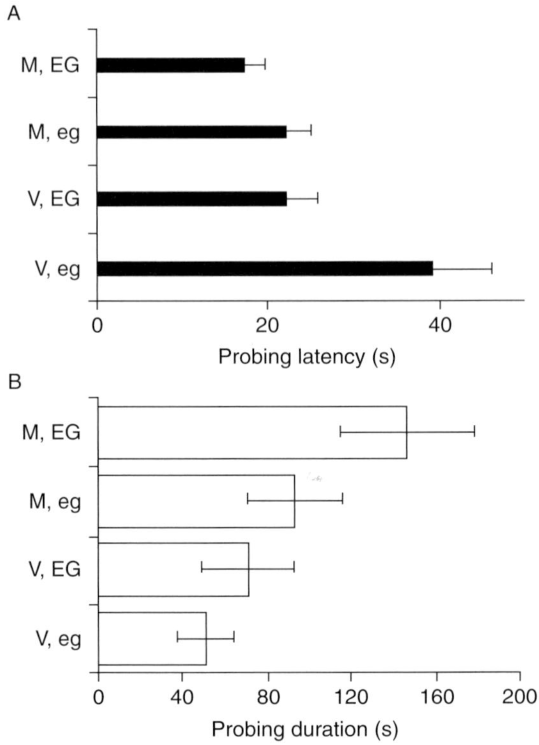

FIGURE 3.5 Mating and egg-load influence parameters of the conditioned probing response to banana odour. (A) Effect on the latency of the response. Females that are mated (M) and have their full egg load (group M, EG) respond the most rapidly, whereas unmated females with decreased egg-load (group V, eg) present the longest latency of response. (B) Effect on the duration of the response. The longest probing duration is observed in mated females with their full egg load, and the shortest in virgin females with decreased egg load. These effects of both factors are significant. *Note: Redrawn from data of Perez-Maluf and Kaiser (1998).*

The probability of conditioned probing response to banana odour was not influenced either by the mating or the egg-load status. However both factors had a significant effect on the latency and duration of the learned response. In the short term, females with full egg load probed for a longer time than egg-depleted females, and this a longer probing was also observed in association with mated status (within each egg-load status). The same effect was maintained in the longer term (Fig. 3.5), where probing responses were solely due to associative memory, without a sensitization effect. In addition, in both the short and long term, both mating and egg-load status affected the latency of the conditioned response. Mated females responded more rapidly than virgin ones (for each egg-load level), as well as females with full egg load compared to females with decreased egg load (for each mating category). A decrease of latency combined with an increase in probing duration can be simply interpreted as an increase in motivation to search for hosts. So here, mated females with full egg load exhibited the highest motivation to search, more than double the lowest one, observed for virgin females with reduced egg load. These results are coherent with prediction of optimal foraging theories, linking behavior to fitness gain. A virgin female lays only male eggs due to haplo-diploid sex-determination, so its motivation to oviposit is expected to be lower than in mated females. A mated female having already laid about 60 eggs is expected to show lower motivation than a female having laid five eggs. This study on *L. boulardi* showed how mating and egg load can modulate the recall of memory for a host-related stimulus according to its physiological need. This could be one mechanism accounting for the recent finding of the relationship between mating status and adjustment of patch-time to host density. Fauvergue et al. (2008) demonstrated in field conditions that in the aphid parasitoid *L. testaceipes*, mated females adjusted their patch time and rate of attack to host density whereas virgin females did not, which could result from differing learning performance while foraging on the patches.

3.6. Genetic Variability of the Learned Searching Response

Next we investigated whether interindividual variability of the conditioned probing response could be partly explained by genotype variability. Genetic differences between individual abilities for chemical learning have been observed in flies (McGuire and Hirsch, 1977; Mery and Kawecki, 2002; Mery et al., 2007; Tully and Hirsch, 1982) and bees (Bhagavan et al., 1994; Brandes, 1991) where bidirectional selection allowed isolation of good and poor learning lines. Here we used the technique of isofemale lines (Carton et al., 1989; Hoffman and Parsons, 1988) to estimate the proportion of variance explained by genetic differences (Perez-Maluf et al., 1998). An isofemale line is the progeny from a single randomly mated female. Significance of genetic variance in a population can be obtained by analysis of variance (ANOVA) on the line effect.

Thirty lines were started at the grandmother generation. Females of the mother generation were mated with their brothers[7] to reduce genetic variability within lines. At the daughter generation, seven or eight females per mother were randomly taken to be conditioned and tested. No variation was

[7] Due to haplo-diploidy, brothers are haploid and share 50% of their genes with their sisters.

found in the probability of conditioned probing to banana odour, but there was a significant line effect on the latency of the response (Fig. 3.6). The probing duration did not differ significantly between lines. Since the test was performed 1 h after conditioning, differences in latency were attributable to genetic variation in sensitization or associative memory.

This genetic variability of a learned response to a host-habitat cue within a laboratory strain encouraged us to question the adaptive character of the trait by comparing two natural populations facing different levels of identified reproductive constraints.

3.7. PROBING IN RESPONSE TO FRUIT ODOUR: WHEN IS IT ADAPTIVE?

3.7.1. Differentiation of innate but not learned responses to host-habitat odours between two genotypes of *L. heterotoma*

Since only the latency of the probing response to a fruit odour has a genetic variability, we compared populations for which activity levels should contribute to fitness and is thus expected to be favored by natural selection. In *L. heterotoma*, a population from the Mediterranean French coast (Antibes) and one from Burgundy (Tailly) have genetically different circadian rhythms, rates of locomotory activity, rates of oviposition activity (Fleury et al., 1995) and fecundity (Ris, 2003; Chapter 1 by Fleury and Allemand). Higher rates of locomotion and oviposition observed in the Mediterranean strain are interpreted as a mechanism to coexist with the competitor species *L. boulardi*, which is absent from Burgundy (Allemand et al., 1999), having mainly a Mediterranean and tropical distribution (Carton et al., 1986). *L. heterotoma* has to oviposit first to win the larval competition when both parasitoids infest the same larva (Fleury et al., 2000), and hence has a limited time window for successful parasitism. We compared both populations for their oviposition behavior, and for their naïve and conditioned probing responses to banana odour (Perez-Maluf et al., 2008). Odour conditioning of probing was done as before (five odour-associated ovipositions) and conditioned responses were tested the next day. Naïve females were treated identically, except that they were not exposed to larvae or odour before being tested.

The results showed a differentiation of innate but not learned responses to banana odour (Fig. 3.7). The more active genotype, Antibes, had a higher probability and a shorter latency of innate probing to the odour than the less active genotype, Tailly, but probing durations were not different. Antibes females also had a higher probability of innate probing to pear flavor, but both populations showed equally low values in response to mushroom flavor. This is coherent with the fact that

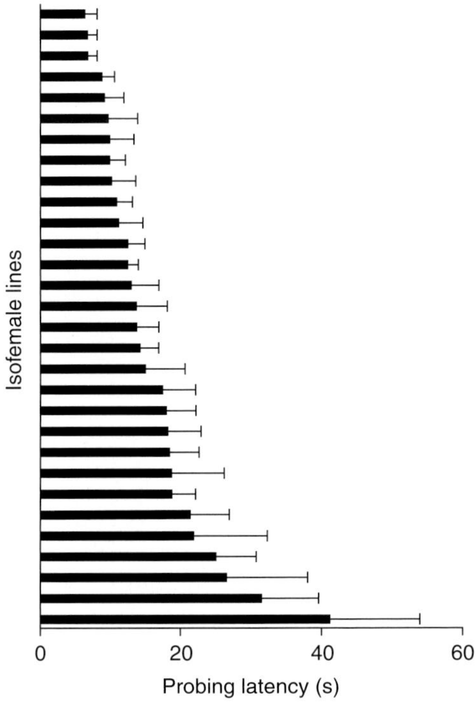

FIGURE 3.6 Variability of the latency of the conditioned probing response to banana odour between isofemales lines. Each bar represents the mean value (+ standard error of the mean) of seven or eight sisters per line. The observed variability from 6 to 42 s is significant. *Note: Redrawn from data of Perez-Maluf* et al. *(1998).*

L. heterotoma is mainly found in frugivorous *Drosophila*, and the studied strains were started from insects caught in fruit-baited traps. Antibes females also found larvae and completed infestations more rapidly. Odour learning equalized the probability and the latency of probing to the odours in both strains, by decreasing the latency of the slow strain. Learning also increased the probing duration of both strains.

These results are in accordance with the previous finding of genetic variability in the latency to probe when *L. boulardi* responds to habitat odours. They additionally indicate a selection of faster innate responses to both host and habitat cues. There was however no differentiation of learned responses between the two studied *L. heterotoma* populations. This suggests that the initial host discovery is more crucial to fitness than subsequent ones. It is consistent with the wasp and host biology. The initial host discovery requires a long distant search because no young larvae can be found on the decayed fruit the wasps emerge from. Once a first larva has been discovered, further host discoveries should be quite probable because host larvae are gregarious.

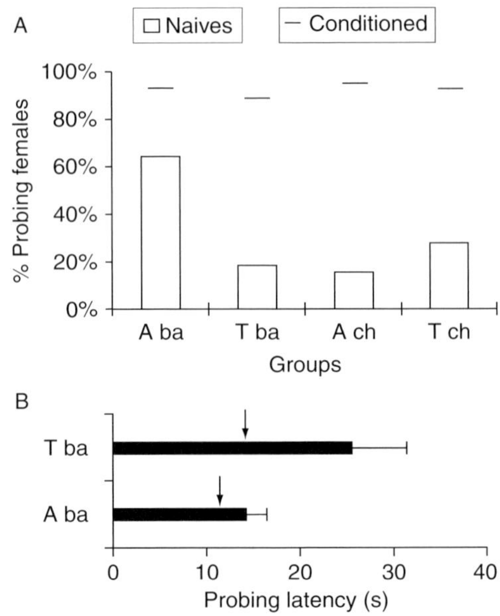

FIGURE 3.7 Variability of naïve and conditioned probing responses between two strains of *L. heterotoma*. One strain originates from Antibes (A) and the other from Tailly (T). (A) A high probability of naïve response to banana odour (ba) is exhibited by Antibes females, but not in response to mushroom odour, and not in females from Tailly. After associative conditioning to banana or mushroom, females from both strains exhibit a high probability of probing to their CS (test performed 24 h after conditioning), without a difference between strains or odour. (B) Naïve females of Tailly present a longer latency of response to banana odour than females from Antibes. After associative conditioning, both strains present equally short latency of response (arrows).
Note: Redrawn from data of Perez-Maluf et al. *(2008).*

3.7.2. Genetic components of innate fruit odour recognition

In a preliminary experiment with two strains of *L. boulardi*, we observed that they differed in their innate frequency of probing to banana odour, just as described above between the two populations of *L. heterotoma*. In the strain originating from Nasrallah (Tunisia), innate responses were rare, whereas they were frequent in the one from Brazzaville (Congo). It was interesting to investigate the genetic basis of this difference in searching behavior, because these strains should face contrasted constraints on reproductive success, linked to their genetic differences in virulence (ability to suppress the encapsulation of their eggs by larvae of *D. melanogaster*; Carton and Nappi, 2001; Dupas et al., 1998; Chapter 6 by Dubuffet et al.). Encapsulation is fully suppressed by the strain from Nasrallah, whereas about half of the Brazzaville progeny is killed by to the host immune reaction (Carton and Nappi, 1991).

We compared innate probing responses of both strains and their hybrids, using a range of fruit aromas (banana, pear, orange and prickly pear) and the mushroom aroma (Campan et al. 2002). The four fruits are breeding sites for *L. boulardi*, and the species is never found on mushrooms. As before, the ovipositor probing into agar in response to an odour was characterized as the frequency of responding females, and the latency and duration of their probing response.

The preliminary observation was confirmed for all fruit odours. Most Brazzaville females exhibited innate probing in response to the four fruit odours, whereas this was rare in the Nasrallah strain. Mushroom odour did not trigger any probing response in either strain.[8] In addition, the latency of response was shorter in Brazzaville females, but not the duration of response, which was not different between strains.

In parallel, the probing behavior was observed in the first (F1) and second (backcrosses: BC) hybrid generations. This was done only in response to banana odour, because the parental difference was found for all fruits. Due to haplo-diploidy, F1 males are not hybridized, having only a maternal origin. The cytoplasm of BC females is only of maternal origin, as well as 75% of their chromosomes. Then, it is possible to estimate the linear regression between the mean value of a trait and the percentage of the nuclear genome from the maternal lineage, equal to 50 in F1 hybrid daughters, 75 in BC daughters, 100 in their mother and 0 in the mother of the other strain. It is also possible to compare the mean value measured in a hybrid group and the expected value under the hypothesis of purely additive inheritance of the studied trait, which is the mid value of the theoretical mid-parent of the hybrid (estimated by the mean value of the hybrid's parents).

Results from hybrid females (F1 and BCs) showed that both probability and latency of the probing response to banana odour were strongly heritable. Regarding the probability to probe (Fig. 3.8A), there was an apparent complete dominance of Nasrallah characteristics, whereas additive inheritance was found in the Brazzaville lineage. The contrast between reciprocal F1 hybrids suggested a strong interaction with nonchromosomal factors. It is possible that a cytoplasmic factor in the Nasrallah strain decreases females' propensity to probe in response to significant odours. The latency of probing appeared to be under a chromosomal influence (Fig. 3.8B).

The fact that innate probing in response to host habitat odour can be strongly genetically determined suggests genetic variability in the

[8] The fact that *L. heterotoma* can probe in response to mushroom odour but not *L. boulardi* reflects their difference in host range.

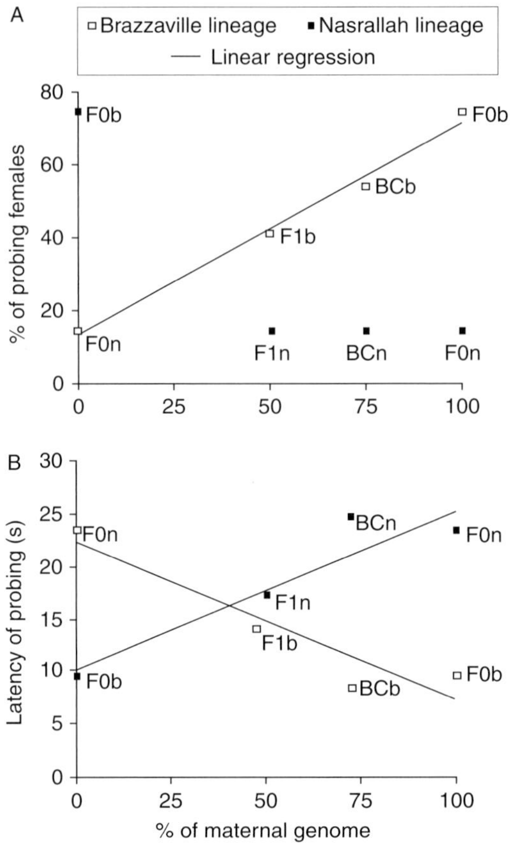

FIGURE 3.8 Heritability of naïve probing response shown by crosses between two strains of *L. boulardi*. Linear regression between the probability of response and the percentage of maternal genome was calculated based on the parental generation (F0), the F1 hybrids and the backcrosses (BC) between Nasrallah (n) and Brazzaville (b). (A) The probability of innate probing to banana odour is high in Brazzaville mothers and low in those of Nasrallah. Low values of hybrids from Nasrallah maternal origin (F1n, BCn) show an apparent complete dominance of Nasrallah characteristics, whereas values of hybrids from Brazzaville maternal origin are proportional to the percentage of their maternal genome, showing additive inheritance in the Brazzaville lineage. (B) The latency of probing follows a significant linear regression with the percentage of maternal genome, showing additive inheritance of this strain for both maternal origin. *Note: Adapted from Campan* et al. *(2002).*

threshold for triggering the ovipositor search. Indeed, Nasrallah females can exhibit a high probing frequency to banana odour if memorized by associative learning, equivalent to the innate response of Brazzaville females. At the population level, the genetically determined high level of ovipositor searching activity in naïve Brazzaville females might contribute to balance the reproductive constraint linked to the death of half the progeny due to the poor ability of the strain to suppress the immune reaction of the host.

3.8. GENERAL DISCUSSION AND CONCLUSIONS

Based on the observation of a searching response to a simple stimulus representing a potential host-habitat, this experimental analysis of behavior documents contributions of the intrinsic factors determining behavioral variability: individual experience (here, learning), physiological state and genotype. It pointed out that the probability of exhibiting a behavioral response is determined by both genotype and individual experience, but not by the physiological state defined by mating and egg load. With regards to the quality of the searching response, its rapidity is determined by the three factors, but its duration does not depend on the genotype (Table 3.1). This may be explained by the high interindividual variability of this trait. The study on inheritance of searching traits in *L. boulardi* showed that in one strain, a nonchromosomal factor had a strong influence on females' propensity to probe in response to fruit odour. This might be due to a symbiotic organism. In *L. boulardi*, a symbiotic virus affects other behaviors (see Chapter 13 by Varaldi et al.). We also documented an interaction between learning and genotype, since the genotype influences naïve responses but not learned responses.

Altogether, these results indicate that naïve but not learned responses are subject to selection, as well as the rapidity but not the duration of naïve responses. This suggests that finding the first host is more crucial to the parasitoid female fitness than finding subsequent ones, which is

TABLE 3.1 Presence or absence of the effects of learning, physiological state and genotype on the probability and the quality (latency, duration) of searching responses to fruit odours

	Probability	Latency	Duration
Learning[a]	YES	YES	YES
Physiological state[b] (mating and egg load)	NO	YES	YES
Genotype[c]	YES	YES	NO

Notes: [a] The effect of learning was assessed in both orientation and probing responses of *L. boulardi,* and in probing responses of *L. heterotoma.*
[b] The effect of physiological state was assessed in probing responses of *L. boulardi.*
[c] The effect of the genotype was assessed in probing responses of *L. boulardi* and *L. heterotoma.*

coherent with the aggregative nature of the host larvae. This also corroborates the often accepted idea that selection is less efficient on more plastic traits, which is the case for those influenced by learning and physiological state. The finding of innate responses to host habitat odour in a particular genotype also represents variation to the general scheme of host selection. In this scheme (Vet et al. 1990; Vinson, 1976), searching responses to host-habitat odours result from learning, and host-habitat odours attract females from a distance whereas subsequent behaviors involved in host searching within the habitat, host examination and oviposition, are triggered mostly by host specific stimuli. Here we document unusual situations where habitat odour triggers innate responses of searching within the habitat. This particularity can be explained by identified constraints on the probability to reproduce, either environmental (*L. heterotoma* in Perez-Maluf et al., 2008) or intrinsic to the genotype (*L. boulardi*, Campan et al., 2002). We documented important genetic variability in searching responses to host-habitat odour, in these two species. This stresses one great interest of *Drosophila* parasitoids, which is the availability of several well-known populations in different species (see Chapter 1 by Fleury and Allemand). This is one key element for the study of the evolution of behavior.

The studies on the ability to learn successively different host-habitat odours pointed out original and remarkable cognitive capacity of the small wasp *L. boulardi*, showing that they are able to track the timing of events they experience. Whether such memory capacity is used in real life or not still needs to be investigated because both the host and the parasitoid are expected to prefer the first host-habitat they have encountered, which limits the probability of shifting to a different substrate. The experiments on the odour-conditioning of the probing behavior revealed an important plasticity determined by experienced events[9] and motivation arising from mating and egg load. This conditioned response, which represents a searching response to the presence of host larvae, is finely tuned to host availability and to individual need. It can be seen as a potential mechanism of Bayesian updating of host resources, and of optimal foraging in general. Considering its cognitive capacities, the adaptive plasticity of its learned responses and its relatively long life, *L. boulardi* is a useful model to address current issues on mechanisms to cope with temporal variations faced by insects.

We failed to find genetic variation in conditioned responses, but we may have missed traits of memory that might be genetically determined. Long-lasting memory requires *de novo* protein synthesis,[10] which corresponds to a physiological cost which has been estimated in estimated in

[9] Number of successful ovipositions, of encounters with the odour in the absence of host larvae.
[10] This has been verified for many animal and insect species, including a parasitic wasp (Collatz et al., 2006).

Drosophila (Mery and Kawecki, 2005). The duration of memory may be one trait subjected to positive selection. More generally, if learning and memory performances have been shaped by evolutionary constraints, it is expected that they will differ between species or even populations. Such constraints can depend on the host range, learning being expected to be more developed in generalist than specialist insects (Poolman-Simons et al., 1992; Vet et al., 1995). Memory capacity could also differ between egg-limited and time-limited species, and according to the frequency of decisions female parasitoids make (Roitberg et al., 1993) and to host resource predictability (Stephens, 1993). Regarding the biological diversity among *Drosophila* parasitoids, they are an invaluable group to address the evolution of cognitive capacities.

In this review, we have presented intrinsic sources of behavioral variability. An extrinsic source, sublethal exposure to neurotoxic insecticides, was also investigated in *L. boulardi* and *L. heterotoma*. It transiently increased the probing activity in response to host larvae and to banana odour, but also without identified stimulus (Rafalimanana et al., 2002). It increased the residence time in agarose patches coated with host-larvae products (Delpuech et al., 2005; Komeza et al., 2001). Changes in activity were coherent with known neuronal effects of the insecticide molecules.[11] It could be also interpreted as an insect response to such a lethal threat, by increasing reproductive activity, or to low host availability in treated areas, by staying in a host-infested patch. This extrinsic trigger represents an additional factor of the great plasticity of the probing behavior.

ACKNOWLEDGEMENTS

The authors express their gratitude to their colleagues and students that participated in the studies presented in this chapter, for their essential contribution: the research was carried out at the former Laboratoire de Neurobiologie des Invertébrés, INRA, Bures-sur-Yvette, under the direction of C. Masson and M. H. Pham-Delègue, and in collaboration with Y. Carton, R. De Jong and L. Vet, E. Wajnberg, F. Fleury and J. C. Sandoz, with the contribution of the master students C. Bartheye, V. Kerguelen, E. Campan and H. Rafalimanana, and with the technical assistance of F. Frey, D. Huet, N. Menay, R. Charles and E. Genecque. We also thank L. Guibouret from Haarmann and Reimer for providing the fruit aromas.

REFERENCES

Allemand, R., Fleury, F., Lemaitre, C., Boulétreau, M., 1999. Dynamique des populations et interactions compétitive chez deux espèces de *Leptopilina*, parasitoïdes de Drosophiles, dans la Vallée du Rhone (Hymenoptera, Figitidae). Ann. Soc. Entomol. France 35, 97–103.

Arthur, A.P., 1966. Associative learning in *Itoplectis conquisitor* Say (Hym., Ichneumonidae). Can. Entomol. 98, 213–223.

[11] Chlorpyrifos ethyl prolongs synaptic activity and can explain the activation of the probing behavior.

Arthur, A.P., 1971. Associative learning by *Nemeritis canescens* (Hym., Ichneumonidae). Can. Entomol. 103, 1137–1141.

Bhagavan, S., Benatar, S., Cobey, S., Smith, B.H., 1994. Effect of genotype but not of age or caste on olfactory learning performance in the honeybee, *Apis mellifera*. Anim. Behav. 48, 1357–1369.

Brandes, C., 1991. Genetic differences in learning behavior in honeybees (*Apis mellifera capensis*). Behav. Genet. 21, 271–274.

Campan, E., Couty, A., Carton, Y., Pham-Delegue, M.H., Kaiser, L., 2002. Variability and genetic components of innate fruit odour recognition in a parasitoid of *Drosophila*. Physiol. Entomol. 27, 243–250.

Carton, Y., Nappi, A., 1991. The *Drosophila* immune reaction and the parasitoid capacity to evade it, genetic and coevolutionary aspects. Acta Oecol. 12, 89–104.

Carton, Y., Nappi, A.J., 2001. Immunogenetic aspects of the cellular immune response of *Drosophila* against parasitoids. Immunogenetics 52, 157–164.

Carton, Y., Boulétreau, M., van Alphen, J.J.M., van Lenteren, J.C., 1986. The *Drosophila* parasitic wasps. In: Ashburner, Carlson, Thompson, (Eds.), The Genetics and Biology of *Drosophila*, Vol. 3. Academic Press, London, pp. 347–394.

Carton, Y., Capy, P., Nappi, A., 1989. Genetic variability of host–parasite relationship traits, utilization of isofemale lines in a *Drosophila simulans* parasitic wasp. Genet. Select. Evol. 21, 437–446.

Chabaud, M.A., Devaud, J.M., Pham-Delègue, M.H., Preat, T., Kaiser, L., 2006. Olfactory conditioning of proboscis activity in *Drosophila melanogaster*. J. Comp. Physiol. A 192, 1335–1348.

Chapman, T., Davies, S.J., 2004. Functions and analysis of the seminal fluid proteins of male *Drosophila melanogaster* fruit flies. Peptides 25, 1477–1490.

Collatz, J., Müller, C., Steidle, J.L.M., 2006. Protein synthesis-dependent long-term memory induced by one single associative training trial in the parasitic wasp *Lariophagus distinguendus*. Learn. Mem. 13, 263–266.

Collins, M.D., Dixon, A.F.G., 1986. The effect of egg depletion on the foraging behaviour of an aphid parasitoid. J. Appl. Entomol. 102, 342–352.

Colomb, J., Kaiser, L., Chabaud, M.A., Preat, T., 2009. Parametric and genetic analysis of *Drosophila* appetitive long-term memory and sugar motivation. Genes Brain Behav. 8, 407–415.

Couty, A., Kaiser, L., Huet, D., Pham-Delègue, M.H., 1999. The attractiveness of different odour sources from the fruit-host complex on *Leptopilina boulardi*, a larval parasitoid of *Drosophila*. Physiol. Entomol. 24, 76–82.

De Jong, R., Kaiser, L., 1991. Odour learning by *Leptopilina boulardi*, a specialist parasitoid (Hymenoptera, Eucoilidae). J. Insect Behav. 4, 743–750.

De Jong, R., Kaiser, L., 1992. Odour preference of a parasitic wasp depends on order of learning. Experientia 48, 902–904.

Delpuech, J.M., Bardon, C., Boulétreau, M., 2005. Increase of the behavioural response to kairomones by the parasitoid wasp *Leptopilina heterotoma* surviving insecticides. Arch. Environ. Contam. Toxicol. 49, 186–191.

Dicke, M., van Lenteren, J.C., Boskamp, G.J.F., van Dongen-van Leeuwen, E., 1984. Chemical stimuli in host-habitat location by *Leptopilina heterotoma* (Thomson) (Hymenoptera, Eucoilidae), a parasite of *Drosophila*. J. Chem. Ecol. 10, 695–712.

Dupas, S., Frey, F., Carton, Y., 1998. A single parasitoid segregating factor controls immune suppression in *Drosophila*. J. Heredity 89, 306–311.

Fauvergue, X., Lo Genco, A., Lo Pinto, M., 2008. Virgins in the wild, mating status affects the behaviour of a parasitoid foraging in the field. Oecologia 156, 913–920.

Fleury, F., Allemand, R., Fouillet, P., Boulètreau, M., 1995. Genetic variation in locomotor activity rhythm among populations of *Leptopilina heterotoma* (Hymenoptera, Eucoilidae), a larval parasitoid of *Drosophila* species. Behav. Genetics 25, 81–89.
Fleury, F., Allemand, R., Vavre, F., Fouillet, P., Boulétreau, M., 2000. Adaptive significance of a circadian clock, temporal segregation of activities reduces intrinsic competitive inferiority in *Drosophila* parasitoids. Proc. R. Soc. Lond. B 267, 1005–1010.
Friedrich, A., Thomas, U., Muller, U., 2004. Learning at different satiation levels reveals parallel functions for the cAMP-protein kinase A cascade in formation of long-term memory. J. Neurosci. 24, 4460–4468.
Gallistel, C.R., 1989. Animal cognition, the representation of space, time and number. Ann. Rev. Psychol. 40, 155–189.
Hedlund, K., Vet, L.E.M., Dicke, M., 1996. Generalist and specialist parasitoid strategies of using odours of adult drosophilid flies when searching for larval hosts. Oikos 77, 390–398.
Hoffmann, A.A., Parsons, P.A., 1988. The analysis of quantitative variation in natural populations with isofemale strains. Genetic Select. Evol. 20, 87–98.
Kaiser, L., De Jong, R., 1993. Multi-odour memory influenced by learning order. Behav. Proc. 30, 175–184.
Kaiser, L., Marion-Poll, F. (in press). Lutte contre les insectes ravageurs des plantes. In: Baudoin, Boissy, Pham-Delégue, (Eds.), Ethologie Appliquée, Editions Quae, Versailles, France.
Kaiser, L., Bartheye, C., Kerguelen, V., Pham-Delègue, M.-H., 1995. Odour conditioning of ovipositor probing in a parasitic wasp. Ethol. Ecol. Evol. 7, 245–255.
Kaiser, L., Perez-Maluf, R., Sandoz, J.C., Pham-Delègue, M.H., 2003. Dynamics of odour learning in *Leptopilina boulardi*, a Hymenopterous parasitoid. Anim. Behav. 66, 1077–1084.
Kester, K.M., Barbosa, P., 1991. Behavioral and ecological constraints imposed by plants on insect parasitoids, implications for biological control. Biol. Control 1, 94–106.
Kolterman, R., 1974. Periodicity in the activity and learning performance of the honeybee. In: Barton-Brown, (Ed.), Experimental Analysis of Insect Behavior. Springer, Heidelberg, New York, pp. 218–227.
Komeza, N., Fouillet, P., Boulétreau, M., Delpuech, J.M., 2001. Modification, by the insecticide chlorpyrifos, of the behavioural response to kairomones of a parasitic wasp, *Leptopilina boulardi*. Arch. Environ. Contamin. Toxicol. 41, 436–442.
Lewis, W.J., Takasu, K., 1990. Use of learned odours by a parasitic wasp in accordance with host and food needs. Nature 348, 635–636.
Lewis, W.J., Tumlinson, J.H., 1988. Host detection by chemically mediated associative learning in a parasitic wasp. Nature 331, 257–259.
Mangel, M., 1993. Motivation, learning and motivated learning. In: Papaj, Lewis, (Eds.), Insect Learning – Ecological and Evolutionary Perspectives. Chapman & Hall, London, pp. 158–173.
McFarland, D., 2001a. In: Le Comportement Animal, Section 2.2. L'Animal et L'Environnement. De Boeck Université, Paris, pp. 261–282.
McFarland, D., 2001b. In: Le Comportement Animal, Section 2.3. L'Apprentissage Animal. De Boeck Université, Paris, pp. 339–352.
McGuire, T.R., Hirsch, J., 1977. Behavior-genetic analysis of *Phormia regina*, conditioning, reliable individual differences, and selection. Proc. Natl. Acad. Sci. USA. 74, 5193–5197.
Menzel, R., 1999. Memory dynamics in the honeybee. J. Comp. Physiol. A 185, 323–340.
Menzel, R., Greggers, U., Hammer, M., 1993. Functional organization of appetitive learning and memory in a generalist pollinator, the honey bee. In: Papaj, Lewis, (Eds.), Insect Learning – Ecological and Evolutionary Perspectives. Chapman & Hall, London, pp. 79–125.

Mery, F., Kawecki, T.J., 2005. A cost of long-term memory in *Drosophila*. Science 308, 1148.
Mery, F., Kawecki, T.J., 2002. Experimental evolution of learning ability in fruit flies. Proc. Natl. Acad. Sci. USA. 22, 14274–14279.
Mery, F., Pont, J., Preat, T., Kawecki, T.J., 2007. Experimental evolution of olfactory memory in *Drosophila melanogaster*. Physiol. Biochem. Zool. 80, 399–405.
Papaj, D., Vet, L.E.M., 1990. Odour learning and foraging success in the parasitoid, *Leptopilina heterotoma*. J. Chem. Ecol. 16, 3137–3150.
Perez-Maluf, R., Kaiser, L., 1998. Mating and oviposition experience influence odour learning in *Leptopilina boulardi* (Hymenoptera, Eucoilidae), a parasitoid of *Drosophila*. Biol. Control 11, 154–159.
Perez-Maluf, R., Kaiser, L., Wajnberg, E., Carton, Y., Pham-Delègue, M.-H., 1998. Genetic variability of conditioned probing responses to a fruit odour in *Leptopilina boulardi* (Hymenoptera, Eucoilidae), a *Drosophila* parasitoid. Behav. Genet. 28, 67–73.
Perez-Maluf, R., Rafalimanana, H., Campan, E., Fleury, F., Kaiser, L., 2008. Differentiation of innate but not learnt responses to host-habitat odours contributes to rapid host finding in a parasitoid genotype. *Physiol. Entomol.* 33, 226–232.
Pettersson, J., 1970. An aphid sex attractant. I. Biological studies. Entomol. Scand. 1, 63–73.
Poolman-Simons, M.T.T., Suverkropp, B.P., Vet, L.E.M., De Moed, G., 1992. Comparison of learning in related generalist and specialist eucoilid parasitoids. Entomol. Exp. Appl. 64, 117–124.
Rafalimanana, H., Kaiser, L., Delpuech, J.-M., 2002. Stimulating effects of the insecticide chlorpyrifos on host searching and infestation efficacy of a parasitoid wasp. Pest Manag. Sci. 58, 321–328.
Ris, N., 2003. In: Spatial heterogeneity, phenotypic plasticity and environmental trade-off, role of host species and temperature in the genetic differentiation of populations of the parasitoid *Leptopilina heterotoma* (Hymenoptera) [PhD thesis]. University Claude Bernard Lyon 1, p. 159 (in French).
Roitberg, B.D., Reid, M.L., Li, C., 1993. Choosing hosts and mates, the value of learning. In: Papaj, Lewis, (Eds.), Insect Learning – Ecological and Evolutionary Perspectives. Chapman & Hall, London, pp. 174–194.
Schwärzel, M., Müller, U., 2006. Dynamic memory networks: dissecting molecular mechanisms underlying associative memory in the temporal domain. Cell. Mol. Life Sci. 63, 989–998.
Slater, P.J.B., 1985. In: An Introduction to Ethology. Cambridge University Press, Cambridge, p. 196.
Smith, B., 1993. Merging mechanisms and adaptation, an ethological approach to learning and generalization. In: Papaj, Lewis, (Eds.), Insect Learning – Ecological and Evolutionary Perspectives. Chapman & Hall, London, pp. 126–157.
Stephens, D.W., 1993. Learning and behavioral ecology, incomplete information and environmental predictability. In: Papaj, Lewis, (Eds.), Insect learning – Ecological and evolutionary perspectives. Chapman & Hall, London, pp. 195–218.
Tanimoto, H., Heisenberg, M., Gerber, B., 2004. Event timing turns punishment to reward. Nature 430, 26.
Tentelier, C., Fauvergue, X., 2007. Herbivore-induced plant volatiles as cues for habitat assessment by a foraging parasitoid. J. Anim. Ecol. 76, 1–8.
Tentelier, C., Lacroix, M.N., Fauvergue, X., 2009. Inflexible wasps, the aphid parasitoid *Lysiphlebus testaceipes* does not track multiple changes in habitat profitability. Anim. Behav. 77, 95–100.
Thiel, A., Driessen, G., Hoffmeister, T.S., 2006. Different habitats, different habits? Response to foraging information in the parasitic wasp *Venturia canescens*. Behav. Ecol. Sociobiol. 59, 614–623.

Tully, T., Hirsch, J., 1982. Behavior-genetic analysis of *Phormia regina*. I. Isolation of lines pure-breeding for a high and low levels of central excitatory state (CES) from an unselected population. Behav. Genet. 12, 395–415.

Turlings, T.C.J., Wäckers, F.L., Vet, L.E.M., Lewis, W.J., Tumlinson, J.H., 1993. Learning of host-finding cues by hymenopterous parasitoid. In: Papaj, Hewis, (Eds.), Insect Learning Ecological and Evolutionary Perspectives. Chapman & Hall, London, pp. 51–78.

Van Lenteren, J.C., 1976. The development of host discrimination and the prevention of superparasitism in the parasite *Pseudochoila bochei* (Hym., Cynipidae). Neth. J. Zool. 26, 1–83.

Vet, L.E.M., Bakker, K., 1985. A comparative functional approach to the host detection behaviour of parasitic wasps. 2. A quantitative study on eight eucoilid species. Oikos 44, 487–498.

Vet, L.E.M., Dicke, M., 1992. Ecology of infochemical use by natural enemies in a tritrophic context. Ann. Rev. Entomol. 37, 141–172.

Vet, L.E.M., Groenewold, A.W., 1990. Semiochemicals and learning in parasitoids. J. Chem. Ecol. 16, 3119–3135.

Vet, L.E.M., Van Alphen, J.J.M., 1985. A comparative functional approach to the host detection behaviour of parasitic wasps. 1. A qualitative study on Eucoilidae and Alysiinae. Oikos 44, 478–486.

Vet, L.E.M., De Jong, A.G., Franchi, E., Papaj, D.R., 1998. The effect of complete versus incomplete information on odour discrimination in a parasitic wasp. Anim. Behav. 55, 1271–1279.

Vet, L.E.M., Lewis, W.J., Cardé, R.T., 1995. Parasitoid foraging and learning. In: Cardé, Bell, (Eds.), Chemical Ecology of Insects. 2nd edn Chapman & Hall, London, pp. 65–104.

Vet, L.E.M., Lewis, W.J., Papaj, D.R., van Lenteren, J.C., 1990. A variable-response model for parasitoid foraging behavior. J. Insect Behav. 3, 471–490.

Vet, L.E.M., Sokolowski, M.B., MacDonald, D.E., Snellen, H., 1993. Responses of a generalist and a specialist parasitoid (Hymenoptera, Eucoilidae) to drosophilid larval kairomones. J. Insect Behav. 6, 615–624.

Vet, L.E.M., van Lenteren, J.C., Heymans, M., Meelis, E., 1983. An airflow olfactometer for measuring olfactory response of *Hymenopterous* parasitoids and other small insects. Physiol. Entomol. 8, 97–106.

Vinson, S.B., 1976. Host selection by insect parasitoids. Ann. Rev. Entomol. 21, 109–134.

Vinson, S.B., Barfield, C.S., Henson, R.D., 1977. Oviposition behaviour of *Bracon mellitor*, a parastoid of the boll weevil (*Anthonomus grandis*). II. Associative learning. Physiol. Entomol. 2, 157–164.

Wardle, A.R., 1990. Learning of host microhabitat colour by *Exeristes roborator* F. (Hymenoptera: Ichneumonidae). Anim. Behav. 39, 914–923.

Wardle, A.R., Borden, J.H., 1985. Age-dependant associative learning by *Exeristes roborator* (F.) (Hymenoptera, Ichneumonidae). Can. Entomol. 117, 605–616.

Wertheim, B., Vet, L.E.M., Dicke, M., 2003. Increased risk of parasitism as ecological costs of using aggregation pheromones, laboratory and field study of *Drosophila–Leptopilina* interaction. Oikos 100, 269–282.

Wiskerke, J.S.C., Dicke, M., Vet, L.E.M., 1993. Larval parasitoid uses aggregation pheromone of adult hosts in foraging behaviour – a solution to the reliability-detectability problem. Oecologia 93, 145–148.

Yang, C.H., Rumpf, S., Xiang, Y., Gordon, M.D., Song, W., Jan, L.Y., et al., 2009. Control of the postmating behavioral switch in *Drosophila* females by internal sensory neurons. Neuron 61, 519–526.

SECTION 2

THE PHYSIOLOGY AND GENETICS OF IMMUNITY RELATIONSHIPS BETWEEN PARASITOIDS AND THEIR *DROSOPHILA* HOSTS

CHAPTER 4

The Role of Melanization and Cytotoxic By-Products in the Cellular Immune Responses of *Drosophila* Against Parasitic Wasps

A. Nappi,* **M. Poirié,**† and **Y. Carton**‡

Contents

Abstract

The cellular innate immune response of several species of *Drosophila* terminates with the encasement of large foreign objects within melanotic capsules comprised of several layers of adhering blood cells or hemocytes. This reaction is manifested by various *Drosophila* hosts in response to infection by endoparasitic wasps (i.e., parasitoids). Creditable assessments of the factor(s) causing, or contributing

* Department of Biology, Loyola University of Chicago, Chicago, IL 60525, USA
† UMR Interactions Biotiques et Santé Végétale, Institut Agrobiotech, 06 903 Sophia Antipolis, France. INRA, UMR 1301/CNRS UMR 6243/ Université Nice Sophia Antipolis, 28, avenue de Valrose, 06103 Nice Cedex 2, France
‡ IRD, UR072 Laboratoire Evolution, Génomes et Spéciation/UPR9034, CNRS 91198 Gif-sur-Yvette Cedex, France/Université Paris-Sud 11, 91405 Orsay Cedex, France

Advances in Parasitology, Volume 70
ISSN 0065-308X, DOI: 10.1016/S0065-308X(09)70004-1

to, parasite mortality have long been considered as cytotoxic elements certain molecules associated with enzyme-mediated melanogenesis. However, observations that warrant additional or alternative considerations are those documenting parasitoid survival despite melanotic encapsulation, and those where parasitoids are destroyed with no evidence of this host response. Recent studies of the production of some reactive intermediates of oxygen and nitrogen during infection provide a basis for proposing that these molecules constitute important components of the immune arsenal of *Drosophila*. Studies of the virulence factors injected by female wasps during oviposition that suppress the host response will likely facilitate identification of the cytotoxic molecules as well as the cell-signaling pathways that regulate their synthesis.

4.1. INTRODUCTION

The outcome of a host–pathogen association depends on a series of complex interactions involving behavioral, genetic, physiological and biochemical components of the two competing species. To combat pathogens, mammals and many other vertebrates benefit from the synergistic interactions of two immune systems, adaptive and innate. Adaptive or anticipatory immunity generates an almost limitless repertoire of pathogen-specific responses, enabled in large part by considerable genetic plasticity that accounts for the development of cell-surface receptors and immune memory. Innate immunity is nonadaptive, being dependent instead on constitutive (i.e., germ-line encoded) and dedicated cell membrane-bound pattern recognition receptors with limited responsiveness to invariant molecular motifs of certain pathogens. In each system, receptor–ligand binding leads to the activation of signal transduction pathways, the transcription of immune genes and the generation of reactive cells and cytotoxic effector molecules. Successful pathogens use counter strategies that include virulence factors that actively suppress host responses, or they avoid immune detection, either by finding sanctuary refuge within host tissues that are inaccessible to immune cells and cytotoxic effector responses, or by molecular mimicry.

Whether or not invertebrates are capable of immune phenotypic plasticity and memory comparable to that of mammals has long been a matter of debate. Recent insights have been made with the characterization of the insect Down syndrome cell adhesion molecule (Dscam) as a key immune surveillance factor with characteristics analogous to antibodies (Kurtz and Armitage, 2006; Watson et al., 2005). Dscam shows extensive somatic diversification in the immune system both in insects and crustaceans (Brites et al., 2008). It is nevertheless still largely admitted that insects and other invertebrates rely exclusively on innate immune responses to combat infections successfully (Beutler, 2004; Carton et al., 2008; Christensen et al., 2005; Hoffmann, 2003; Royet, 2004; Siva-Jothy et al., 2005; Stanley, 2006; Stanley et al., 2009; Tafalla et al., 2003). Immune effector responses elicited by prokaryotic infections may include phagocytosis, hemolymph coagulation, and the synthesis of proinflammatory cytokines, antimicrobial peptides, reactive intermediates of oxygen (ROI) and nitrogen (RNI), and stress-related proteins (Bettencourt et al., 2004; Bidla et al., 2005; Bodian et al., 2004; Foley and O'Farrell, 2003; Hoffmann, 2003; Kanost et al., 2004; Krishnan et al., 2006; Lemaitre and Hoffmann, 2007; Molina-Cruz et al., 2008; Nappi et al., 2000b; Novas et al., 2004; Ottaviani et al., 2004; Pacelli et al., 1995; Rivero, 2006; Sharma et al., 2008; Shrestha and Kim, 2008; Stanley et al., 2009; Tafalla et al., 2003; Terland et al., 2006; Wright et al., 2006; Fig. 4.1). Eukaryotic parasites that are too large to be phagocytosed provoke an encapsulation response mediated by macrophage-like blood cells (hemocytes) that rapidly form multilayer capsules around the foreign organisms (Carton et al., 2008; Lavine and Strand, 2002). In insects and other arthropods, hemocyte-mediated encapsulations characteristically are accompanied by melanogenesis, a feature that has long been viewed as evidence that the process constitutes an essential component of the defense response of these animals (Bidla et al., 2007; Cerenius and Soderhall, 2004; Christensen and Soderhall, 2005; De Gregorio et al., 2002a; Ligoxygakis et al., 2002; Liu et al., 2007; Nappi and Christensen, 2005). However, the precise role of melanogenesis in insect immunity remains to be established, and there exists little information as to the identity of the killing molecules and their target-specific mode of action. Observations that question the role of melanin in defense against endoparasites are those where they succumb early during capsule formation or when there is no visible evidence of a melanotic capsule, and reports that document parasite survival despite extensive melanogenic activity (Henter and Via, 1995; Nappi and Streams, 1970; Schnitger et al., 2007; Tardieu and Rabasse, 1988; Vernick et al., 1995).

The interactions of various species of endoparasitic wasps with their *Drosophila* hosts provide exceptionally good models for investigating various aspects of insect cellular innate immunity and mechanisms of parasite virulence (Carton and Nappi, 1997; Carton et al., 2005; Dubuffet et al., 2007; Dupas et al., 2003; Fleury et al., 2004; Nappi et al., 1991). The availability of well-defined resistant and susceptible host lines, and virulent and avirulent parasitoid lines (see Chapter 6 by Dubuffet et al.), made possible comparative investigations of the genetic and biochemical aspects of the cellular immune reactions of *Drosophila* larvae, and the way parasitoids deal with such reactions.

This review focuses on the role of melanization and of the cytotoxic molecules generated during host hemocyte-mediated melanotic

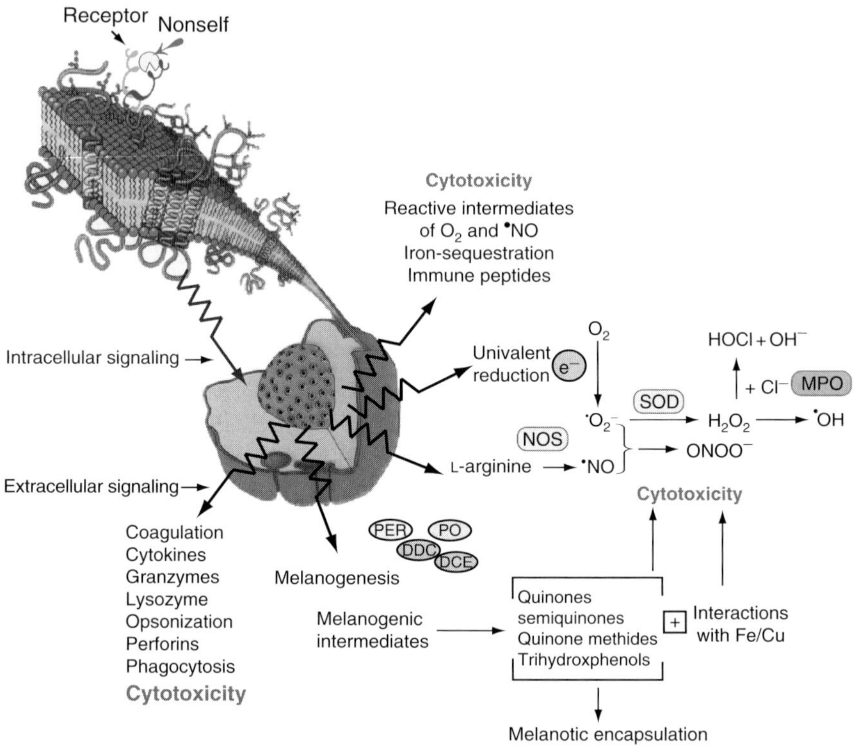

FIGURE 4.1 Overview of some innate immune responses manifested by various invertebrates. Melanotic encapsulation is a characteristic defense reaction made by arthropods infected with eukaryotic pathogens. Nonself recognition may involve plasma membrane receptors independently functioning, or cooperatively engaging nonself binding molecules in the host's hemolymph. *Notes: DCE, dopachrome conversion enzyme; DDC, dopa decarboxylase; MPO, myeloperoxidase; NOS, nitric oxide synthase; PER, peroxidase; PO, phenoloxidase; SOD, superoxide dismutase.*

encapsulation responses to endoparasitic wasps, and the ability of certain of them to circumvent these potentially damaging molecules. However, in rare cases, markedly different immune responses occur in some *Drosophila* spp., such as the species belonging to the *melanica* group (Nappi and Streams, 1970), where eggs or young parasitoid larvae succumb to the host's immune response but are never enveloped by melanotic capsules. Recent studies of the virulence factors injected by female wasps during oviposition that suppress the host response will likely facilitate identification of the cytotoxic molecules and the cell-signaling pathways regulating their synthesis (Poirié et al., 2009).

4.2. HEMOCYTE-MEDIATED ENCAPSULATION

Various aspects of the cellular immune reaction of several species of *Drosophila* against the parasitoids *Leptopilina boulardi*, *L. heterotoma*, *Asobara tabida and A. citri* have been reported in previous reviews (Carton and Kitano, 1981; Carton et al., 1997, 2005, 2008; De Gregorio et al., 2002a,b; Russo et al., 1996; Vass et al., 1993a; Williams, 2007). Typically, eggs of the endoparasitic wasps provoke a rapid host hemocyte-mediated melanotic encapsulation response, with pigment appearing early on the surface of the eggs, just before or at the time the hemocytes begin to adhere to form a capsule around the dead parasitoid (Fig. 4.2). This response terminates with the dead parasite sequestered in the melanotic capsule formed by the collaborative interactions of two types of hemocytes, plasmatocytes and lamellocytes (Fig. 4.2D and E; Williams, 2007). Comparative analyses of hemocyte profiles during melanotic encapsulation responses characteristically document a precocious elevation in the total number of both cell categories (Russo et al., 2001), lamellocytes rarely being observed in noninfected larvae. Capsule-forming hemocytes were considered to be those in circulation at the time of infection, and others recruited in large numbers from the hemocytopoietic glands (i.e., lymph glands) situated along the dorsal vessel (Fig. 4.2A–C; Russo et al., 2001; Sorrentino et al., 2002). However, a subepidermal population of sessile blood cells is now known to be released into the circulation in response to a parasitoid infection, and sessile hemocytes might be considered as a novel hematopoietic compartment and the main source of lamellocytes (Markus et al., 2009; Zettervall et al., 2004).

Melanin is believed to be derived in large part from pigment precursors primarily associated with a third cell type, the crystal cell, numbers of which are significantly diminished in host larvae (Bidla et al., 2007; Crozatier and Meister, 2007; Irving et al., 2005; Lanot et al., 2001; Meister, 2004; Meister and Lagueux, 2003; Rizki et al., 1985; Sorrentino et al., 2002; Williams, 2007), but recent studies also point to lamellocytes as sources for melanogenesis (Irving et al., 2005; Nam et al., 2008). The single account of capsule formation in *D. melanogaster* without pigmentation (Rizki and Rizki, 1990) appears to represents a mistaken identification of a single layer of spherical cells in the proximity of the parasite surface as capsule-forming hemocytes (Russo et al., 1996).

4.3. MELANIZATION DURING THE *DROSOPHILA* CELLULAR IMMUNE REACTION

Two basic types of melanin are found in animals, brownish-black eumelanin and yellow to reddish-brown pheomelanin. Both cytoprotective and cytotoxic roles have been attributed to these pigments, given their capacity to

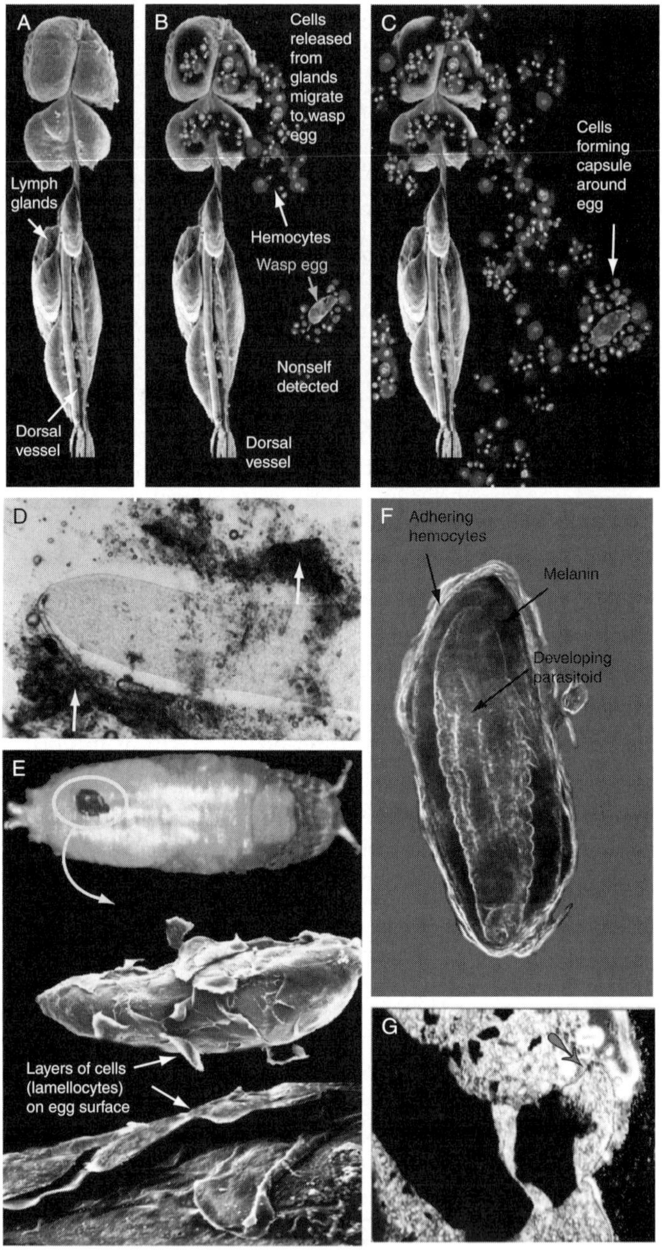

FIGURE 4.2 (A–C) Graphic representation of the early events occurring during a typical hemocyte-mediated melanotic encapsulation response in *D. melanogaster* larvae infected by the endoparasitoid *L. boulardi*. Cells released from the disintegrating lymph glands enter into the body cavity of the host and specifically adhere to the surface of

scavenge potentially toxic organic and inorganic cations and free-radical species, engage in metal-binding and sequestering responses, initiate redox reactions, cross-link proteins and mediate detoxification processes.

Much of our current understanding of melanogenesis is derived from studies of mammalian systems where a combination of enzyme-catalyzed and chemical reactions common to both pigment pathways have been characterized. An initial enzyme-mediated reaction involves the hydroxylation of *L*-phenylalanine to *L*-tyrosine. This reaction is catalyzed by phenylalanine hydroxylase (PAH), an enzyme that requires the cofactor 6(*R*)-*L*-erythro-5,6,7,8-tetrahydrobiopterin (BH_4). Ensuing oxidations of *L*-tyrosine and/or *L*-DOPA, either by the copper-containing monoxygenase tyrosinase (i.e., insect phenoloxidase, PO; Terland et al., 2006), or the heme protein peroxidase (PER; Kasraee, 2002; Okun, 1996) generate dopaquinoine, a reactive intermediate essential for the formation of both eumelanin and pheomelanin (Fig. 4.3). Following the formation of dopaquinone, a series of enzyme-regulated and/or spontaneous oxidoreductions occur at rates that vary considerably depending upon the presence and concentration of hydrogen ion, metal ions (e.g., manganese, copper, zinc and iron) and reducing compounds. In the absence of thiol compounds, dopaquinone can then undergo two distinct reactions; an intramolecular 1,4- addition of the side chain amino group to the benzene ring (i.e., cyclization) and/or a water addition reaction. The latter yields such reactive and potentially cytotoxic molecules as quinone methide, the trihydroxyphenol TOPA and its derived *o*- and *p*-quinones (Q) and semiquinones (QH^-). The cyclization reaction produces leukodopachrome, which is quickly oxidized in redox reactions with dopaquinone to form dopachrome, and in the process dopaquinone is reduced back to *L*-DOPA. When formed in mammalian systems, dopachrome can either be converted by dopachrome tautomerase (DT) to 5,6-dihydroxyindole-2-carboxylic acid (DHICA) and then oxidized by DHICA oxidase (DO), or undergo a nonenzymatic decarboxylation to form the less stable 5,6-dihydroxyindole (DHI). Derivatives of DHICA and DHI, which include their representative indolequinones, QH^-, methides and imines, eventually polymerize to produce a brown-black heteropolymer. In insects, which apparently do not produce DHICA, dopachrome conversion enzyme (DCE) accelerates the formation of the less stable DHI by

---

the parasite, which characteristically shows melanin deposits during the early stage of infection. (D) Early response involving the deposition of melanin on the surface of the wasp egg. (E) Scanning electron micrographs showing layers of lamellocytes adhering to the surface of a fully formed melanotic capsule removed from the host. (F) Larva of parasite developing despite a host melanotic encapsulation response. (G) Parasite larva escaping from a fully formed melanotic capsule. Arrow indicates mouth area of the emerging parasite. (See Page 1 in Color Section at the back of the book.)

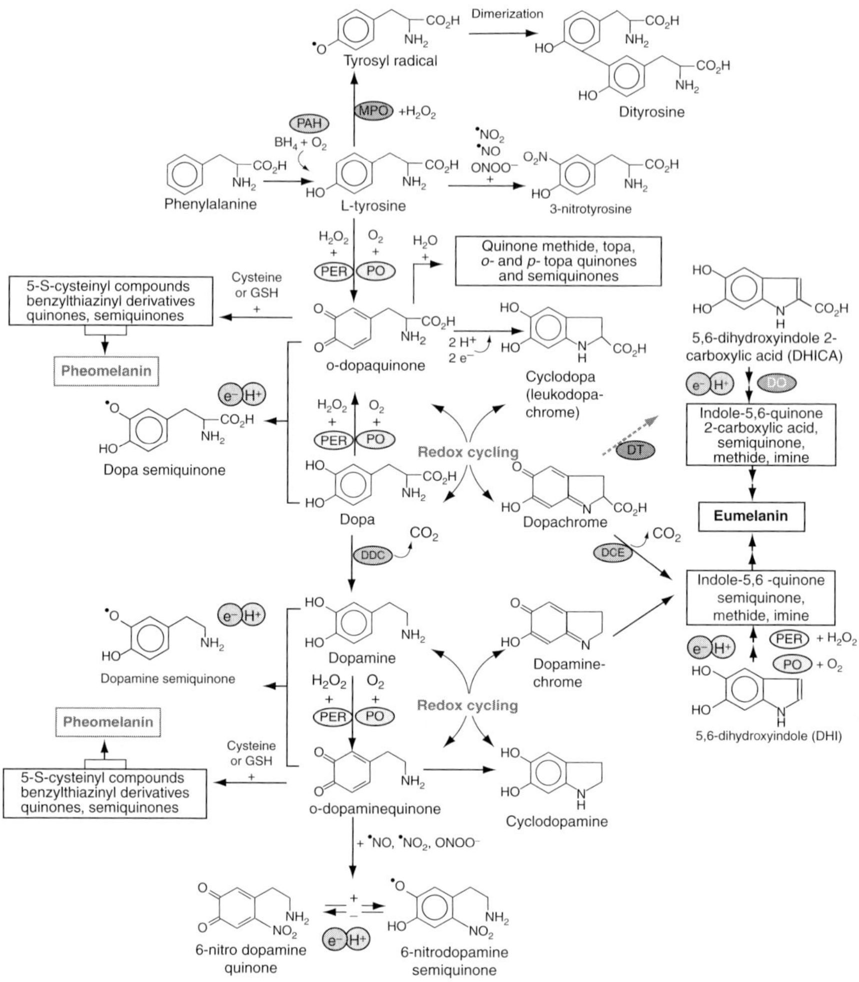

FIGURE 4.3 Overview of the pathways involved in the formation of eumelanin and pheomelanin, and some their reactive intermediates, including quinones and semiquinones. Redox cycling and univalent transfers, which represent important mechanisms for generating cytotoxic molecules, also occur with DHI-derived noncyclized indolequinone and indolesemiquine (not illustrated). Insects apparently are incapable of forming DHICA. *Notes: DCE, dopachrome conversion enzyme; DDC, dopa decarboxylase; DO, 5,6-dihydroxyindole-2-carboxylic acid (DHICA) oxidase; GSH, glutathione; MPO, myeloperoxidase; PAH, phenylalanine hydroxylase; PER, peroxidase; PO, phenoloxidase.*

promoting the decarboxylation and rearrangement of dopachrome to DHI (Li and Nappi, 1991). The pathway to DHI also can be achieved by the decarboxylation of *L*-DOPA to dopamine, a reaction catalyzed by dopa decarboxylase (DDC).

When the concentrations of cysteine and reduced glutathione (GSH) are high, these sulfhydryl compounds conjugate with dopaquinone to form cysteinyldopa and glutathionyldopa, respectively, and initiate pheomelanogenesis (Fig. 4.3). GSH, which is rejuvenated from oxidized glutathione by action of glutathione reductase, the latter obtaining the requisite reducing equivalents from the nicotinamide adenine dinucleotide phosphate (NADPH)/nicotinamide adenine dinucleotide phosphate (reduced form) (NADH) recycling system is essential for the formation of glutathionyldopa.

Although the detection of DHI unequivocally established the synthesis of eumelanin in the defense response of wasp-infected *Drosophila* larvae (Nappi et al., 1992a; Vass et al., 1993a), the contribution of pheomelanin intermediates in capsule formation have yet to be examined. The presence of *N*-acetylarterenone, a sclerotizing agent, in infected larvae also suggests the cellular capsules formed in infected larvae are most likely comprised of both melanin and sclerotin (Nappi et al., 1992a; Vass et al., 1993b).

Thus, the initiation of a successful melanotic response of *Drosophila* larvae immediately following infection by a parasitoid may be dependent on the rapid mobilization and catabolism of existing levels of *L*-tyrosine by either PO or PER, an equally timely PAH response, availability of the cofactor BH_4, and on the NADPH/NADP recycling system, components that have yet to be collectively and comprehensively studied to ascertain the precise involvement of melanogenesis in insect immunity. The DDC-mediated pathway to DHI may be a principal route for production of pigment precursors in infected *Drosophila*, as the melanotic encapsulation response against eggs of the parasitic wasp *Leptopilina boulardi* is severely compromised in temperature-sensitive DDC-deficient mutants (Nappi et al., 1992b). Accordingly, it was recently shown that silencing the genes for DDC and DCE significantly reduced melanization of foreign objects implanted in the mosquito *Anopheles gambiae* (Paskewitz and Andreev, 2008). In the medfly *Ceratitis capitata*, DDC-dependent pathways have been shown to regulate such immune functions as phagocytosis, nodulation and melanization by hemocytes (Sideri et al., 2008).

4.4. CYTOTOXIC MOLECULES ASSOCIATED WITH MELANIZATION

The oxidoreduction reactions that generate Q and QH^- during melanogenesis represent sites of critical electron transfers that can be employed by insects for cytotoxic reactions against pathogens. However, melanization is not a universal feature of insect cellular immunity. In some species of *Drosophila*, as well as in other insects, the fate of the pathogen appears to be unrelated to host melanotic encapsulation reactions (Figs. 2.2E and 4.4; Henter and Via, 1995; Nappi, 1973; Nappi and Streams, 1970; Tardieu and Rabasse, 1988; Vernick et al., 1995). In the *Drosophila*–parasitoid systems, such as the *D. paramelanica*–*L. heterotoma* system, eggs of the endoparasite succumb with no evidence of blood cell-mediated encapsulation and no pigment reaction (Carton et al., 2009; Nappi and Streams, 1970). Although the identity of the cytotoxic molecules remains unknown, attention has focused on ROI and RNI, given that elevated levels of some of these molecules have been found in immune responsive hosts (Foley and O'Farrell, 2003; Luckhart and Li, 2001; Molina-Cruz et al., 2008; Novas et al., 2004; Whitten et al., 2001), including those in which hemocyte-mediated melanotic encapsulation reactions are typically formed, that is, *Drosophila* spp. belonging to the subgroup *melanogaster* (Nappi and Vass, 1998, 2001a,b; Nappi et al., 1995, 2000b), and in wasp-infected *D. paramelanica* where parasites are destroyed but melanotic capsules are not produced (Carton et al., 1992).

Potentially damaging ROI are readily produced during cellular aerobic metabolism as a result of successive univalent reductions of molecular oxygen (O_2) to form water (Fig. 4.5). Major cellular sources for ROI include the mitochondrial respiratory chain, peroxisomes and the activity of several enzyme systems (e.g., cyclooxygenase, NADPH-oxidase, NADPH dehydrogenase, ubiquinone-cytochrome C reductase, cytochrome P-450 reductase and xanthine dehydrogenase/oxidase). Electrons derived from these sources can reduce molecular oxygen to superoxide anion ($^{\bullet}O_2^-$). The latter is removed from tissues by spontaneous or enzyme-mediated (superoxide dismutase, SOD) dismutations to hydrogen peroxide (H_2O_2). H_2O_2 is then readily detoxified to water by catalase and glutathione peroxidase. However, $^{\bullet}O_2^-$ and H_2O_2 can react with each other or with certain transition metal ions, chloride (Cl^-) and nitric oxide ($^{\bullet}NO$), to generate highly reactive and potentially cytotoxic molecules such as the hydroxyl radical ($^{\bullet}OH$), peroxynitrite ($ONOO^-$) and hypochlorous acid (HOCl; Fig. 4.5). Elevated levels of ROI cause tissue damage by various mechanisms, including lipid peroxidation, deoxyribonucleic acid (DNA) damage, protein cross-links, sulfhydryl oxidation, stimulation of proinflammatory cytokine release and inappropriate activation of nuclear factor κB. Interestingly, the univalent oxidations of redox active o-diphenols (QH_2) such as *L*-DOPA and dopamine by PO and/or PER form $^{\bullet}QH^-$ and Q, with the unpaired electrons transferred to O_2 and H_2O_2 to generate $^{\bullet}O_2^-$ and $^{\bullet}OH$, respectively, thereby initiating cytotoxic reactions (Fig. 4.5). The process of melanogenesis in response to infection must by tightly regulated to avoid the generation of cytotoxic molecules at nonspecific sites within the host hemocoel.

The most reactive free radical produced in biological systems is $^{\bullet}OH$, a highly electrophilic molecule that reacts with virtually any organic

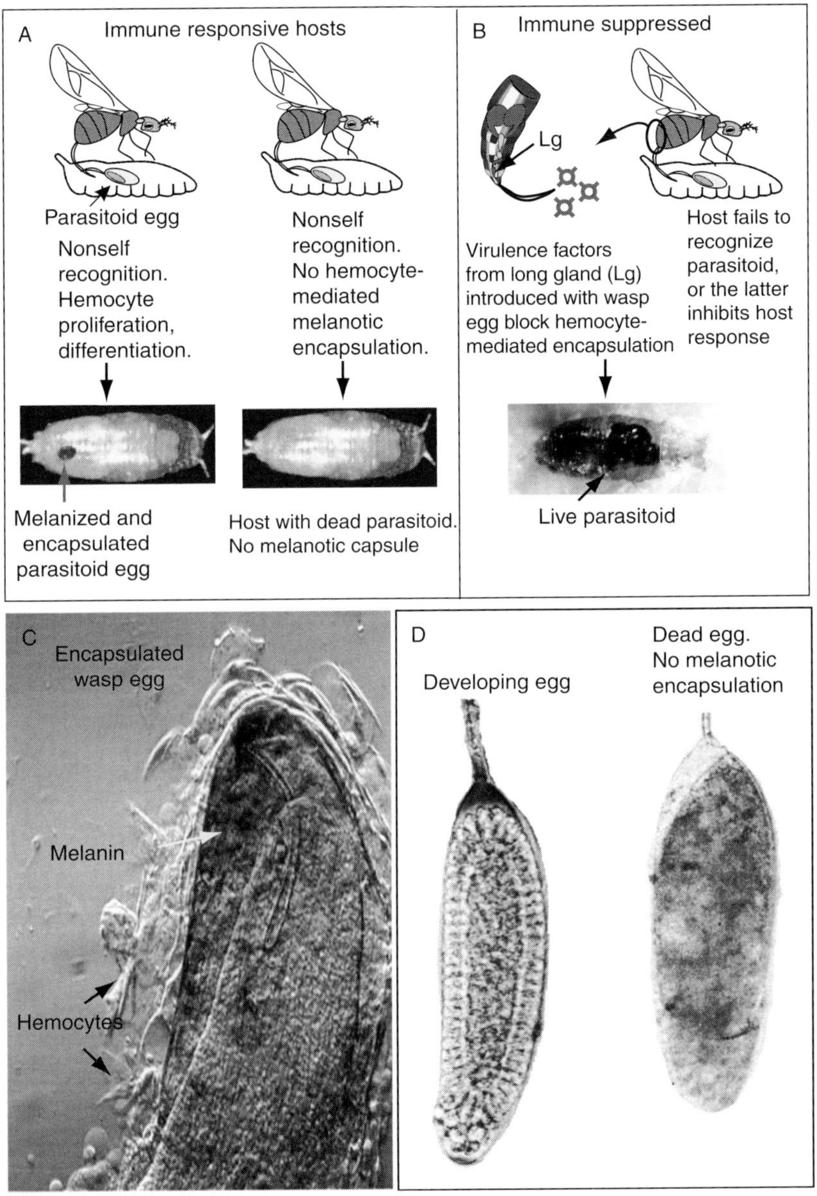

FIGURE 4.4 (A) Illustrations of the possible outcomes of *Drosophila* host–wasp parasite interactions. Immune competent hosts manifest either a melanotic encapsulation response, or a response that does not involve hemocyte-mediated encapsulation. The cytotoxic elements in each case have not been characterized, but some evidence suggests they may be one or more reactive intermediates of oxygen and or nitrogen. (B) Successful wasp species introduce virulence factors into the host that suppress hemocyte-mediated encapsulation. (C) Typical melanotic encapsulation response. (D) Example of a *Leptopilina* egg destroyed by *D. paramelanica* and lacking melanotic encapsulation. (See Page 2 in Color Section at the back of the book.)

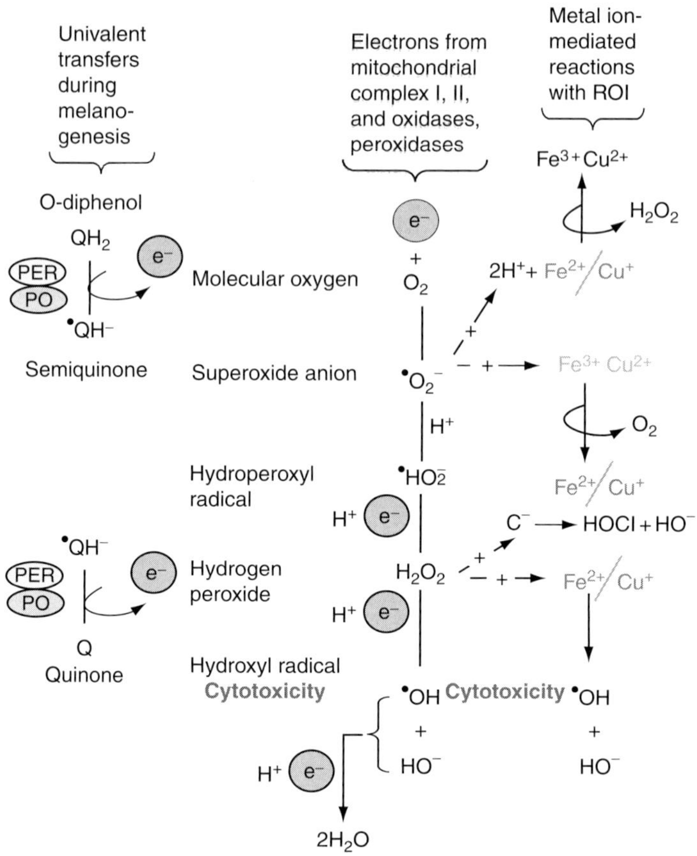

FIGURE 4.5 Outline of the sequence of univalent reductions of oxygen that lead to the formation of water. Interactions of the intermediates hydrogen peroxide (H_2O_2) and superoxide anion ($^{\bullet}O_2^-$), either with each other or with nitric oxide ($^{\bullet}NO$) and chloride (Cl^-), represent major sources of hydroxyl radical ($^{\bullet}OH$)-mediated damage. Univalent transfers initiated by the copper- and/or iron-containing melanogenic enzymes can augment $^{\bullet}OH$ damage and also serve to localize such response to sites of infection. *Notes: PER, peroxidase; PO, phenoloxidase; ROI, reactive intermediates of oxygen.*

compound at diffusion-limited rates (Halliwell and Gutteridge, 1997). An essential mechanism involved in the production of this radical is the reduction of certain transition metals, such as Fe^{3+} and/or Cu^{2+}, a reaction that can be initiated by $^{\bullet}O_2^-$ (Fig. 4.5). Whether $^{\bullet}O_2^-$ mediates the reduction or the oxidation of these metal ions depends in large part on the rate constants and concentrations of reactants. Because the concentration of O_2 is normally higher than that of $^{\bullet}O_2^-$, the favored reaction is the oxidation of Fe^{2+}/Cu^{+}. If the concentration of $^{\bullet}O_2^{-}$ is higher than that of O_2, as might occur during a host immune response, the overriding response will be the reduction of Fe^{3+}/Cu^{2+} to Fe^{2+}/Cu^{+}. Interactions of Fe^{2+}/Cu^{+} with H_2O_2 generate $^{\bullet}OH$ (Fig. 4.5). Thus, the binding of the copper-containing PO or heme-containing PER to pathogens, or the reactions of these enzymes in the vicinity of an infection, would provide a localization for metal ion-mediated $^{\bullet}OH$ killing. Because of its intrinsic coordination properties, copper can induce a more site-specific $^{\bullet}OH$ cytotoxicity to bound ligands than can iron (Berthon, 1993). In addition to transmission metal-mediated reactions, $^{\bullet}OH$ can be generated from the interactions of H_2O_2 with $^{\bullet}NO$ (Fig. 4.6; Nappi and Ottavani, 2000a; Nappi and Vass, 2001a,b). Recent studies in our laboratory (Carton et al., 2009) support earlier reports that document the involvement of $^{\bullet}NO$ in mediating various toxic responses (Alterton et al., 2001; Colasanti and Venturini, 1998; Conte and Ottaviani, 1995; Franchini et al., 1995; Pech and Strand, 2000; Rivero, 2006; Sharma et al., 2008; Wink and Mitchell, 1998), in *Drosophila* (Foley and O'Farrell, 2003; Nappi et al., 2000b) and in other invertebrates (Novas et al., 2004; Pacelli et al., 1995; Sharma et al., 2008; Whitten et al., 2001, 2007). In *D. paramelanica* where elevated levels of $^{\bullet}NO$ are produced in response to infection by *L. heterotoma*, immune capacity is diminished when a specific nitric oxide synthase (NOS) inhibitor is introduced in host larvae immediately

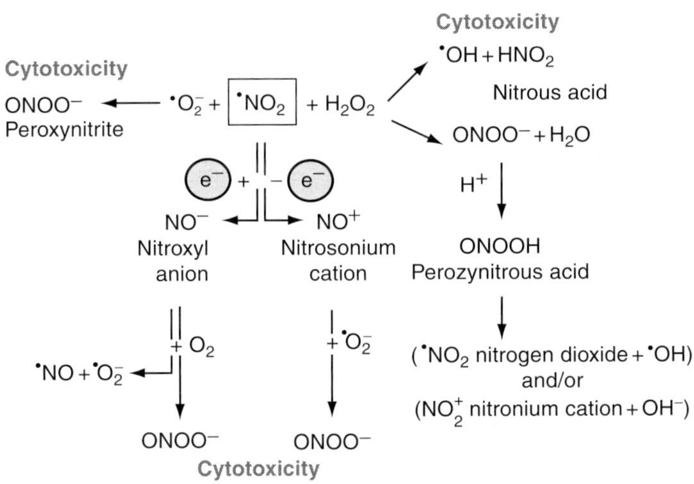

FIGURE 4.6 Overview of some cytotoxic responses generated by the interactions of $^{\bullet}NO$ with H_2O_2 and $^{\bullet}O_2^{-}$. Potentially damaging molecules include peroxinitrite ($OONO^{-}$), $^{\bullet}OH$ and nitrogen dioxide radical ($NO_2^{\bullet}$).

following infection (Carton et al., 2009). Nitric oxide has been shown to enhance peroxidase/H_2O_2-mediated oxidations of *L*-DOPA and dopamine, and diminish the tyrosinase/O_2-mediated oxidations of these melanin precursors (Nappi and Vass, 2001b). Thus, $^{\bullet}NO$ can influence cellular integrity and function directly by altering the rates of conversion of melanin precursors to reactive Q and QH^- and by forming noncyclized nitrosyl complexes with these molecules. Various derivatives of $^{\bullet}NO$, including nitrogen dioxide radical ($NO_2^{\bullet}$) and peroxinitrite ($ONOO^-$) are nitrating species that can interact with tyrosine and other phenols and thereby represents additional mechanisms of toxicity (Fig. 4.6; Casella et al., 2002).

To assess the role of $^{\bullet}O_2^-$ and H_2O_2 in *Drosophila* hosts, strains of *D. melanogaster* with differing immune capabilities against the wasp parasitoid *L. boulardi* were examined for the production of these ROI during infection, and compared to reactions made by strains deficient in SOD and catalase (CAT; Nappi et al., 1995). Elevated levels of $^{\bullet}O_2^-$ were produced by immune reactive hosts during melanotic encapsulation of the parasitoid, but not by susceptible hosts in which the parasitoid developed unmolested. Both SOD- and CAT-deficient strains also produced melanotic capsules and manifested elevated levels of $^{\bullet}O_2^-$ when infected, but these reactions were unsuccessful and the parasitoids survived inside the capsule, indicating that neither the quinoid precursors of melanin nor $^{\bullet}O_2^-$ *per se* were cytotoxic. Immune incompetence in SOD-deficient and CAT-deficient hosts was attributed in part to defects in H_2O_2 metabolism, and/or the inability of the enzyme-deficient strains to initiate the metal-mediated reductive cleavage of H_2O_2 required for the production of the cytotoxic $^{\bullet}OH$. The role proposed for $^{\bullet}O_2^-$ in *Drosophila* cellular immunity is one of initiating the formation of $^{\bullet}OH$. Melanin, which contains both oxidizing and reducing components, may serve a dual role in producing $^{\bullet}O_2^-$ and sequestering redox-active metal ions, thereby confining the production of ROI to sites of infection.

No study has yet identified the killing components produced in conjunction with melanotic encapsulation responses, or explained how cytotoxic molecules generated in the open circulatory system of an insect can selectively destroy only foreign tissues. It would be important to learn if PO-initiated reactions occurring at or near foreign surfaces not only initiate melanogenic activity but also enhance the production of cytotoxic molecules by univalent transfers that generate reactive Q and $^{\bullet}QH^-$. Melanogenic activity in such cases would also provide a mechanism for localizing and sequestering the cytotoxic molecules, preventing them from dispersing in the open circulatory system of the insect host. Parasite success in *Drosophila* hosts may be determined by the ability of the parasitoid to modulate hemocyte activity and prevent effective melanotic encapsulation and/or the generation of cytotoxic levels of ROI (Nappi et al., 1995).

4.5. THE PREVENTION OF PHENOLOXIDASE ACTIVITY BY PARASITOID VIRULENCE FACTORS

It is well known that activated PO functions in cuticular melanization and sclerotization, and it is generally believed that at least some of the enzymes and products of these reactions play a critical role in the defense reactions of insects against invaders, although the latter issue still remains a matter of debate (Cerenius and Soderhall, 2004; Schnitger et al., 2007). If melanogenesis, ROI or RNI represent critical components of the cytotoxic arsenal of insects, one would expect successful parasites to have evolved strategies that either target and actively suppress such potentially biochemically hostile reactions, or that effectively avoid host detection (passive immune evasion; Vinson, 1990).

What little is known of immune suppression of host melanogenesis by insect parasitoids has mainly come from investigations of effects of parasitism or parasite-derived factors *in vivo* and *in vitro*. Diminished host PO activity and lack of melanotic encapsulation has been reported in parasitoid-infected *Lepidoptera* (Beck et al., 2000; Beckage et al., 1990; Strand and Noda, 1991) and in *Drosophila* larvae infected by *Asobara* spp. (Moreau et al., 2003). Successful development of *L. boulardi* in *Drosophila* has been attributed to immune-suppressive factors introduced in host larvae by the female wasp during oviposition (Labrosse et al., 2003; Rizki and Rizki, 1991, 1994). Precisely how these parasitoid-derived factors function remains largely unexplored. These molecules may inhibit host cell proliferation and differentiation, interfere with their adhesion to form multicellular capsules (Rizki and Rizki, 1994), inhibit PO and other melanogenic enzymes, or interfere with the production of cytotoxic ROI and RNI.

In larvae of certain melanotic tumor mutants of *Drosophila*, a hemocyte-mediated autoimmune response that is virtually morphologically indistinguishable from melanotic encapsulation of parasitoids targets instead endogenous tissues (Nappi, 1984). Interestingly, varying degrees of suppression of melanotic tumors by different parasitoid species and strains have been reported (Labrosse et al., 2003; Nappi, 1975; Walker, 1959). More recent studies have shown that components of the *L. boulardi* venom (ISy strain, see description in Chapter 6 by Dubuffet et al.) inhibit activation of PO in *D. yakuba* larvae hemolymph by affecting some step (s) of the cascade leading to PO activation, but not PO activity by itself. In another study designed to determine if venom factors from *L. boulardi* targeted the principal oxidation pathways leading to synthesis of eumelanin in larvae of *D. melanogaster*, sensitive electrochemical detection methods showed that venom factors diminished the oxidations of the two diphenol eumelanin precursors, dopamine and DHI, while oxidations of the monophenol tyrosine, and two other related diphenols, dopa and DHICA, were not significantly inhibited (Kohler et al., 2007). Collectively, these related studies (Colinet et al., 2007; Labrosse et al., 2005) suggest that, in addition to the targeting specific hemocytes, the oxidation pathways synthesizing certain pigment precursors, most likely decarboxylated pigment precursors derived from DHI, constitute some of the specific host responses suppressed by *L. boulardi*.

The cascade leading to activation of PO from the enzymatically inactive zymogen form of PO and its regulation are still largely undescribed in *Drosophila*, most data having been obtained from analyses in Lepidopteran species such as *Manduca sexta*. In the *Drosophila melanogaster* genome, the proPO isoforms identified to date include proPO1 and proPO2, which are expressed in crystal cells, and proPO3, which was predominantly found in the lamellocytes. A recent study by Nam et al. (2008) showed that proPO3 is enzymatically active as a zymogen, and that its expression leads to high melanin production, while its absence drastically reduces melanization. Since parasitoid egg encapsulation involves surrounding layers of lamellocytes as well as melanin deposit, the biochemistry and role of this proenzyme are now important to decipher. The melanization reaction induced by activated phenoloxidase must be tightly controlled because of the risk of systemic melanization damage to the hosts. Among the known regulators are PO itself (Kan et al., 2008), which can act as a competitive inhibitor of melanization complex formation, and serine protease inhibitors that control serine proteases involved in the successive steps of proteolytic cleavage leading to activation of proPO into PO. The *Drosophila* genome encodes 29 serpins (Spn), of which only Spn27A, necrotic and recently Spn28D have been analyzed in detail. Flies deficient for Spn27A, a negative regulator of PO activation, exhibit spontaneous melanization in larvae and adults (Ligoxygakis et al., 2002). Spn28D regulates hemolymph PO activity in both larvae and adults, but at a different level than Spn27A. Data from Scherfer et al. (2008) indeed suggest that Spn28D confines PO availability by controlling its initial release, while Spn27A limits the melanization reaction to the wound site. Interestingly, injection of Serpin 27A into normally highly immune-competent *D. melanogaster* larvae reduces their ability to form melanotic capsules around the eggs of the parasitoid *L. boulardi* (Nappi et al., 2005), which confirms implication of the PO cascade in successful encapsulation of parasitoids. Of considerable interest was the observation that Spn27A not only inhibited PO-mediated melanization, but also prevented capsule formation by plasmatocytes and lamellocytes. It may be that Spn27A, by limiting the melanization reaction to the parasitoid egg, plays an essential role in localizing hemocyte responses to sites of injury or infection. Some support for this proposal derives from studies showing hemocoelic melanization reactions in parasitized larvae of a Spn27A-deficient mutant

(Spn27A1) to be uncontrolled and diffuse, and thus not specifically directed against eggs of *L. boulardi* (Nappi et al., 2005). Together with the inhibitory effect of *L. boulardi* venom on PO activation, these data indicated that factors somehow mimicking the role of Spn27A or other regulating serpins of the PO cascade might be contained in wasp venom. Accordingly, a serpin domain-containing protein, LbSPNy was recently identified in the venom of the ISy strain of this parasitoid species (Colinet et al., 2009). Besides its high abundance in the venom, LbSPNy contains a serpin domain whose sequence and structural characteristics suggest it might inhibit serine proteases with trypsin-like specificity, as all insect serpins demonstrated so far to be involved in regulating the PO cascade. The inhibitory effect of LbSPNy on the *Drosophila* PO cascade was confirmed *in vitro* and *in vivo* using the recombinant protein produced in bacteria. Together with the 460-fold overexpression of the LbSPNy-encoding gene in venom glands compared to the rest of the body, this pointed to LbSPNy as the first serpin described as a parasitoid virulence factor that target host melanization. The number of genes encoding serine proteases in *D. melanogaster*, and the fact that LbSPNy serpin domain is not closely related to that of any particular insect serpin, makes it difficult to hypothesize on the possible target(s) of LbSPNy. However, interestingly, the residues known to determine protease specificity in the hypervariable reactive center loop (RCL) region of LbSPNy are identical with those of Spn6 of *M. sexta*, which was already demonstrated to inhibit PAP-3 (Zou and Jiang, 2005), a ProPO activating protein, as well as other components of the melanization complex. Current studies to characterize LbSPNy targets in *D. melanogaster* should provide new insights into how the *Drosophila* PO cascade is regulated, and how parasitoids might interfere with this regulation process for their survival.

How some of *L. boulardi*-derived factors function has now been established (Colinet et al., 2007, 2009) but the complete arsenal of immune suppressive molecules remains largely unexplored. Still-undescribed molecules may indeed inhibit host cell proliferation and differentiation, interfere with their adhesion to form multicellular capsules (Rizki and Rizki, 1994), inhibit other melanogenic enzymes, or interfere with the production of cytotoxic ROI and RNI.

4.6. CONCLUSIONS

Host–parasitoid interactions represent coevolved adaptations of great complexity. Insects and other arthropod hosts typically manifest a unique defense response against metazoan parasites that involves hemocyte-mediated melanotic encapsulation. The use of melanin for protection from foreign insult is a fascinating process that minimally involves a multifaceted biochemistry and an equally complex genetic regulation that we have yet to comprehend fully. An intriguing challenge for future investigations is the assessment of the role of ROI and RNI in the immune arsenal of insects, most notably in hosts that do not form melanotic capsules but nevertheless kill parasitoids. Studies that merely correlate host melanogenesis with immune competence, and those that define host susceptibility or parasite virulence on the lack of a pigment reaction, do not provide substantive information about the actual cytotoxic mechanism involved.

A fascinating component of insect host–parasitoid combative relationships is the ability of some wasp species and strains to develop unmolested within otherwise immune-competent hosts. Either such parasitoids evolve with one or more passive immune evasion strategies that effectively preclude host detection, or with the capacity to actively combat and render ineffective host defenses (Eslin and Prevost, 2000; Strand and Pech, 1995; Vinson, 1990). It is anticipated that future proteomic and transcriptomic studies of parasitoid virulence proteins will facilitate identification of the cytotoxic molecules, the cell-signaling pathways that regulate their synthesis, and their mode of target-specific engagement with foreign organisms.

REFERENCES

Alterton, W.K., Cooper, C.E., Knowles, R.G., 2001. Nitric oxide synthase: structure, function and inhibition. Biochem. J. 357, 593–615.

Beck, M., Theopold, U., Schmidt, O., 2000. Evidence for serine protease inhibitor activity in the ovarian calyx fluid of the endoparasitoid *Venturia canescens*. J. Insect Physiol. 46, 1275–1283.

Beckage, N., Metcalf, J., Nesbit, D., Schleifer, K., Zetlan, S.R., De Buron, I., 1990. Host hemolymph monophenoloxidase activity in parasitized *Manduca sexta* larvae and evidence for inhibition by wasp polydnavirus. Insect Biochem. 20, 285–294.

Berthon, G., 1993. Is copper pro- or anti-inflammatory? A reconciling view and a novel approach for the use of copper in the control of inflammation. Agents Actions 39, 210–217.

Bettencourt, R., Asha, H., Dearolf, C., Ip, Y.T., 2004. Hemolymph-dependent and -independent responses in *Drosophila* immune tissue. J. Cell Biochem. 92, 849–863.

Beutler, B., 2004. Innate immunity: an overview. Mol. Immunol. 40, 845–859.

Bidla, G., Dushay, M.S., Theopold, U., 2007. Crystal cell rupture after injury in *Drosophila* requires the JNK pathway, small GTPases and the TNF homolog Eiger. J. Cell Sci. 120, 1209–1215.

Bidla, G., Lindgren, M., Theopold, U., Dushay, M.S., 2005. Hemolymph coagulation and phenoloxidase in *Drosophila* larvae. Dev. Comp. Immunol. 29, 669–679.

Bodian, D.L., Leung, S., Chiu, H., Govind, S., 2004. Cytokines in *Drosophila* hematopoiesis and cellular immunity. Prog. Mol. Subcell. Biol. 34, 27–46.

Brites, D., McTaggart, S., Morris, K., Anderson, J., Thomas, K., Colson, I., et al., 2008. The Dscam homologue of the crustacean *Daphnia* is diversified by alternative splicing like in insects. Mol. Biol. Evol. 25, 1429–1439.

Carton, Y., Kitano, H., 1981. Evolutionary relationships to parasitism by seven species of the *Drosophila melanogaster* subgroup. Biol. J. Lin. Soc. 16, 227–241.

Carton, Y., Nappi, A.J., 1997. *Drosophila* cellular immunity against parasitoids. Parasitol. Today 13, 218–227.

Carton, Y., Frey, F., Nappi, A.J., 1992. Inheritance of cellular immune resistance in *Drosophila melanogaster*. Heredity 69, 393–399.

Carton, Y., Frey, F., Nappi, A.J., 2009. Parasite-induced changes in nitric oxide levels in *Drosophila paramelanica*. J. Parasitol. 95(4), in press.

Carton, Y., Nappi, A.J., Poirie, M., 2005. Genetics of anti-parasite resistance in invertebrates. Dev. Comp. Immunol. 29, 9–32.

Carton, Y., Poirie, M., Nappi, A.J., 2008. Insect resistance to parasitoids. Insect Sci. 15, 67–87.

Casella, L., Monzani, E., Roncone, R., Nicolis, S., Sala, A., De Riso, A., 2002. Formation of reactive nitrogen species at biologic heme centers: a potential mechanism of nitrogen-dependent toxicity. Environ. Health Perspect. 110, 709–711.

Cerenius, L., Soderhall, K., 2004. The prophenoloxidase-activating system in invertebrates. Immunol. Rev. 198, 116–126.

Christensen, B.M., Li, J., Chen, C.-C., Nappi, A.J., 2005. Melanization immune responses in mosquito vectors. Trends Parasitol. 21, 192–199.

Colasanti, M., Venturini, G., 1998. Nitric oxide in invertebrates. Mol. Neurobiol. 17, 157–174.

Colinet, D., Dubuffet, A., Cazes, D., Moreau, S., Drezen, J.M., Poirie, M., 2009. A serpin from the parasitoid wasp *Leptopilina boulardi* targets the *Drosophila* phenoloxidase cascade. Dev. Comp. Immunol. 33, 681–689.

Colinet, D., Schmitz, A., Depoix, D., Crochard, D., Poirié, M., 2007. Convergent use of RhoGAP toxins by eukaryotic parasites and bacterial pathogens. PLoS Pathog. 3, e203.

Conte, A., Ottaviani, E., 1995. Nitric oxide synthase activity in molluscan hemocytes. FEBS Lett. 365, 120–124.

Crozatier, M., Meister, M., 2007. *Drosophila* haematopoiesis. Cell Microbiol. 9, 1117–1126.

De Gregorio, E., Han, S.J., Lee, W.J., Baek, M.J., Osaki, T., Kawabata, S.I., et al., 2002a. An immune-responsive serpin regulates the melanization cascade in *Drosophila*. Dev. Cell 3, 581–592.

De Gregorio, E., Spellman, P.T., Tzou, P., Rubin, G.M., Lemaitre, B., 2002b. The Toll and Imd pathways are major regulators of the immune response in *Drosophila*. EMBO J. 21, 2568–2579.

Dubuffet, A., Dupas, S., Frey, F., Drezen, J.M., Poirié, M., Carton, Y., 2007. Genetic interactions between the parasitoid wasp *Leptopilina boulardi* and its *Drosophila* hosts. Heredity 98, 21–27.

Dupas, S., Carton, Y., Poirie, M., 2003. Genetic dimension of the coevolution of virulence-resistance in *Drosophila*–parasitoid wasp relationships. Heredity 90, 84–89.

Eslin, P., Prevost, G., 2000. Racing against host's immunity defenses: a likely strategy for passive evasion of encapsulation in *Asobara tabida* parasitoids. J. Insect Physiol. 46, 1161–1167.

Fleury, F., Ris, N., Fouillet, P., Carton, Y., Boulétreau, M., 2004. Genetic and ecological interactions in *Drosophila*–Parasitoids communities: a case study with *Drosophila melanogaster*, *D. simulans* and their *Leptopilina* species. Genetica 120, 181–194.

Foley, E., O'Farrell, P.H., 2003. Nitric oxide contributes to induction of innate immune responses to gram-negative bacteria in *Drosophila*. Genes Dev. 17, 115–125.

Franchini, A., Conte, A., Ottaviani, E., 1995. Nitric oxide — an ancestral immunocyte effector molecule. Adv. Neuroimmunol. 5, 463–478.

Halliwell, B., Gutteridge, J.M.C., 1997. Lipid peroxidation in brain homogenates — the role of iron and hydroxyl radicals. J. Neurochem. 69, 1330.

Henter, H., Via, S., 1995. The potential for coevolution in a host–parasitoid system. I. Genetic variation within an aphid population in susceptibility to a parasitic wasp. Evolution 49, 427–438.

Hoffmann, J.A., 2003. The immune response of *Drosophila*. Nature 426, 33–38.

Irving, P., Ubeda, J.M., Doucet, D., Troxler, L., Lagueux, M., Zachary, D., et al., 2005. New insights into *Drosophila* larval haemocyte functions through genome-wide analysis. Cell Microbiol. 7, 335–350.

Kan, H., Kim, C.H., Kwon, H.M., Park, J.W., Roh, K.B., Lee, H., et al., 2008. Molecular control of phenoloxidase-induced melanin synthesis in an insect. J. Biol. Chem. 283, 25316–25323.

Kanost, M.R., Jiang, H.B., Yu, X.Q., 2004. Innate immune responses of a lepidopteran insect, *Manduca sexta*. Immunol. Rev. 198, 97–105.

Kasraee, B., 2002. Peroxidase-mediated mechanisms are involved in the melanocytotoxic and melanogenesis-inhibiting effects of chemical agents. Dermatology 205, 329–339.

Kohler, L.J., Carton, Y., Mastore, M., Nappi, A.J., 2007. Parasite suppression of the oxidations of eumelanin precursors in *Drosophila melanogaster*. Arch. Insect Biochem. Physiol. 66, 64–75.

Kurtz, J., Armitage, S.A., 2006. Alternative adaptive immunity in invertebrates. Trends Immunol. 27, 493–496.

Krishnan, N., Hyrsl, P., Simek, V., 2006. Nitric oxide production by hemocytes of larva and pharate prepupa of *Galleria mellonella* in response to bacterial lipopolysaccharide: cytoprotective or cytotoxic? *Comp*. Biochem. Physiol. Toxicol. Pharmacol. 142, 103–110.

Labrosse, C., Carton, Y., Dubuffet, A., Drezen, J.M., Poirié, M., 2003. Active suppression of *D melanogaster* immune response by long gland products of the parasitic wasp *Leptopilina boulardi*. J. Insect Physiol. 49, 513–522.

Labrosse, C., Stasiak, K., Lesobre, J., Grangeia, A., Huguet, E., Drezen, J.M., et al., 2005. A RhoGAP protein as a main immune suppressive factor in the *Leptopilina boulardi* (Hymenoptera, Figitidae)–*Drosophila melanogaster* interaction. Insect Biochem. Mol. Biol. 35, 93–103.

Lanot, R., Zachary, D., Holder, F., Meister, M., 2001. Postembryonic hematopoiesis in *Drosophila*. Dev. Biol. 230, 243–257.

Lavine, M.D., Strand, M.R., 2002. Insect hemocytes and their role in immunity. Insect Biochem. Mol. Biol. 32, 1295–1309.

Lemaitre, B., Hoffmann, J., 2007. The host defense of *Drosophila melanogaster*. Ann. Rev. Immunol. 25, 697–743.

Li, J., Nappi, A.J., 1991. Electrochemical identification of dopachrome isomerase in *Drosophila melanogaster*. Biochem. Biophys. Res. Commun. 180, 724–729.

Ligoxygakis, P., Pelte, N., Ji, C.Y., Leclerc, V., Duvic, B., Belvin, M., et al., 2002. A serpin mutant links Toll activation to melanization in the host defence of *Drosophila*. EMBO J. 21, 6330–6337.

Liu, H., Jiravanichpaisal, P., Cerenius, L., Lee, B.L., Soderhall, I., Soderhall, K., 2007. Phenoloxidase is an important component of the defense against *Aeromonas hydrophila* Infection in a crustacean, *Pacifastacus leniusculus*. J. Biol. Chem. 282, 33593–33598.

Luckhart, S., Li, K., 2001. Transcriptional complexity of the *Anopheles stephensi* nitric oxide synthase gene. Insect Biochem. Mol. Biol. 31, 249–256.

Markus, R., Laurinyecz, B., Kurucz, E., Honti, V., Bajusz, I., Sipos, B., et al., 2009. Sessile hemocytes as a hematopoietic compartment in *Drosophila melanogaster*. Proc. Natl Acad. Sci. USA 106, 4805–4809.

Meister, M., 2004. Blood cells of *Drosophila*: cell lineages and role in host defence. Curr. Opin. Immunol. 16, 10–15.

Meister, M., Lagueux, M., 2003. *Drosophila* blood cells. Cell Microbiol. 5, 573–580.

Molina-Cruz, A., De Jong, R.J., Charles, B., Gupta, L., Kumar, S., Jaramillo-Gutierrez, G., et al., 2008. Reactive oxygen species modulate *Anopheles gambiae* immunity against bacteria and Plasmodium. J. Biol. Chem. 283, 3217–3223.

Moreau, S.J.M., Eslin, P., Giordanengo, P., Doury, G., 2003. Comparative study of the strategies evolved by two parasitoids of the genus *Asobara* to avoid the immune response of the host, *Drosophila melanogaster*. Dev. Comp. Immunol. 27, 273–282.

Nam, H.J., Jang, I.H., Asano, T., Lee, W.J., 2008. Involvement of pro-phenoloxidase 3 in lamellocyte-mediated spontaneous melanization in *Drosophila*. Mol. Cells 26, 606–610.

Nappi, A.J., 1973. The role of melanization in the immune reaction of larvae of *Drosophila algonquin* against *Pseudocoila bochei*. Parasitology 66, 23–32.

Nappi, A.J., 1975. Inhibition by parasites of melanotic tumour formation in *Drosophila melanogaster*. Nature 255, 402–404.

Nappi, A.J., 1984. Hemocyte reactions and early cellular changes during melanotic tumor formation in *Drosophila melanogaster*. J. Invertebrate Pathol. 43, 395–406.

Nappi, A.J., Christensen, B.M., 2005. Melanogenesis and associated cytotoxic reactions: applications to insect innate immunity. Insect Biochem. Mol. Biol. 35, 443–459.

Nappi, A.J., Frey, F., Carton, Y., 2005. *Drosophila* Serpin 27A is a likely target for immune suppression of the blood cell-mediated melanotic encapsulation response. J. Insect Physiol. 51, 197–305.

Nappi, A.J., Ottavani, E., 2000a. BioEssays 22, 469–480.

Nappi, A.J., Streams, F.A., 1970. Abortive development of the cynipid parasite *Pseudocoila bochei* (Hymenoptera) in species of the *Drosophila melanica* group. Ann. Entomol. Soc. Am. 63, 321–327.

Nappi, A.J., Vass, E., 1998. Hydrogen peroxide production in immune-reactive *Drosophila melanogaster*. J. Parasitol. 84, 1150–1157.

Nappi, A.J., Vass, E., 2001a. Cytotoxic reactions associated with insect immunity. In: Beck, Sugumaran, Cooper, (Eds.), Phylogenetic Perspectives on the Vertebrate Immune System. Kluwer Academic-Plemum, New York, pp. 329–348.

Nappi, A.J., Vass, E., 2001b. The effects of nitric oxide on the oxidations of L-dopa and dopamine mediated by tyrosinase and peroxidase. J. Biol. Chem. 276, 11214–11222.

Nappi, A.J., Carton, Y., Frey, F., 1991. Parasite-induced enhancement of hemolymph tyrosinase activity in a selected immune reactive strain of *Drosophila melanogaster*. Arch. Insect Biochem. Physiol. 18, 159–168.

Nappi, A.J., Carton, Y., Vass, E., 1992b. Reduced cellular immune competence of a temperature-sensitive dopa decarboxylase mutant strain of *Drosophila melanogaster* against the parasite *Leptopilina boulardi*. Comp. Biochem. Physiol. Comp. Biochem. 101, 453–460.

Nappi, A.J., Vass, E., Carton, Y., Frey, F., 1992a. Identification of 3,4-dihydroxyphenylalanine, 5,6-dihydroxyindole, and *N*-acetylarterenone during eumelanin formation in immune reactive larvae of *Drosophila melanogaster*. Arch. Insect Biochem. Physiol. 20, 181–191.

Nappi, A.J., Vass, E., Frey, F., Carton, Y., 1995. Superoxide anion generation in *Drosophila* during melanotic encapsulation of parasites. Eur. J. Cell Biol. 68, 450–456.

Nappi, A.J., Vass, E., Frey, F., Carton, Y., 2000b. Nitric oxide involvement in *Drosophila* immunity. Nitric Oxide 4, 423–430.

Novas, A., Cao, A., Barcia, R., Ramos-Martinez, J.I., 2004. Nitric oxide release by hemocytes of the mussel *Mytilus galloprovincialis* Lmk was provoked by interleukin-2 but not by lipopolysaccharide. Int. J. Biochem. Cell Biol. 36, 390–394.

Okun, M.R., 1996. The role of peroxidase in mammalian melanogenesis: a review. Physiol. Chem. Phys. Med. NMR 28, 91–101.

Ottaviani, E., Malagoli, D., Franchini, A., 2004. Invertebrate humoral factors: cytokines as mediators of cell survival. Prog. Mol. Subcell. Biol. 34, 1–25.

Pacelli, R., Wink, D.A., Cook, J.A., Krishna, M.C., DeGraff, W., Friedman, N., et al., 1995. Nitric oxide potentiates hydrogen peroxide-induced killing of *Escherichia coli*. J. Exp. Med. 182, 1469–1479.

Paskewitz, S.M., Andreev, O., 2008. Silencing the genes for dopa decarboxylase or dopachrome conversion enzyme reduces melanization of foreign targets in *Anopheles gambiae*. Comp. Biochem. Physiol. Biochem. Mol. Biol. 150, 403–408.

Pech, L.L., Strand, M.R., 2000. Plasmatocytes from the moth *Pseudoplusia includens* induce apoptosis of granular cells. J. Insect Physiol. 46, 1565–1573.

Poirié, M., Carton, Y., Dubuffet, A., 2009. Virulence strategies in parasitoid Hymenoptera as an example of adaptative diversity. C. R. Biol. 334, 311–320.
Rivero, A., 2006. Nitric oxide: an antiparasitic molecule of invertebrates. Trends Parasitol. 22, 219–225.
Rizki, T.M., Rizki, R.M., 1990. Encapsulation of parasitoid eggs in phenoloxidase-deficient mutants of *Drosophila melanogaster*. J. Insect Physiol. 36, 523–529.
Rizki, R.M., Rizki, T.M., 1991. Effects of lamellolysin from a parasitoid wasp on *Drosophila* blood cells *in vitro*. J. Exp. Zool. 257, 236–244.
Rizki, T.M., Rizki, R.M., 1994. Parasitoid-induced cellular immune deficiency in *Drosophila*. Ann. N. Y. Acad. Sci. 712, 178–194.
Rizki, T.M., Rizki, R.M., Bellotti, R.A., 1985. Genetics of a *Drosophila* phenoloxidase. Mol. Gen. Genet. 201, 7–13.
Royet, J., 2004. Infectious nonself recognition in invertebrates: lessons from *Drosophila* and other insect models. Mol. Immunol. 41, 1063–1075.
Russo, J., Brehelin, M., Carton, Y., 2001. Haemocyte changes in resistant and susceptible strains of *Drosophila melanogaster* caused by virulent and avirulent strains of the parasitic wasp *Leptopilina boulardi*. J. Insect Physiol. 47, 167–172.
Russo, J., Dupas, S., Frey, F., Carton, Y., Brehelin, M., 1996. Insect immunity — early events in the encapsulation process of parasitoid (*Leptopilina boulardi*) eggs in resistant and susceptible strains of *Drosophila*. Parasitology 112, 135–142.
Scherfer, C., Tang, H., Kambris, Z., Lhocine, N., Hashimoto, C., Lemaitre, B., 2008. *Drosophila* Serpin-28D regulates hemolymph phenoloxidase activity and adult pigmentation. Dev. Biol. 323, 189–196.
Schnitger, A.K., Kafatos, F.C., Osta, M.A., 2007. The melanization reaction is not required for survival of *Anopheles gambiae* mosquitoes after bacterial infections. J. Biol. Chem. 282, 21884–21888.
Sharma, A., Raghavendra, K., Adak, T., Dash, A.P., 2008. Determination of nitric oxide metabolites, nitrate and nitrite, in *Anopheles culicifacies* mosquito midgut and haemolymph by anion exchange high-performance liquid chromatography: plausible mechanism of refractoriness. Malaria J. 7, 71.
Shrestha, S., Kim, Y., 2008. Eicosanoids mediate prophenoloxidase release from oenocytoids in the beet armyworm *Spodoptera exigua*. Insect Biochem. Mol. Biol. 38, 99–112.
Sideri, M., Tsakas, S., Markoutsa, E., Lampropoulou, M., Marmaras, V.J., 2008. Innate immunity in insects: surface-associated dopa decarboxylase-dependent pathways regulate phagocytosis, nodulation and melanization in medfly haemocytes. Immunology 123, 528–537.
Siva-Jothy, M.T., Moret, Y., Rolff, J., 2005. Insect immunity: an evolutionary ecology perspective. Adv. Insect Physiol. 32, 1–48.
Sorrentino, R.P., Carton, Y., Govind, S., 2002. Cellular immune response to parasite infection in the *Drosophila* lymph gland is developmentally regulated. Dev. Biol. 243, 65–80.
Stanley, D., 2006. Prostaglandins and other eicosanoids in insects: biological significance. Ann. Rev. Entomol. 51, 25–44.
Stanley, D., Miller, J., Yunaz, H., 2009. Eicosanoid actions in insect immunity. J. Innate Immun. 1, 281–290.
Strand, M.R., Noda, T., 1991. Alterations in the haemocytes of *Pseudoplusia includens* after parasitism by *Microplitis demolitor*. J. Insect Physiol. 37, 839–850.
Strand, M.R., Pech, L.L., 1995. Immunological basis for compatibility in parasitoid host relationships. Ann. Rev. Entomol. 40, 31–56.
Tafalla, C., Gomez-Leon, J., Novoa, B., Figueras, A., 2003. Nitric oxide production by carpet shell clam (*Ruditapes decussatus*) hemocytes. Dev. Comp. Immunol. 27, 197–205.
Tardieu, I., Rabasse, M., 1988. Some aspects of host immunity and physiological suitability in aphids attacked by *Aphidius colemani*. In: Niemczyk, Anderson, (Eds.), Ecology and Effectiveness of Aphophaga. Academic Publishing, The Hague, pp. 311–315.

Terland, O., Almas, B., Flatmark, T., Andersson, K.K., Sorlie, M., 2006. One-electron oxidation of catecholamines generates free radicals with an *in vitro* toxicity correlating with their lifetime. Biol. Med. 41, 1266–1271.

Vass, E., Nappi, A.J., Carton, Y., 1993a. Comparative study of immune competence and host susceptibility in *Drosophila melanogaster* parasitized by *Leptopilina boulardi* and *Asobara tabida*. J. Parasitol. 79, 106–112.

Vass, E., Nappi, A.J., Carton, Y., 1993b. Alterations in the activities of tyrosinase, *N*-acetyltransferase, and tyrosine aminotransferase in immune reactive larvae of *Drosophila melanogaster*. Dev. Comp. Immunol. 17, 109–118.

Vernick, K.D., Fujioka, H., Seeley, D.C., Tandler, B., Aikawa, M., Miller, L.H., 1995. *Plasmodium gallinaceum*: a refractory mechanism of ookinete killing in the mosquito, *Anopheles gambiae*. Exp. Parasitol. 80, 583–595.

Vinson, S.B., 1990. How parasitoids deal with the immune system of their hosts: an overview. Arch. Insect Biochem. Physiol. 13, 3–27.

Walker, I., 1959. Die Abwehrreaktion des Wirtes *Drosophila melanogaster* gegen die zoophage Cynipidae *Pseudocoila bochei* Weld. Rev. suisse Zool. 66, 570–632.

Watson, F.L., Puttmann-Holgado, R., Thomas, F., Lamar, D.L., Hughes, M., Kondo, M., et al., 2005. Extensive diversity of Ig-superfamily proteins in the immune system of insects. Science 309, 1874–1878.

Whitten, M., Sun, F., Tew, I., Schaub, G., Soukou, C., Nappi, A., et al., 2007. Differential modulation of *Rhodnius prolixus* nitric oxide activities following challenge with *Trypanosoma rangeli, T. cruzi* and bacterial cell wall components. Insect Biochem. Mol. Biol. 37, 440–452.

Whitten, M.M., Mello, C.B., Gomes, S.A., Nigam, Y., Azambuja, P., Garcia, E.S., et al., 2001. Role of superoxide and reactive nitrogen intermediates in *Rhodnius prolixus* (Reduviidae)/*Trypanosoma rangeli* interactions. Exp. Parasitol. 98, 44–57.

Williams, M.J., 2007. *Drosophila* hemopoiesis and cellular immunity. J. Immunol. 178, 4711–4716.

Wink, D.A., Mitchell, J.B., 1998. Chemical biology of nitric oxide: insights into regulatory cytotoxic, and cytoprotective mechanisms of nitric oxide. Free Radic. Biol. Med. 25, 434–456.

Wright, B., Lacchini, A.H., Davies, A.J., Walker, A.J., 2006. Regulation of nitric oxide production in snail (*Lymnaea stagnalis*) defence cells: a role for PKC and ERK signalling pathways. Biol. Cell 98, 265–278.

Zettervall, C.-J., Anderl, I., Williams, M.J., Palmer, R., Kurucz, E., Ando, I., et al., 2004. A directed screen for genes involved in *Drosophila* blood cell activation. Proc. Natl Acad. Sci. USA 101, 14192–14197.

Zou, Z., Jiang, H., 2005. *Manduca sexta* serpin-6 regulates immune serine proteinases PAP-3 and HP8: cDNA cloning, protein expression, inhibition kinetics and function elucidation. J. Biol. Chem. 280, 14341–14348.

CHAPTER 5

Virulence Factors and Strategies of *Leptopilina* spp.: Selective Responses in *Drosophila* Hosts

Mark J. Lee,*[*] Marta E. Kalamarz,[*,†] **Indira Paddibhatla,**[*,†] **Chiyedza Small,**[*,†] **Roma Rajwani,**[*] and **Shubha Govind**[*,†]

Contents

[*] Department of Biology, The City College of New York, New York, NY 10031, USA
[†] The Graduate Center of the City University of New York, New York, NY 10016, USA

Advances in Parasitology, Volume 70
ISSN 0065-308X, DOI: 10.1016/S0065-308X(09)70005-3

Abstract

To ensure survival, parasitic wasps of *Drosophila* have evolved strategies to optimize host development to their advantage. They also produce virulence factors that allow them to overcome or evade host defense. Wasp infection provokes cellular and humoral defense reactions, resulting in alteration in gene expression of the host. The activation of these reactions is controlled by conserved mechanisms shared by other invertebrate and vertebrate animals. Application of genomics and bioinformatics approaches is beginning to reveal comparative host gene expression changes after infection by different parasitic wasps. We analyze this comparison in the context of host physiology and immune cells, as well as the biology of the venom factors that wasps introduce into their hosts during oviposition. We compare virulence strategies of *Leptopilina boulardi* and *L. heterotoma*, in relation to genome-wide changes in gene expression in the fly hosts after infection. This analysis highlights fundamental differences in the changes that the host undergoes in its immune and general physiology in response to the two parasitic wasps. Such a comparative approach has the potential of revealing mechanisms governing the evolution of pathogenicity and how it impacts host range.

5.1. INTRODUCTION

Parasitic wasps act as foundation species in natural ecosystems (LaSalle and Gauld, 1993). *Leptopilina, Asobara* and *Ganaspis* are part of a large group of insects making up more than 20% of all insect species. Thus, even though the parasitoid lifestyle is somewhat atypical, it is by no means unusual. Parasites hijack the body of the developing larva, taking over its resources and reprogramming the development of a single parasite within the host's body. In large measure, the natural success of parasitic wasps is due to the presence of virulence factors in the venom glands of female wasps. Virulence factors can be proteins, or in some cases, mutualists, microparasites or microbial symbionts. The nature of the virulence factors in parasitic wasps of *Drosophila*, as well as the relationships between virulence factors and their insect hosts (wasps), or wasp hosts (*Drosophila*), remain largely unknown. Because virulence factors represent the interface of host–pathogen interactions and are subject to natural selection, they are likely to shape the genetic structures of both the host and parasite populations. In this chapter, we review the immune competence of the host, and the nature of putative virulence factors and virus-like particles produced in venom glands of *L. boulardi, L. heterotoma* and *L. victoriae*. We also address the relationship of virulence strategies to cellular and molecular changes, especially as they relate to host defense in *Drosophila*.

5.1.1. Parasitism by *Leptopilina* spp.

The cosmopolitan genus *Leptopilina* (Nordlander, 1980) consists of three species groups: *heterotoma* (five described species, including *L. heterotoma* and *L. victoriae*), *boulardi* (three described species) and *longipes* (five described species; Allemand et al., 2002; Schilthuizen et al., 1998). As evident from considerable coverage in accompanying chapters, the biology of *L. boulardi* and *L. heterotoma* is particularly well studied. These parasitic wasps infect (or superinfect) second-to-early third-instar stages of *Drosophila* larvae. Infection causes a slight delay in host development (Kopelman and Chabora, 1984; Schlenke et al., 2007).

One of the clearest responses to wasp infection is encapsulation of the wasp egg or the early embryo. This innate immune reaction has been documented in many invertebrates (Brehélin, 1985), and is likely to be a universal defense mechanism in animals to combat large foreign bodies. Encapsulation of wasp eggs is relatively easy to observe in whole *Drosophila* larvae because the larval cuticle is transparent, and encapsulation is often accompanied by melanization (see Chapter 4 by Nappi et al.). While the dead, dark encapsulated wasp in the host hemocoel has fascinated biologists for a number of years, only recently we have learned that wasp infection also activates the humoral pathways in fly larvae (Schlenke et al., 2007; Wertheim et al., 2005). Humoral responses in insects include localized melanization and systemic induction of antimicrobial peptides (AMPs). These innate immune responses in *Drosophila* have thus far been characterized mainly in the context of microbial (bacterial and fungal) infections of the adult fly. Many molecules that are involved in nonself recognition, core components of the NF-κB pathways (Toll and IMD) and the effector molecules that limit infection *in vivo* are now known (see Lemaitre and Hoffmann, 2007 for a recent comprehensive review). However, the way in which wasp eggs are recognized in the host hemocoel, and various functions of antimicrobial peptides or other effector molecules, specifically in host defense against wasps, are not known. In the three species of *Leptopilina* discussed here (*L. boulardi* and the sister species *L. heterotoma* and *L. victoriae*), these immune responses are avoided, blocked, or suppressed, complicating the understanding of specific host responses.

We and others have studied host physiology, genetics and genomics in response to infections by *L. heterotoma/L. victoriae* and *L. boulardi* to probe similarities and differences in their host range and virulence strategies. Because these species present different virulence strategies, host responses differ dramatically: *L. heterotoma/L. victoriae* produce 300-nm wide virus-like particles (VLPs) within specialized long glands (also called venom glands). These particles are deposited along with the eggs. VLPs bind to host lamellocytes, become internalized and promote lamellocyte lysis (Morales et al., 2005; Rizki and Rizki, 1984, 1990, 1992, 1994). *L. heterotoma/L. victoriae* infection also leads to apoptosis of hemocytes in the circulation and in the lymph gland, respectively (Chiu and Govind, 2002). Different strains of *L. heterotoma* tested on *Drosophila melanogaster* are consistently highly virulent. Furthermore, morphologically identical 300-nm VLPs have been reported from at least three independently isolated strains of this wasp.

L. boulardi, in contrast, exhibits substantial intraspecific variation with respect to virulence on *D. melanogaster* spp. (Dupas et al., 1996; see also Chapter 6 by Poirié et al. and Chapter 11 by Dupas et al.) and venom content. Even though venom glands of *L. boulardi-17* and *G486* strains produce VLPs (Dupas et al., 1996; Schlenke et al., 2007), their ability to provoke encapsulation is different, the former strain being more virulent. In addition, filamentous viruses (Varaldi et al., 2006; see Chapter 13 by Varaldi et al.) are also found in venom glands of some *L. boulardi* strains. Regardless, unlike *L. heterotoma/L. victoriae* venom, lytic or apoptotic effects of *L. boulardi* venom on host hemocytes has not been reported.

A second major difference between *heterotoma/L. victoriae* and *L. boulardi* has to do with the location of the wasp eggs soon after oviposition: while eggs of *L. heterotoma* or *L. victoriae* are found floating freely in host hemolymph, *L. boulardi* eggs are often found attached to host tissues. This difference suggests that, in addition to active suppression affecting hemocytes, *L. boulardi* also employs a passive or evasive method of protecting its eggs from complete hemocyte encapsulation (Melk and Govind, 1999; Rizki et al., 1990). Together, these differences in the proteins, particles and other secretions of the venom gland appear to confer unique properties on *Leptopilina* spp. (and perhaps even to individual strains), resulting in unique host responses.

5.2. THE HOST RANGE OF *L. BOULARDI* AND *L. HETEROTOMA*

A host range or spectrum is the total variety of species that a parasite infects in nature. Factors contributing to this spectrum include shared ecosystem, behavioral compatibility and the ability of the host to mount a robust immune response. The strength of the immune response depends on the genetic factors that directly control production of immune effector cells and molecules. The success of this response also rests on the host's general health and physiology (tolerance to pathogens) of the healthy host (Fig. 5.1A; e.g., Ayres et al., 2008). The complex interactions of the genetic and physiological factors within the host and with the parasite dictate the

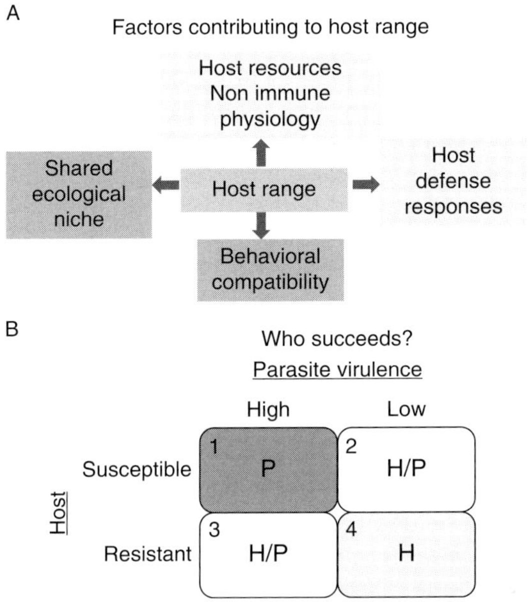

FIGURE 5.1 Host range and susceptibility. (A) A confluence of ecological, behavioral and host factors contributes to host range. (B) In the event of infection, success of a parasite will depend on the combination of specific individual host and parasite factors, as shown. Host resistance (immune mechanisms), host resources (nonimmune physiological factors that enable host defense) versus mechanisms of evasion and/or virulence on the part of the parasitoid. Clear outcomes of either the parasite [quadrant 1; "high virulence" parasite (e.g., *L. heterotoma*) infecting "low resistance" host] or the host [(quadrant 4; "low virulence" parasites (e.g., *A. tabida*) infecting a "high resistance" host)] can be predicted. However, in case of high virulence/high resistance or low virulence/low resistance (quadrants 2 and 3), specific outcomes may be more difficult to predict.

outcome in each infection (Fig. 5.1B). Thus, a parasite is a generalist if it successfully infects a number of related or unrelated hosts. In contrast, a specialist parasite succeeds on one or few host species in nature.

Infection by wasps is known to modify host development: a parasite can take control of host development, generally by delaying it. This delay allows the parasite to utilize host resources optimally. For example, the polydnavirus of the ichneumonid parasitoid, *Campoletis sonorensis* (CsV) was found to be the only component of its calyx fluid responsible for causing developmental arrest of its host *Heliothis virescens* (Dover et al., 1987). This developmental delay is attributed to a reduction of ecdysteroid titers that actually occurs because of partial degeneration of the prothoracic glands. Thus, infection introduces factors that modify aspects of host physiology, other than host immunity, that facilitate parasite success.

In laboratory infection experiments involving 18 *Drosophila* spp., *L. boulardi* strain-*17* was found to be far less infectious than *L. heterotoma* strain-*14* (Schlenke et al., 2007). *L. boulardi-17* succeeded (>90% wasp emergence) only on *D. melanogaster*, and was only moderately successful (~ 50% wasp emergence) on *D. mauritiana, D. sechellia* and *D. simulans*, all of which are closely related species of the melanogaster group. *L. heterotoma-14* showed higher success on these hosts.

The difference in infection outcomes on *D. yakuba, D. santomea* and *D. teissieri* was stark: whereas *L. boulardi-17* failed (0–1% wasp emergence) on these three closely related hosts, *L. heterotoma-14* succeeded with greater than 60% wasp emergence. *L. boulardi-17* infection induced clear encapsulation in these hosts (Schlenke et al., 2007; Fig. 5.2A); the capsules are similar to those formed in *D. melanogaster*, showing typical aggregation and melanization of hemocytes (Schlenke et al., 2007).

Of the remaining 11 species tested, *L. heterotoma-14* succeeded on six species (>30% wasp emergence), whereas *L. boulardi-17* infection was successful on only two species.

The ability of *L. heterotoma-14* to infect diverse hosts of different sizes and developmental times successfully is quite remarkable (Schlenke et al., 2007). It implies that the egg and/or the factors that are introduced with the egg, interfere with host development and defense in a "species-nonspecific" manner. Part of this strategy is likely to involve the deadly effects that *L. heterotoma* infection unleashes on host hemocytes and hematopoiesis, a strategy not shared by *L. boulardi*. The host range of *L. victoriae* on laboratory-raised fly hosts has not been studied. Because *L. heterotoma* and *L. victoriae* share similar virulence mechanisms, especially their lethal effects on host hemocytes, it will be interesting to compare the ability of *L. victoriae* to succeed on these same species that *L. heterotoma* infects.

Because of the dramatic differences in host range in laboratory-raised hosts (Schlenke et al., 2007) and hosts in natural habitat (Carton et al., 1986), *L. boulardi* is characterized as a "specialist," whereas. *L. heterotoma* is a "generalist." It is worth noting that *L. boulardi-17* infection induces encapsulation in many species (Fig. 5.2A and B; Schlenke et al., 2007), whereas *L. heterotoma-14* infection does not. This difference underscores the importance of using the optimal wasp/host pair to study encapsulation. More importantly, it highlights the contribution of the encapsulation reaction in host defense across different *Drosophila* spp.

5.2.1. Infection by *L. boulardi-G486* triggers stage-specific hematopoiesis in third-instar hosts

The encapsulation reaction (Fig. 5.2A and B) is complex, highly controlled and quite effective. Successful encapsulation requires coordinated interactions of three hemocyte types, namely plasmatocytes, crystal cells

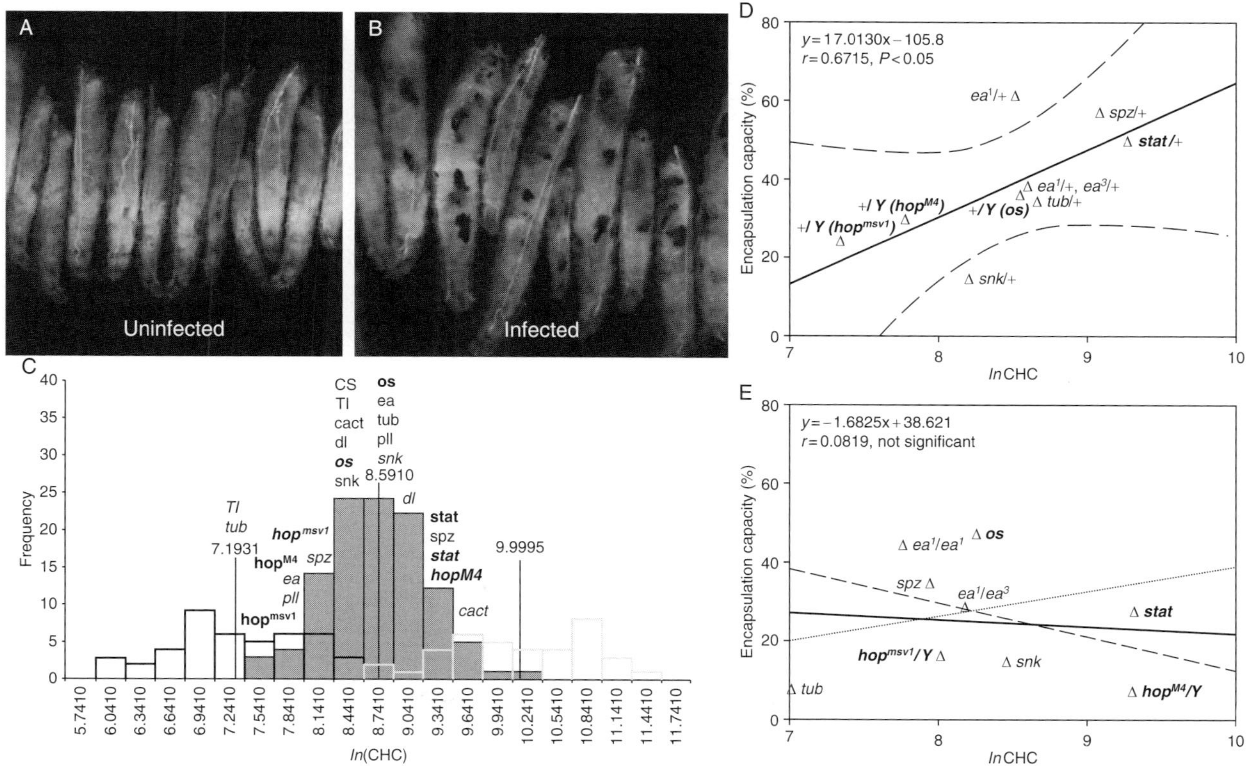

FIGURE 5.2 Hemocyte-mediated encapsulation depends on both hemocyte concentrations and hemocyte differentiation. (A) and (B) *L. boulardi-17* infection induces massive encapsulation in *D. yakuba*. (A) Uninfected and (B) infected third-instar larvae. (C) Hemocyte concentration in *D. melanogaster* larvae (log normal transformation of hemocyte concentrations) shows a normal distribution. Empirically

and lamellocytes. While the encapsulation reaction is triggered by oviposition, it is not known what factors limit the hemocytic response and what determines its timely termination.

The hemocoel of normal uninfected third-instar larvae is populated mainly by plasmatocytes. These small, phagocytic cells (Fig. 5.3H and I) consume bacteria, scavenge dead cells and secrete antibacterial peptides and extracellular matrix components (Fessler et al., 1994; Rizki and Rizki, 1984). Crystal cells make up less than 5% of all hemocytes in circulation and synthesize substrates and enzymes for melanization reactions.

The lymph gland is a small organ at the anterior dorsal region (behind the brain through the ring gland, in the anterior abdominal segments) of the larva. It houses hematopoietic progenitors of the adult fly (Govind, 2008; Holz et al., 2003; Lemaitre and Hoffmann, 2007; Martinez-Agosto et al., 2007; Meister and Lagueux, 2003).

When faced with parasitization by *L. boulardi-G486* the fly larva triggers differentiation of lamellocytes and crystal cells (Lanot et al., 2001; Sorrentino et al., 2002). Furthermore, a limited burst of mitosis follows shortly after infection (Sorrentino et al., 2002), suggesting that both cell division and differentiation of lymph gland progenitors are required for encapsulation. These changes, observed in the lymph glands of third instar, but not of second-instar hosts, are almost always accompanied by dispersal of the anterior lobes themselves. In this dispersal response, the continuous basement membrane that lines (and possibly holds) the cells of the anterior lobes is disrupted (Sorrentino et al., 2004). Lamellocytes and their precursors are also present in the posterior hematopoietic compartment. Cells from this location also contribute to host defense (Markus et al., 2009).

A link between host development and immune competence was confirmed in genetic experiments using mutant hosts in which development was blocked during mid-to-late larval stages. *Drosophila* strains where ecdysone levels are low (*ecdysoneless*) or ecdysone signaling is

---

determined range of normal hemocyte concentration in staged third-instar larvae is shown (vertical lines). Values of hemocyte concentration from heterozygotes and mutants were derived from outcrossed animals. Genotype of mutants is italicized (allele name, where used, is superscripted), heterozygotic genotypes are not. Genotypes shown in bold pertain to JAK-STAT signaling components (os, outstretched; hop, hopscotch; STAT, STAT92E), whereas those that are not bold relate to Toll signaling (snk, snake; ea, easter; spz, spatzle; Tl, Toll; tub, tube; pll, pelle; cact, cactus. CS, Canton S is a wild-type strain). (D,E) Direct correlation of encapsulation capacity with hemocyte concentration in wild-type and control values (D). This correlation is violated in mutant animals. Genotypes are referred to as in legend above. *Note: Figures modified from* Sorrentino et al. (2004) *with permission from the publisher.* (See Page 3 in Color Section at the back of the book.)

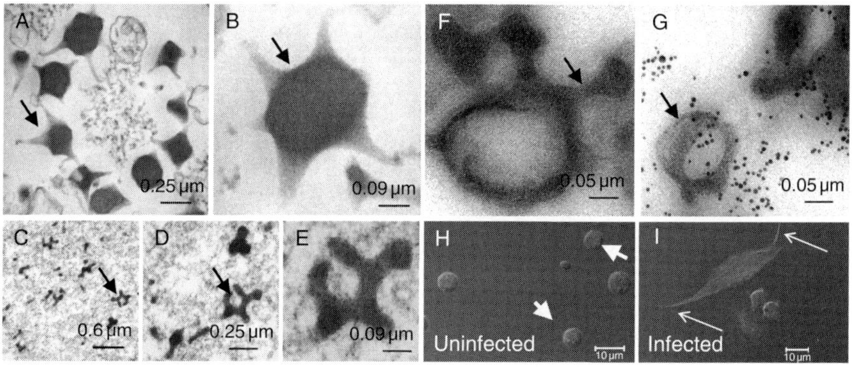

FIGURE 5.3 Virus-like particle (VLP) structure and biogenesis. (A,B) Sections of gradient-purified VLPs from *L. victoriae*. Notice the somewhat irregular pentagonal/hexagonal organization of the VLP core that extends into projections. (C–E) Section of a region from the lumen of *L. victoriae* long gland with immature VLPs at different magnifications. (F,G) Association of p40 antigen within and around immature VLPs of *L. heterotoma* (arrows point to immuno-gold particles linked to secondary antibody (G)). Sample in panel (F) was not treated with the primary antibody. (H and I) *L. victoriae* infection induces lamellocytes to alter morphology. Hemocytes from hemolymph of normal (H) or infected hosts (I) stained with rhodamine-phalloidin (F-actin) and Hoechst (blue). Normal hemolymph has 10-μm wide spherical plasmatocytes (arrowheads) but very few, if any, lamellocytes. None are present in this panel. Infected lamellocytes exhibit altered, spindle-shaped morphology.

blocked (nonpupariating allele of the transcription factor *broad*), were infected by *L. boulardi-G486*. The encapsulation response in such hosts was severely compromised: (1) the postinfection mitotic amplification in the lymph glands of third-instar *ecdysoneless* hosts was absent; (2) there was a reduction in crystal cell maturation in the lymph gland; and (3) there was also a reduction in the postinfection circulating lamellocyte concentration (Sorrentino et al., 2002). These results suggested that development of the precursors continues through larval instars, a prediction that was confirmed in subsequent studies of lymph gland cell populations in which precursor population was distinguished from differentiating cells (Jung et al., 2005). It appears then that these parasitic wasps are able to infect larval hosts while the hosts are still immune incompetent, while giving themselves almost the entire third larval instar to program their own development. The infection cycle is therefore developmentally aligned with the development of the host. Conversely, lymph gland precursors for lamellocytes are available to divide and differentiate at precisely the time that these wasps are able to recognize and infect their hosts.

Sorrentino et al. (2002) also predicted the existence of an ecdysone-activated pathway that potentiates precursors of effector cell types to respond to parasitization by proliferation and differentiation. This prediction regarding hormonal control of immune cell development remains to be explored at the molecular level.

5.2.2. Does the concentration of circulating hemocytes have a bearing on successful encapsulation?

While virulence factors interfere with encapsulation, variabilities in the host's immune physiology and cells are also likely to contribute to the wasp/fly outcome. Hemocytes normally circulate freely in the body cavity and do not clump or spread inside the hemocoel. Yet, upon introduction of parasites into the hemocoel, some of these cells flatten and become adhesive to form a capsule around the nonself entity. Sorrentino et al. (2004) characterized circulating hemocyte concentration (CHC) from wild-type third-instar hosts to examine if hemocyte density has a bearing on wasp encapsulation. They found that: (1) the control mean raw CHCs exhibit a nearly sevenfold range of values; (2) the distribution of wild-type Canton-S or control ($n = 110$) CHC values does not follow a normal distribution. Instead, when the CHC values were converted to natural logarithms of raw CHC values (ln CHC), the frequency distribution showed a normal distribution (Fig. 5.2C). Using *L. boulardi-G486* to infect developmentally staged animals, they reported a significant correlation between hemocyte concentration and encapsulation capacity among wild-type larvae and larvae heterozygous for mutations in the JAK-Stat92E and Toll-NF-κB pathways (Fig. 5.2D).

5.2.3. Toll-NF-κB and JAK-STAT signaling in encapsulation

JAK-STAT92E and Toll-NF-κB signaling control cellular physiology, proliferation and/or differentiation. Activated by extracellular cytokines, these signaling pathways mediate cellular and systemic responses to infection. Many core components of these pathways (that are responsible for relaying information from the membrane to the nucleus) have been identified (see Fig. 5.4). Highly conserved in metazoan animals, JAK kinases, transcription factors NF-κB/Rel or STAT proteins and their inhibitors (IκB and PIAS, respectively) also regulate aspects of mammalian hematopoiesis (Baker et al., 2007; Bottero et al., 2006; Martinez-Agosto et al., 2007).

Heterozygous animals carrying loss-of-function (recessive) mutations in *Toll*, *tube* or *pelle* show CHC in the normal range. However, homozygous larvae carrying loss-of-function mutations in these "dorsal group" genes (*Toll*^{5BRE}, *tube*^{238} and *pelle*^{rm8}) have significantly reduced circulating hemocyte concentrations (Qiu et al., 1998). These mutants are severely compromised in their ability to mount an effective encapsulation

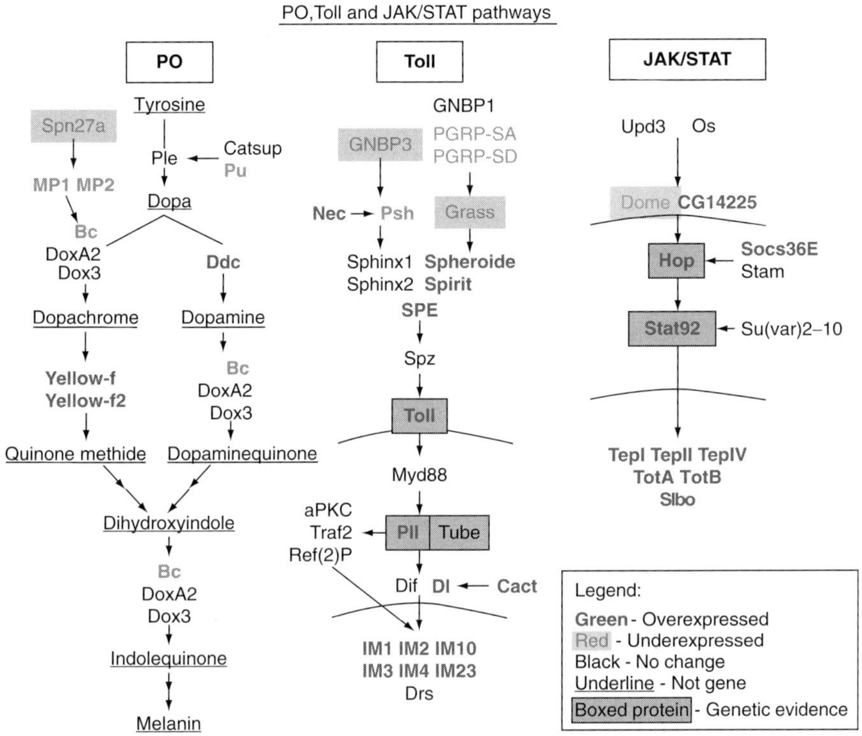

FIGURE 5.4 Components and targets of pro-phenol oxidase, Toll-Dorsal/Dif and JAK/STAT pathways. Genes whose expression is modulated by *L. boulardi*-17 infection (up- or downregulated) are shown. The Toll-Dorsal/Dif and JAK/STAT pathway components that were tested in genetic experiments (Sorrentino *et al.*, 2004) for their requirement for a robust encapsulation response are also shown (genetic evidence). *Note: Figure modified from* Schlenke et al. (2007).

response (Sorrentino et al., 2004; Fig. 5.2E). Loss of function in IκB/*cactus* results in the opposite effect reflected in increased CHC outside of the normal range (Qiu et al., 1998). Affected mutants show microtumors. The genetic lesion results in overproliferation and constitutive lamellocyte differentiation of the hematopoietic tissue, which in turn encapsulates self tissue.

The situation with components of the JAK-STAT signaling is different. Here heterozygous or mutant larvae deficient in Hopscotch-Stat92E signaling (*outstretched*^{0}, *hopscotch*^{M4}, *hopscotch*^{msv1}, *stat92E*^{HJ}) exhibit ln CHC in the control range (Fig. 5.2E). Yet infection of loss-of-function mutant animals affecting *hop* or *STAT* genes affects encapsulation; mutant lymph gland progenitors are unable to differentiate into lamellocytes (Sorrentino et al., 2004).

The genetic regulation of lamellocyte differentiation is not entirely clear. The JAK-STAT signal appears to be essential for holding hematopoietic progenitors in their immature state within the lymph gland (Krzemień et al., 2007).

While *L. boulardi-G486* provokes a substantial and measurable cellular immune response in *D. melanogaster*, this is not true for *L. heterotoma*. Oviposition by *L. heterotoma* has different effects on the hematopoietic system of *D. melanogaster* (see Section 5.3). Encapsulation can be observed in *L. victoriae*-infected *D. melanogaster* hosts, presumably because the *L. victoriae* venom acts at a slower rate *in vivo* than that of *L. heterotoma* (Chiu and Govind, 2002; see Section 5.3).

5.3. ORIGIN OF *L. HETEROTOMA/L. VICTORIAE* VLPs AND THEIR EFFECTS ON HOST HEMOCYTES

5.3.1. An actin-lined canal system controls biogenesis and release of virulence factors

The venom apparatus in the female wasp is associated with the reproductive tract and produces some of the factors that accompany the egg. The gland has three main regions: the venom (or long) gland, the reservoir and the ovipositor (Fig. 5.5A and B). The venom gland is the site of production of VLP precursors and other secreted factors. These factors are secreted into the gland lumen, pass through a connecting duct, into the reservoir, where they mature further. The venom fluid is also stored in the reservoir until the female injects it with its eggs. The ovipositor is structurally sharp and chitinous, and capable of extending outside the wasp abdomen.

The venom gland is composed of a peripheral layer of large secretory cells and an internal intimal layer of narrow cells. Both cell layers are concentric to the gland lumen. We recently discovered a specialized system of canals in venom glands of five parasitoid wasps that are quite different in their infection strategies (Ferrarese et al., 2009). This supracellular system of canals is made up of individual secretory units lined with bundles of filamentous actin. Each unit has two (continuous) parts: (a) the proximal rough canal, which originates within the secretory cell and is organized into brush border morphology; and (b) the smooth canal, which is narrower and passes through the intimal cells opening into the gland lumen. Three-dimensional reconstruction of fluorescently labeled canals and cell nuclei reveals that the canal system occupies the whole organ. Each secretory cell has one canal that is oriented roughly perpendicular to the venom gland lumen (Fig. 5.5C). This analysis, at the light microscopy level is reinforced by observations at higher magnification using

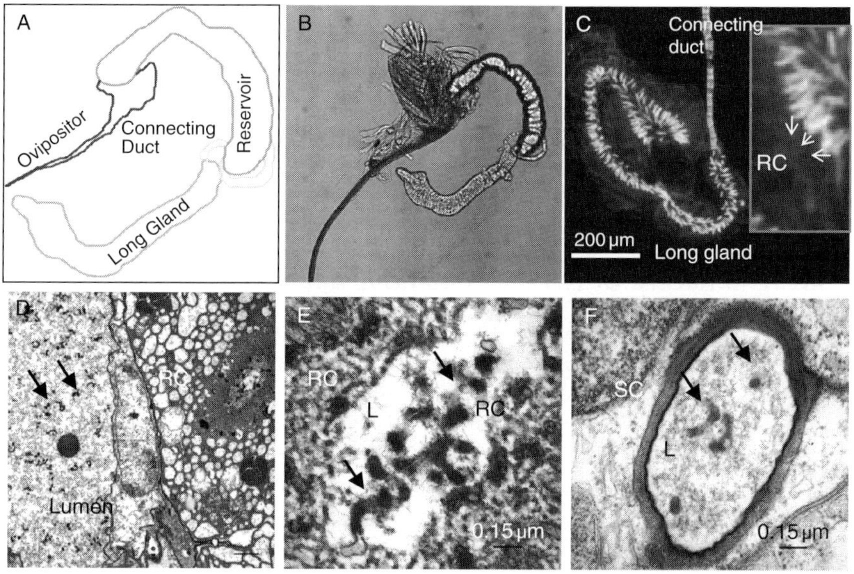

FIGURE 5.5 Venom apparatus, canals and contents. (A) and (B) A schematic (A) that corresponds to the whole mount (phase, B) of a dissected sample of a venom apparatus from *L. heterotoma*. Different parts of the organ are labeled. (C) Phalloidin staining of the venom (long) gland from *L. heterotoma* reveals the scope and organization of a supra-cellular canal system. Made of individual secretory units (one per secretory cell), VLP precursors and other constituent proteins make their way, initially through the rough portion of the canal system, present in the cytoplasm of the large secretory cell itself. This portion (panel inset) is composed of membranous folds of actin-rich microvilli that give the structure a rough appearance. Rough canal (RC) loses the folds and narrows into the smooth canal, which opens into the gland lumen. (D–F) Transmission electron micrographs of the venom gland. (D) Rough canal in cross section, a cell of the intimal layer and immature VLPs are seen in this low magnification view of the gland. (E and F) Cross-sections of the rough (E) and smooth (F) canals of *L. victoriae* venom glands. Both these structures contain VLP precursors (arrows). (See Page 4 in Color Section at the back of the book.)

transmission electron micrographic (TEM) methods (Chiu et al., 2006; Ferrarese et al., 2009; Morales et al., 2005; Fig. 5.5D–F). Based on localization of p40 to the microvilli of the rough canals, within the smooth canals and in the gland lumen, and its close association with VLP precursors (Fig. 5.3F and G) in immunostaining experiments, we have proposed that the canal system is adapted for efficient trafficking of the molecular components from secretory cells to the lumen (Ferrarese et al., 2009). p40 is a putative virulence factor of *L. heterotoma* VLPs (Chiu et al., 2006, see below).

Remarkably, structures with a very similar organization of actin-lined canals were observed in three *Leptopilina* spp. and *Ganaspis xanthopoda*, parasitoids of *Drosophila* spp., as well as in *Campoletis sonorensis*, an ichneumonid parasitoid of *Heliothis virescens*. These observations suggest that the novel supracellular canal system may be a shared trait of venom glands in parasitic wasps. This system appears to be essential for efficient biogenesis and delivery of virulence factors (Ferrarese et al., 2009).

5.3.2. The nature of *L. victoriae* and *L. heterotoma* VLPs

Wasp venom is fluid-like, composed of a variety of proteins and microscopic entities (bacteria, viruses or virus-like particles). Mature VLPs of *L. victoriae* and *L. heterotoma* (Fig. 5.3A and B) are pentagonal and hexagonal in shape, with varying numbers of spike-like appendages (Chiu et al., 2006; Morales et al., 2005). Through a silver stain gel analysis of purified VLPs obtained from *L. heterotoma* and *L. victoriae*, we showed that mature VLPs of these two closely related wasps are composed of at least four major proteins. Of these, p40 and p47.5 are the most abundant in the respective species (Chiu et al., 2006). An antibody raised against purified *L. heterotoma* VLPs recognizes p40 and cross reacts with p47.5 of *L. victoriae*.

Using fluorescence light microscopy and immuno-EM methods, we could track biogenesis of VLPs from secretory cells of the venom gland, all the way into the host hemocytes using the anti-p40 antiserum (Chiu and Govind, 2002; Chiu et al., 2006; Ferrarese et al., 2009): (1) VLP precursors (p40) are produced in the perinuclear region of secretory cells; (2) p40 moves from the perinuclear region toward the membrane microvilli region, also known as the "rough" canal; (3) It is delivered into the venom gland lumen via the smooth canal (Fig. 5.5D–F). VLP precursors and partially assembled particles continue to undergo assembly within the venom gland lumen (e.g., compare morphologies in Fig. 5.3C–E with Fig. 5.3A and B). Additional morphological changes are observed in sections from regions adjacent to the connecting duct and from within the reservoir, where they eventually mature into VLPs (Chiu et al., 2006; Morales et al., 2005).

The presence of VLPs has also been reported in different *L. boulardi* strains including *L. boulardi-G486*, although their biogenesis, structure and mechanism of action are not well understood. *L. boulardi* venom induces cytoskeletal changes affecting function of lamellocytes (Dupas et al., 1996; Labrosse et al., 2005). Regardless, it is clear that the population of symbiotic/microbial structures formed and residing within parasitoid venom gland can profoundly modulate parasitoid–host interaction. Their characterization will clarify the nature of this interaction and shed light on the evolution of *Drosophila* spp.

5.3.3. The lethal effects of *L. heterotoma/L. victoriae* VLPs on host hemocytes

The VLPs of both *L. heterotoma* and *L. victoriae* have spikes with knobs at the end extending from the center core (Chiu et al., 2006; Rizki and Rizki 1994; Fig. 5.3A and B). Immuno-EM staining of mature VLPs from both species shows that p40 and p47.5 proteins are largely located in the periphery and along the spike-like structures of VLPs (Fig. 5.3F and G; Chiu et al., 2006). In scanning electron microscope (SEM) preparations, VLPs aggregate with each other via VLP spikes. VLPs also attach to the lamellocyte membrane via these extensions.

Infection by *L. heterotoma* and *L. victoriae* (but not *L. boulardi*) results in a concerted and active deletion of larval hemocytes: first, infection leads to the apoptosis of the larval lymph glands (Chiu and Govind, 2002). This effect was observed in an *in situ* terminal deoxynucleotidyltransferase-mediated 2′-deoxyuridine 5′-triphosphate nick-end labeling (TUNEL) assay. Factors that trigger these changes in the lymph gland are also not fully known.

Second, few cells that are able to differentiate into lamellocytes undergo rapid lysis, but first assume a bipolar morphology (Fig. 5.3H and I). *In vitro*, venom extracted from either *L. heterotoma* or *L. victoriae* induces lamellocytes (from microtumor-bearing $hop^{Tum\text{-}l}$ larvae) to assume bipolar morphology. Such bipolar cells have p40/p47.5 localized within them. Significantly, anti-p40 antibody specifically neutralizes this cellular transformation (almost completely for *L. heterotoma* VLPs and by more than half for *L. victoriae* VLPs) implicating these proteins in VLP-lamellocyte recognition or binding (Chiu et al., 2006). Bipolar cells remain TUNEL negative when induced *in vitro* (Chiu and Govind, 2002).

Finally, incubation of *L. heterotoma* or *L. victoriae* venoms induces apoptosis of mature circulating plasmatocytes in short-term cultures. These changes occur when circulating hemocytes from $hop^{Tum\text{-}l}$ mutant larvae were treated with venom from either *L. heterotoma* or *L. victoriae, in vitro*. In such cultures, approximately 30% of the hemocyte population becomes TUNEL positive (2–4 days after infection). Interestingly, the overwhelming majority of the TUNEL-positive cells were plasmatocytes. Furthermore, immunofluorescence experiments revealed that VLPs are actually localized inside the cytoplasm of the TUNEL-positive hemocytes. These findings suggest that VLPs may play an important role in inducing apoptosis in circulating hemocytes of host larvae (Chiu and Govind, 2002). The exact mechanisms or effectors involved in either apoptosis of plasmatocytes or bipolar cell lysis of lamellocytes are not fully understood.

The lytic and apoptotic effects of *L. heterotoma* and *L. victoriae* venom were compared *in vivo* in which tumor-bearing $hop^{Tum\text{-}l}$ larvae were infected by the wasps. The effect of *L. heterotoma* venom was stronger where the development of most tumors in the mutant animals was inhibited. In contrast, *L. victoriae* infection resulted in encapsulation of wasp eggs in larvae of the same genetic background. This difference in "strength" of the venom was also observed in *in vitro* bipolar lysis assays (Morales et al., 2005).

5.4. HOST GENE EXPRESSION CHANGES AFTER *L. BOULARDI* AND *L. HETEROTOMA* INFECTION

One approach to understanding the effects of different infection strategies on host physiology is to examine differences in the global gene expression patterns after *L. boulardi-17* and *L. heterotoma-14* infections. Schlenke et al. (2007) infected second-instar hosts for 2 h and harvested ribonucleic acid (RNA) from animals at 2–5, 9–12 or 21–24 h postinfection by either wasp for microarray analysis (Fig 5.6A).

First, of the classes of genes that are most significantly upregulated by both wasp species are proteolysis and energy generation (mitochondrial electron transport, oxidative phosphorylation), while the gene functional class most significantly downregulated by both wasp species is development (Fig. 5.6B). Upregulation of a wide range of proteolytic genes is indicative that one common response to wasp parasitism is the activation of proteolytic cascades, which are believed to be important in extracellular signaling, hemolymph coagulation and humoral immune signaling. The dynamic between upregulation in energy generation and downregulation of development genes points to the host's response to conserve energy by slowing down normal physiological activities and devote molecular machinery to mount an effective immune response. These changes in gene expression coincide with delayed pupariation of infected *D. melanogaster* hosts (2 days later than controls; see Schlenke et al., 2007).

Second, far smaller numbers of *Drosophila* genes are highly differentially regulated by *L. heterotoma-14* than by *L. boulardi-17* infection. In fact, the analysis of differentially expressed genes at all three timepoints reveals that hosts infected with *L. boulardi-17* expressed twice as many genes compared to *L. heterotoma-14* (Fig. 5.6B). Furthermore, this discrepancy is also evident by analyzing the number of genes that are differentially expressed at individual timepoints (Fig. 5.6D). The number of genes differentially regulated after *L. boulardi-17* infection increase over time, while it remains the same after *L. heterotoma-14* infection, widening the gap between two infections. Thus, gene expression in *D. melanogaster* is more robustly regulated in response to infection by the specialist *L. boulardi-17* than to the generalist *L. heterotoma-14*.

Third, *L. boulardi-17* infection leads to differential regulation of host gene expression, not only for those genes related to immune responses,

A

1. Larval hosts
2. Control or infect
3. Prepare RNA
4. cDNA, labeling reaction
5. Hybridize to chip
6. Intensity of fluorescent cDNA
7. Replicates
8. Statistics; compute signal for each gene in experimental sample relative to corresponding gene in control sample
9. Fold change
10. Analysis of data: gene ontology, GenMapp, etc.

B

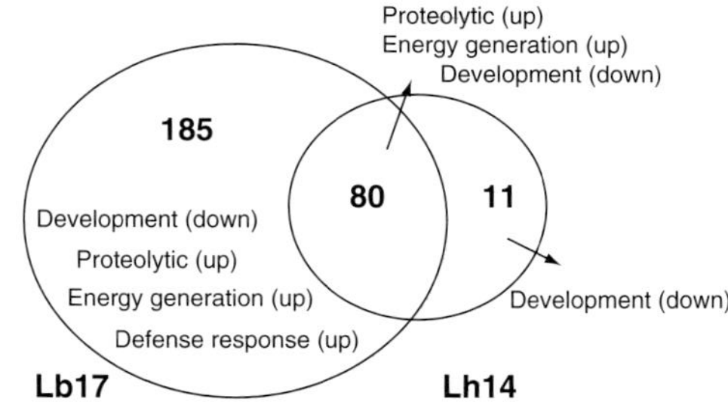

C

Biofunction	Lb17	Lh14	Overlap
Defense response	43	1	20
Structural constituent of cuticle	33	0	15
Proteolysis and peptidolysis	59	3	16
Development	41	7	26
Generation of energy	9	0	3
Total	185	11	80

D Differential gene regulation in *Drosophila* larvae upon wasp infection

	5h					
	Upregulated			Downregulated		
	Lb17	Lh14	Overlap	Lb17	Lh14	Overlap
Defense response	16	0	3	5	8	2
Structural constituent of cuticle	0	0	0	30	2	0
Proteolysis and peptidolysis	27	1	2	7	2	1
Development	0	0	0	13	6	13
Generation of energy	5	0	1	0	1	0
	12h					
	Upregulated			Downregulated		
	Lb17	Lh14	Overlap	Lb17	Lh14	Overlap
Defense response	22	1	2	8	9	0
Structural constituent of cuticle	33	0	0	5	12	0
Proteolysis and peptidolysis	40	1	0	9	12	0
Development	1	0	0	10	1	2
Generation of energy	1	0	0	3	0	0
	24h					
	Upregulated			Downregulated		
	Lb17	Lh14	Overlap	Lb17	Lh14	Overlap
Defense response	8	0	0	27	4	6
Structural constituent of cuticle	20	0	0	0	0	1
Proteolysis and peptidolysis	13	0	2	12	2	1
Development	0	0	0	44	4	21
Generation of energy	4	0	1	2	0	1

FIGURE 5.6 (Continued)

but for a variety of other functional classes (Schlenke et al., 2007; also see Fig. 5.6C). Gene Ontology/GenMapp analysis of over 400 genes revealed differential expression of genes encoding recognition proteins, proteolytic enzymes, antimicrobial peptides and components of the Toll/NF-κB, JAK/STAT and the melanization cascade (Fig. 5.4). Gene activation of Toll/NF-κB and JAK/STAT pathway components is consistent with the genetic requirement of pathway components as discussed in Section 5.2.3. These results suggest that activation of cellular and humoral arms in the larval immune system is linked via these signaling pathways. Surprisingly, however, hosts infected by *L. heterotoma-14* did not substantially modulate gene expression; with fewer than half as many genes affected.

Fourth, in contrast to the Toll, JAK/STAT and phenol oxidase (PO) pathways regulation of IMD, JNK and other described *Drosophila* immune pathways appears largely unaffected by wasp infection (Schlenke et al., 2007).

These results indicate that the Toll pathway is fundamentally important not only in regulating the antimicrobial response, but may also be the central regulator of the antiparasite response of insects. The systemic induction of AMP genes in the fat body cells was confirmed in an *in vivo* reporter assay. A host strain carrying a reporter transgene (*Drosomycin* promoter fused to GFP reporter, see schematic in Fig. 5.7A) was infected. *Drosomycin* is a specific target of the Toll pathway and its activation is easily detected in living animals using fluorescence microscopy (Ferrandon et al., 1998). Strong and uniform activation of this promoter was observed after *L. boulardi-17* infection (Fig. 5.7B–E), but not after *L. heterotoma-14* infection (Schlenke et al., 2007). The upregulation of several of the same effector molecules that are activated by microbial infections suggests that these effectors may also play a role in the antiparasite response or that they provide secondary protection from microbial infection.

FIGURE 5.6 Host gene expression changes after *L. boulardi-17* and *L. heterotoma-14* infections. (A) Outline of the design and analysis of the microarray experiment as described in Schlenke *et al.* (2007). (B) Venn diagram showing number of genes whose expression is modulated at any of the three points after *L. heterotoma-14* or *L. boulardi-17* infections. Fold change in gene expression was calculated from the published data and analysis threshold was arbitrarily set (fold change $\leq$0.75 and $\geq$1.5). (C) Differentially expressed genes identified in panel (B) can be classified into five functional classes as shown. *L. boulardi-17*-infected larvae differentially express 265 genes compared to 91 genes in *L. heterotoma-14*-infected larvae. Eighty genes are expressed in the host after either infection. (D) The five functional categories of differentially expressed genes organized by timepoints. Total genes that are differentially expressed at each timepoint after *L. boulardi-17* infection are 125, 136 and 163, and after *L. heterotoma-14* infection are 42, 40 and 43, respectively. Not all genes are differentially regulated at all three timepoints for the same wasp infection, and therefore, the list of differentially regulated genes at one timepoint is different than the list of genes at another timepoint.

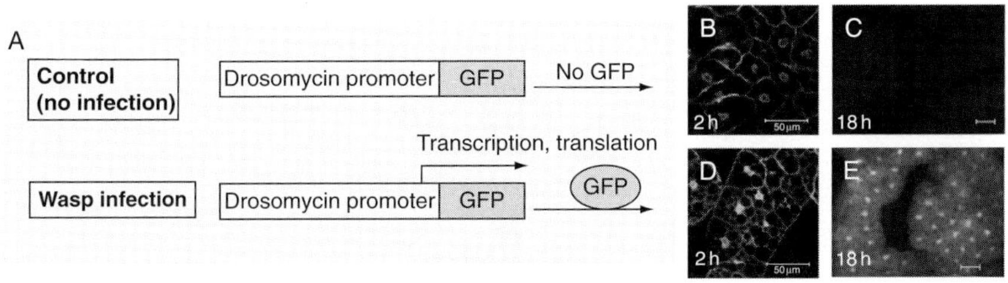

FIGURE 5.7 Wasp-induced fat body expression of *Drosomycin-GFP* reporter *in vivo*. (A) Reporter construct of *Drosomycin-GFP* designed to assay *in vivo* activation of the promoter (Ferrandon *et al.*, 1998; Tzou *et al.*, 2000). (B–E) Upon *L. boulardi* infection, *Drosomycin-GFP* expression is activated through the 24-h period of third-instar larval stages. Hosts were exposed to wasps for 24 h. Fat body samples were dissected 2 or 18 h after infection. *Drosomycin-GFP* was not expressed in the absence of infection (B, C), but was clearly expressed after infection (D, E). (See Page 4 in Color Section at the back of the book.)

The mechanism by which Toll pathway is activated is not known. However, genes encoding specific peptidoglycan recognition proteins are activated by *L. boulardi-17* and *L. heterotoma-14* infections and these might play a role in wasp egg recognition.

Differential activation of melanization by the two wasps was also observed *in vitro* from larval extracts after infection. Cytotoxic quinones, semiquinones and reactive oxygen species have been implicated in melanization and death of *Drosophila* parasites (Nappi and Vass, 1993; Nappi et al., 1995; see Chapter 4 by Nappi et al.).

5.5. CONCLUDING REMARKS

Parasitic wasps of *Drosophila* have evolved various strategies to maximize survival of their progeny. In this chapter, we have reviewed strategies of two wasps that are equally highly successful on *D. melanogaster*. Intriguingly, however, their ability to succeed on other *Drosophila* spp. is quite different. A key difference appears to lie in the lethal effects of venom components on the host hematopoietic system. While *L. heterotoma-14* infection activates immune cells (hemocytes), it quickly disables all arms of the larval innate immune response. The immune-suppressive effect of *L. boulardi-17* is more specific, where only wasp encapsulation is thwarted, while melanization and systemic production of antimicrobial peptide production both continue (Fig. 5.8). The ISm strain of *L. boulardi* produces Rac GTPase (called LbGAP) in its venom that affects lamellocyte morphology by targeting the host Rac1 and Rac2 proteins (Colinet et al., 2007; see Chapter 6 by Poirié et al.). It is not known whether *L. boulardi-17* synthesizes LbGAP.

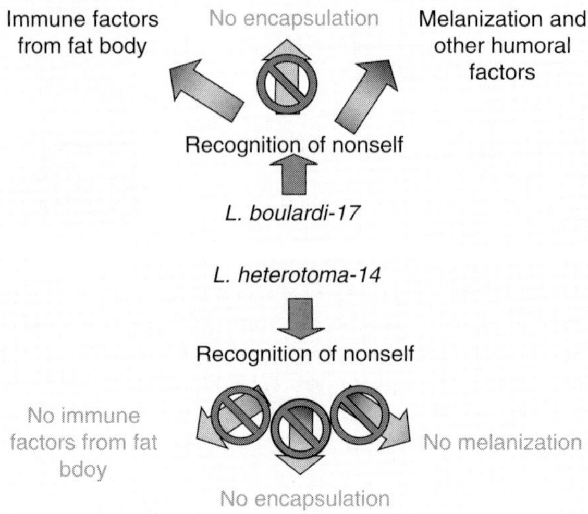

FIGURE 5.8 Effects of *L. boulardi* and *L. heterotoma* on *D. melanogaster*. Differences in the activities of virulence strategies (active vs. passive) and factors (proteins affecting hemocyte morphology vs. viability) may account for differences in the activation of immune pathways after infection by *L. boulardi-17* and *L. heterotoma-14*. In the former case, only encapsulation is abolished in *D. melanogaster* hosts. In the latter, all three aspects of immune responses are compromised.

A comparative approach applied to this three-part (symbiotic virus-like particles/wasp/fly) host–pathogen model system, that combines morphologic, molecular and genomics methods is beginning to provide a more comprehensive view of how this pathogen class succeeds in nature. Our knowledge of natural pathogens of *Drosophila* is still very limited. With application of molecular and proteomic methods, it is now feasible to explore the relationship of this well-characterized host with its natural parasites to understand how NF-κB signaling is activated, or remains repressed and how certain parasites of *Drosophila* have evolved to become highly virulent. The mechanisms of activation and immune suppression in related specific host–parasite pairs are expected to be shared. A systematic analysis of these mechanisms should provide insight into the nature and evolution of virulence.

ACKNOWLEDGMENTS

Financial support from USDA (NRI/USDA CSREES 2006-03817 and 2009-35302-05277), NIH NIGMS S06 GM08168, GM056833-08 and G12-RR03060) and PSC-CUNY is gratefully acknowledged. We thank Felix Castellanos, Roberto Ferrarese and Jorge Morales for help with figures and colleagues for discussions. The *L. heterotoma* strain whose origin is not specified in the text was obtained from P. Chabora. *G. xanthapoda* and *L. victoriae* were obtained from P. Chabora and Jacques J.J.M. van Alphen, respectively.

REFERENCES

Allemand, R., Lemaitre, C., Frey, F., Bouletreau, M., Vavre, F., Nordlander, G., et al., 2002. Phylogeny of six African *Leptopilina* species (Hymenoptera: Cynipoidea, Figitidea), parasitoids of *Drosophila*, with description of three new species. Ann. Soc. Entomol. Fr. 38, 319–332.

Ayres, J.S., Freitag, N., Schneider, D.S., 2008. Identification of *Drosophila* mutants altering defense and endurance of to *Listeria monocytogenes* infection. Genetics 178, 1807–1815.

Baker, S.J., Rane, S.G., Reddy, E.P., 2007. Hematopoietic cytokine receptor signaling. Oncogene 26, 6724–6737.

Bottero, V., Withoff, S., Verma, I.M., 2006. NF-kappaB and the regulation of hematopoiesis. Cell Death Diff. 13, 785–797.

Brehélin, M., 1985. Immunity in Invertebrates: Cells, Molecules, and Defense Reactions. Proceedings in Life Sciences. Springer-Verlag, Berlin; New York.

Carton, Y., Bouletreau, M., van Alphen, J.J.M., van Lenteren, J.C., 1986. The *Drosophila* parasitic wasps. In: Ashburner, Carson, Thompson, (Eds.), The Genetics and Biology of *Drosophila*. Academic Press, London, pp. 347–394.

Chiu, H., Govind, S., 2002. Natural infection of *D. melanogaster* by virulent parasitic wasps induces apoptotic depletion of hematopoietic precursors. Cell Death Diff. 12, 1379–1381.

Chiu, H., Morales, J., Govind, S., 2006. Identification and immuno-localization of p40, a protein component of immune-suppressive virus-like particles from *Leptopilina heterotoma*. J. Gen. Virol. 87, 461–470.

Colinet, D., Schmitz, A., Depoix, D., Chrochard, D., Poirié, M., 2007. Convergent use of RhoGAP toxins by eukaryotic parasites and bacterial pathogens. PLoS Pathog. 3, 2029–2037.

Dover, B.A., Davies, D.H., Vinson, S.B., 1987. Dose-dependent influence of *Campoletis sonorensis* polydnavirus on the development and ecdysteroid titers of last-instar *Heliothis virescens* larvae. Arch. Insect Biochem. Physiol. 8, 113–126.

Dupas, S., Brehelin, M., Frey, F., Carton, Y., 1996. Immune suppressive virus-like particles in a *Drosophila* parasitoid: significance of their intraspecific morphological variations. Parasitology 113, 207–212.

Ferrandon, D., Jung, A.C., Criqui, M., Lemaitre, B., Uttenweiler-Joseph, S., Michaut, L., et al., 1998. A drosomycin-GFP reporter transgene reveals a local immune response in *Drosophila* that is not dependent on the Toll pathway. EMBO J. 17, 1217–1227.

Ferrarese, R., Morales, J., Fimiarz, D., Webb, B.A., Govind, S., 2009. A supracellular system of actin-lined canals controls biogenesis and release of virulence factors in parasitoid venom glands. J. Exp. Biol. 212, 2261–2268.

Fessler, L.I., Nelson, R.E., Fessler, J.H., 1994. *Drosophila* extracellular matrix. Methods Enzymol. 245, 271–294.

Govind, S., 2008. Innate immunity in *Drosophila*: pathogens and pathways. Insect Sci. 15, 29–43.

Holz, A., Bossinger, B., Strasser, T., Janning, W., Klapper, R., 2003. The two origins of hemocytes in *Drosophila*. Development 130, 4955–4962.

Jung, S.H., Evans, C.J., Uemura, C., Banerjee, U., 2005. The *Drosophila* lymph gland as a developmental model of hematopoiesis. Development 132, 2521–2533.

Kopelman, A.H., Chabora, P.C., 1984. Immature stages of *Leptopilina boulardi* (Hymenoptera: Eucoilidae), a protelean parasite of *Drosophila* spp. (Diptera: Drosophilidae). Ann. Ent. Soc. Am. 77, 264–269.

Krzemień, J., Dubois, L., Makki, R., Meister, M., Vincent, A., Crozatier, M., 2007. Control of blood cell homeostasis in *Drosophila* larvae by posterior signaling centre. Nature 446, 325–328.

Labrosse, C., Eslin, P., Doury, G., Drezen, J.M., Poirie, M., 2005. Haemocyte changes in *D. melanogaster* in response to long gland components of the parasitoid wasp *Leptopilina boulardi*: a Rho-GAP protein as an important factor. J. Insect Physiol. 51, 161–170.

Lanot, R., Zachary, D., Holder, F., Meister, M., 2001. Post-embryonic hematopoiesis in *Drosophila*. Dev. Biol. 230, 243–257.

LaSalle, J., Gauld, I.D., 1993. Hymenoptera, their diversity and their impact on the diversity of other organisms. In: LaSalle, Gauld, (Eds.), Hymenoptera and Biodiversity. CAB International, Wallingford, United Kingdom., pp. 1–26.

Lemaitre, B., Hoffmann, J.A., 2007. The host defense of *Drosophila melanogaster*. Ann. Rev. Immunol. 25, 697–743.

Markus, R., Laurinyecz, B., Kurucz, E., Honti, V., Bajusz, I., Sipos, B., et al., 2009. Sessile hemocytes as a hematopoietic compartment in *Drosophila melanogaster*. Proc. Natl. Acad. Sci. USA 106, 4805–4809.

Martinez-Agosto, J.A., Mikkola, H.K., Hartenstein, V., Banerjee, U., 2007. The hematopoietic stem cell and its niche: a comparative view. Genes Dev. 21, 3044–3060.

Meister, M., Lagueux, M., 2003. *Drosophila* blood cells. Cell. Microbiol. 5, 573–580.

Melk, J.P., Govind, S., 1999. Developmental analysis of *Ganaspis xanthopoda*: a larval parasitoid of *Drosophila*. J. Exp. Biol. 202, 1885–1896.

Morales, J., Chiu, H., Plaza, R., Oo, T., Hoskins, S., Govind, S., 2005. Biogenesis, structure, and immune suppressive effects of virus-like particles of a *Drosophila* parasitoid, *Leptopilina victoriae*. J. Insect Physiol. 51, 181–195.

Nappi, A.J., Vass, E., 1993. Melanogenesis and the generation of cytotoxic molecules during insect cellular immune reactions. Pigment Cell Res. 6, 117–126.

Nappi, A.J., Vass, E., Frey, F., Carton, Y., 1995. Superoxide anion generation in *Drosophila* during melanotic encapsulation of parasites. Eur. J. Cell Biol. 68, 450–456.

Nordlander, G., 1980. Revision of the genus Leptopilina Forster, 1869, with notes on the status of some other genera (Hymenoptera, Cynipoidea: Eucoilidae). Entomol. Scand. 11, 428–453.

Qiu, P., Pan, P.C., Govind, S., 1998. A role for the *Drosophila* Toll/Cactus pathway in larval hematopoiesis. Development 125, 1909–1920.

Rizki, R.M., Rizki, T.M., 1984. Selective destruction of a host blood cell type by a parasitoid wasp. Proc. Natl. Acad. Sci. USA 81, 6154–6158.

Rizki, R.M., Rizki, T.M., 1990. Parasitoid virus-like particles destroy *Drosophila* cellular immunity. Proc. Natl. Acad. Sci. USA 87, 8388–8392.

Rizki, T.M., Rizki, R.M., 1992. Lamellocyte differentiation in *Drosophila* larvae parasitized by *Leptopilina*. Dev. Comp. Immunol. 16, 103–110.

Rizki, T.M., Rizki, R.M., 1994. Parasitoid-induced cellular immune deficiency in *Drosophila*. Primordial immunity: foundations for the vertebrate immune system. Ann. N. Y. Acad. Sci. 712, 178–194.

Rizki, T.M., Rizki, R.M., Carton, Y., 1990. *Leptopilina heterotoma* and *L. boulardi*: strategies to avoid cellular defense responses of *Drosophila melanogaster*. Exp. Parasitol. 70, 466–475.

Schilthuizen, M., Nordlander, G., Stouthamer, R., van Alphen, J.J.M., 1998. Morphological and molecular phylogenetics in the genus *Leptopilina* (Hymenoptera: Cynipoidea: Eucoilidae). Syst. Ent. 23, 253–264.

Schlenke, T.A., Morales, J., Govind, S., Clark, A.G., 2007. Contrasting infection strategies in generalist and specialist wasp parasitoids of *Drosophila melanogaster*. PLoS Pathog. 3, 1486–1501.

Sorrentino, R.P., Carton, Y., Govind, S., 2002. Cellular immune response to parasite infection in the *Drosophila* lymph gland is developmentally regulated. Dev. Biol. 243, 65–80.

Sorrentino, R.P., Melk, J.P., Govind, S., 2004. Genetic analysis of contributions of dorsal group and JAK-Stat92E pathway genes to larval hemocyte concentration and the egg encapsulation response in *Drosophila*. Genetics 166, 1343–1356.

Tzou, P., Ohresser, S., Ferrandon, D., Capovilla, M., Reichhart, J.M., Lemaitre, B., et al., 2000. Tissue-specific inducible expression of antimicrobial peptide genes in *Drosophila* surface epithelia. Immunity 13, 737–748.

Varaldi, J., Petit, S., Boulétreau, M., Fleury, F., 2006. The virus infecting the parasitoid *Leptopilina boulardi* exerts a specific action on superparasitism behavior. Parasitology 132, 747–756.

Wertheim, B., Kraaijeveld, A.R., Schuster, E., Blanc, E., Hopkins, M., 2005. Genome-wide gene expression in response to parasitoid attack in *Drosophila*. Genome Biol. 6, R94.

CHAPTER 6

Variation of *Leptopilina boulardi* Success in *Drosophila* Hosts: What is Inside the Black Box?

A. Dubuffet,* D. Colinet,† C. Anselme,† S. Dupas,§,¶ Y. Carton§ and **M. Poirié‡**

Contents

* Institute of Integrative and Comparative Biology, Faculty of Biological Sciences, University of Leeds, United Kingdom
† Institut National de la Recherche Agronomique, INRA Sophia Antipolis, UMR 1301; Centre National de la Recherche Scientifique, CNRS, UMR 6243; Université Nice Sophia Antipolis, UFR Sciences, France
‡ UMR Interactions Biotiques et Santé Végétale, Institut Agrobiotech, 06 903 Sophia Antipolis, France. INRA, UMR 1301/CNRS UMR 6243/Université Nice Sophia Antipolis, 28, avenue de Valrose, 06103 Nice Cedex 2, France
§ IRD, UR072 Laboratoire Evolution, Génomes et Spéciation/UPR9034, CNRS 91198 Gif-sur-Yvette cedex, France/Université Paris-Sud 11, 91405 Orsay cedex, France
¶ Pontificia Universidad Católica del Ecuador, Facultad de Ciencias Exactas y Naturales, Quito, Ecuador

Advances in Parasitology, Volume 70
ISSN 0065-308X, DOI: 10.1016/S0065-308X(09)70006-5

Abstract

Interactions between *Drosophila* hosts and parasitoid wasps are among the few examples in which occurrence of intraspecific variation of parasite success has been studied in natural populations. Such variations can originate from three categories of factors: environmental, host and parasitoid factors. Under controlled laboratory conditions, it is possible to focus on the two last categories, and, using specific reference lines, to analyze their respective importance. Parasitoid and host contributions to variations in parasite success have largely been studied in terms of evolutionary and mechanistic aspects in two *Drosophila* parasitoids, *Asobara tabida* and, in more details, in *Leptopilina boulardi*. This chapter focuses on the physiological and molecular aspects of *L. boulardi* interactions with two *Drosophila* host species, while most of the evolutionary hypotheses and models are presented in Chapter 11 of Dupas et al.

6.1. INTRODUCTION

As for many parasites, the success of parasitoids in the host they infect is not guaranteed. In the first place, the suitability of different host species can vary for a given parasitoid species (host species specificity; Brodeur and Vet, 1995; Mohamed et al., 2003). Additionally, the outcome of a parasitoid species–host species combination can also be quite variable. For instance, the host resists in some cases the infestation through an immune response that kills the parasitoid, while in others the parasitoid escapes this immune response, resulting then in the death of the host. Against endoparasitoids, which develop inside the body cavity of their hosts, the immune response of insects is generally the encapsulation response, which consists in the elaboration of a multicellular and melanized capsule around the parasitoid egg. This encapsulation response can affect the overall parasitoid success significantly, as shown for *Drosophila* hosts (Carton and Kitano, 1981). The physiological and molecular basis of encapsulation is reasonably well characterized, due to numerous studies in lepidopteran and *Drosophila* spp. (Carton et al., 2008; Kanost et al., 2004). The virulence strategies and tactics used by parasitoids (namely the means employed to escape encapsulation) have also been investigated in many models and some of the molecular factors used to achieve these strategies (virulence factors) have been characterized (Carton et al., 2008; Glatz et al., 2004; Moreau and Guillot, 2005; Pennacchio and Strand, 2006;

Poirié et al., 2009). By comparison, less is known about the mechanisms affecting the ultimate outcome of host–parasitoid interactions. In other words, why some interactions resolve in the encapsulation of the parasitoid while others lead to the parasitoid success?

It is generally recognized that the variation in the outcome of any host–parasite interaction can originate from three sources: the variation in host resistance, the variation in the parasite ability to escape host resistance (parasite virulence), which are both genetically determined and the environmental factors. Moreover, there is an increasing evidence for complex interactions between host and parasite genotypes (G_H x G_P interactions; Carius et al., 2001; Lambrechts et al., 2005), which themselves can interact with the environment (G_H x G_P x E interactions; Lazzaro and Little, 2009). In host–parasitoid interactions, the role of environmental factors on the overall variation of success has been studied in various biological models (Bensadia et al., 2006; Calatayud et al., 2002; Oliver et al., 2003). However, most studies on the contribution of host resistance and parasitoid virulence have been restricted to the interactions between *Drosophila* hosts and parasitoid wasps and concern variations in parasitoid encapsulation exclusively. In particular, the parasitoids *Asobara tabida* and *Leptopilina boulardi* have been thoroughly studied. Extensive variation in host resistance and parasitoid virulence in natural populations have been evidenced in these models, and the coevolutionary outcomes largely discussed (Dupas et al., 2003; Kraaijeveld et al., 1998). Recently, significant progress has been made in understanding the genetic and molecular mechanisms underlying variations in immune interactions between *L. boulardi* and *Drosophila* hosts. Here, we review these mechanisms, while the evolutionary hypotheses and models concerning *Drosophila*–parasitoid interactions are presented in Chapter 11 by Dupas et al. First, we show how to "dissect" the variation in parasitoid success in order to identify the factors that influence the outcome of the host–parasitoid interaction (presence or absence of encapsulation). We then review recent data obtained for host resistance and parasitoid virulence. Finally, we discuss the diversity of virulence mechanisms in *Drosophila*–parasitoid interactions, and highlight how the progress in molecular comprehension of host–parasite interactions may help to understand the evolution of pairwise host–parasitoid interactions as well as the evolution of a parasitoid's host range.

6.2. DISSECTION OF THE NATURAL VARIATION OF ENCAPSULATION

Various kinds of environmental factors are known to influence the outcome of host–parasitoid interactions. Abiotic factors, such as temperature (Blumberg and Van Driesche, 2001), presence of insecticides (Delpuech et al., 1996) or host diet (Karimzadeh and Wright, 2008; Ojala et al., 2005), can considerably influence the presence and efficiency of the encapsulation response. Moreover, the host immune response can be affected by biotic factors. The presence of another parasitoid in the host, either from the same species (superparasitism) or from another species (kleptoparasitism) can impair the immune response, eventually increasing the success of a given parasitoid (Kraaijeveld, 1999; Sagarra et al., 2000). More recently, symbionts were shown to influence the success of some parasitoids considerably, impairing or increasing host resistance ability as well as parasitoid virulence (Fytrou et al., 2006; Haine, 2008). Working under controlled laboratory conditions, it is possible to reduce environmental variation and focus on the genetic contribution of hosts and parasitoids.

To assess the occurrence of genetic variation in host resistance or parasitoid virulence within populations, two methods can be used. The first consists of performing selection experiments. If genetic variation exists in the studied trait (resistance or virulence), then its frequency is expected to change as a response to selection. Using this method in *D. melanogaster*, increases in encapsulation rates from less than 5% to more than 40% were obtained for the parasitoids *L. boulardi* and *A. tabida* in less than 10 generations (Fellowes et al., 1998; Kraaijeveld and Godfray, 1997). The second method consists of comparing the resistance or virulence abilities of different host or parasitoid isofemale lines obtained from a population under the same conditions of parasitism. Isofemale lines are each derived from a female that has been inseminated once, and whose progeny inbred during several generations until most loci are homogeneous. Heritability can then be measured by analyzing resistance or virulence of these isofemale lines over two successive generations (Carton and Boulétreau, 1985; Carton et al., 1989). Advantage of founding a series of isofemale lines is that while variation within a line will be lost, a series of independent lines will maintain heritable variation from within the population of interest, and mixing the lines will reconstitute the majority of variation (David et al., 2005).

Between-population variations in resistance and virulence can be assessed either by comparing freshly collected populations or by using the isofemale line method. In this case, several isofemale lines are constituted, thus allowing a ''snapshot'' of the genetic diversity occurring in this population to be taken. Between-population variation in resistance is then tested by comparing the encapsulation rate of a ''reference parasitoid line'' in host populations coming from different geographical areas. Similarly, between-population variation in virulence is then tested by comparing the encapsulation rate of parasitoid populations coming from different geographical areas in a ''host reference line'' (Fig. 6.1A). This method has been largely used in *Drosophila*–parasitoid models (Table 6.1). Substantial variation for both resistance and virulence has been shown in

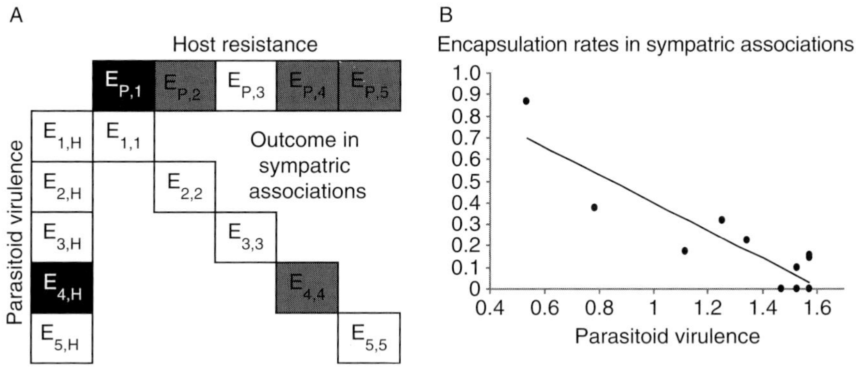

FIGURE 6.1 (A) Assessment of host resistance and parasitoid virulence in populations from different areas, and the outcome of the interaction in sympatric associations. $E_{i,j}$: encapsulation rate of parasitoid from locality *i* in host from locality *j*. Intensity of coloration of each square represents the level of encapsulation rates. Black: high, gray: medium, white: low. H: host reference strain, P: parasitoid reference strain. Resistance of each host population ($E_{P,j}$) is evaluated through the encapsulation rates of a reference parasitoid strain by the host population. Virulence of each parasitoid population ($1\text{-}E_{i,H}$) is evaluated through the encapsulation rates of this parasitoid population by a host reference strain. (B) Impact of parasitoid virulence on encapsulation rates in sympatric associations. Encapsulation rates measured in sympatric associations ($E_{i,j}$) are plotted against parasitoid virulence, which is tested using a host reference strain ($1\text{-}E_{i,H}$). Variables are arcsine transformed. *Note: Drawn from* Dupas et al. (2003).

populations from various geographical areas, and their ecological and evolutionary consequences are discussed elsewhere (Dubuffet et al., 2007; Dupas and Boscaro, 1999; Dupas et al., 2003; Kraaijeveld and Godfray, 1999; see also Chapter 10 by Kraaijeveld and Godfray and Chapter 11 by Dupas et al.).

Interestingly, the evidence for variation is strongly dependent on the line used for the experiment, since some lines fail to reveal variation. The strain ISm of *L. boulardi*, for example, is encapsulated by all the *D. yakuba* strains tested so far, but always escapes encapsulation in *D. melanogaster* (Carton, unpublished data; Dubuffet et al., 2007; Table 6.1). Similarly, the susceptible strain 1088 of *D. melanogaster* encapsulates none of the strains of *L. boulardi* we tested (Carton, unpublished data; Table 6.2). These strains are thus unsuitable to study variations in resistance and virulence in natural populations. Other laboratory lines can evidence genetic variations in the tested natural populations, but they might fail to reveal the whole range of responses. In theory, encapsulation rates obtained using a "good" reference line should range from 0% to 100%. For example, the encapsulation rates obtained using the parasitoid line ISy of *L. boulardi* range from less than 5% to more than 85% using laboratory lines of

TABLE 6.1 List of parasitoid and host species and strains used to demonstrate variations in resistance

Host species tested	Geographical origin of strains tested	Parasitoid species	Strain used for the test	Variation (YES/NO) range of encapsulation rates obtained (mean)	References
D. melanogaster	Europe	*A. tabida*	Sospel	YES 0–63.5 (26.7)	Kraaijeveld and van Alphen (1995)
D. melanogaster	Europe	*L. boulardi*	Tasagil	YES 0–24.2 (5.0)	Kraaijeveld and van Alphen (1995)
D. melanogaster	Worldwide	*L. boulardi*	ISy (G486)	YES 9.6–68.8 (55.2)	Dupas et al. (2003)
D. melanogaster	Worldwide	*L. boulardi*	ISm (G431)	NO (<5)	Carton and Frey (unpublished)
D. yakuba	Africa	*L. boulardi*	ISy (G486)	YES 6–97.9 (65.0)	Dubuffet et al. (2007)
D. yakuba	Africa	*L. boulardi*	ISm (G431)	NO (100)	Dubuffet et al. (2007)

Variation in resistance has been tested in host populations from various geographical origins (Worldwide, European or African distribution) using a single parasitoid strain. "Parasitoid reference strains" (shown in gray) are those which allow to evidence variation in resistance of host populations.

TABLE 6.2 List of parasitoid and host species and strains used to demonstrate variations in virulence

Host species	Strain used for the test	Parasitoid species tested	Geographical origin of strains tested	Variation (YES/NO) range of encapsulation rates (mean)	References
D. melanogaster	R (940)	*L. boulardi*	Worldwide	YES 0–74.2 (12.3)	Dupas and Boscaro (1999), Dupas et al. (2003)
D. melanogaster	S (1088)	*L. boulardi*	Worldwide	NO (<5)	Carton and Frey (unpublished)
D. yakuba	R_1 (1880-D)	*L. boulardi*	Worldwide	YES 10–100	Dupas and Boscaro (1999)
D. simulans	Ds1448	*L. boulardi*	Worldwide	Yes 10–40	Dupas and Boscaro (1999)
D. melanogaster	InHam	*A. tabida*	Europe	Yes <25–100	Kraaijeveld and van Alphen (1994)

Variation in virulence has been evidenced by comparing the encapsulation rate of parasitoid populations from various geographical origins (Worldwide, European or African distribution) on a single host strain. "Reference strains" (shown in gray) are host strains which allow to evidence variation of virulence.

D. melanogaster or *D. yakuba* (Carton et al., 1992; Dubuffet et al., 2007). In contrast, the parasitoid strain "Tasagil" of *L. boulardi* used by Kraaijeveld and van Alphen (1995) does not cover the whole range of encapsulation rates in *D. melanogaster* since the maximum encapsulation rate obtained using this strain is only 50% (Fellowes et al., 1999). The partial virulence of this strain might thus explain the low encapsulation rates measured by Kraaijeveld and van Alphen (1995) in European populations of *D. melanogaster* in comparison to the ones measured by Dupas et al. (2003) using the ISy line, and it might also hide part of the genetic variation of resistance. The choice of the "reference line" is thus critical for those wishing to reveal the genetic variations in resistance and virulence and investigate rationally these genetic interactions. Many well-characterized laboratory lines have been called "reference lines," but we suggest, at least for the present chapter, that the term "reference line" should be restricted to the lines that allow the detection of genetic variation in natural populations of the antagonistic species. We will as well use the term "resistance" as the encapsulation rate of a parasitoid reference line measured in a host population or line, and the term "virulence" as one minus the encapsulation rate of a parasitoid population or line by a host reference strain (see Box 6.1).

Since the amount of genetic variation observed is strongly dependent on the choice of laboratory lines, the use of the reference lines could be questionable for the study of natural variation. Can the outcome of a host–parasitoid interaction be predicted by separate estimation of resistance and virulence levels of each partner using these reference lines? Fortunately, measurements of resistance and virulence using these reference strains actually give good predictions of levels of encapsulation measured in sympatric conditions (hosts and parasitoids coming from the same area, but tested in controlled laboratory conditions; Dupas et al., 2003; Kraaijeveld and Godfray, 2001). Interestingly, in the *L. boulardi*–*D. melanogaster* model, in which host resistance, parasitoid virulence and sympatric outcome were all evaluated, it appears that most of the variation in sympatric host–parasitoid associations comes from the variation in parasitoid virulence (Dupas et al., 2003; Fig. 6.1A and B). It explains 81% of the variance in encapsulation (calculated from Dupas et al., 2003 using the method from Kraaijeveld and Godfray, 2001; $F_{1,11}$= 47.4; $P < 0.001$). The addition of host resistance to the regression does not increase the variance explained ($F_{1,4} = 3.045$, $P = 0.156$), due to the fact that most parasitoid populations are highly virulent on *D. melanogaster* (Fig. 6.2). As a result, resistance variation in *D. melanogaster* only accounts for the overall host–parasitoid outcome in tropical Africa, where parasitoid virulence is low. By comparison, both virulence and resistance explain the overall variance in the *A. tabida*–*D. melanogaster* model, but with

BOX 6.1 Definition of terms used in this chapter

As stressed in this chapter, the presence or absence of encapsulation of parasitoids by their hosts can depend on the interaction between the host and parasitoid genotypes (Dubuffet *et al.*, 2007). This means that a host genotype that is resistant to one parasitoid genotype is not necessarily resistant to all parasitoid genotypes. Similarly, virulence is relative to the antagonistic partner. The terms "resistance" and "virulence" have then to be considered carefully, because they do not design the overall outcome of the host or parasitoid, but their genetic potential toward one particular genotype of the interacting partner. In order to clarify all the terms related to the topic of this chapter, we give their definitions below.

Host reference line: Host line that is used to evidence the genetic variation of virulence in parasitoid populations.

Parasitoid reference line: Parasitoid line that is used to evidence the genetic variation of resistance in host populations.

Resistance: Ability of a host to encapsulate a parasitoid reference line.

Variation in virulence strategy: Genetic variation in the means used by parasitoids to successfully overcome encapsulation by hosts.

Variation of resistance: Genetic variation in the ability of hosts to encapsulate a reference parasitoid line. Individuals from the "**resistant**" line encapsulate the parasitoid while the "**susceptible**" ones do not.

Variation of virulence: Genetic variation in the ability of parasitoids to overcome encapsulation by a reference host line. Individuals from the "**virulent**" line escape encapsulation while the "**avirulent**" ones do not.

Virulence: Ability of a parasitoid to overcome encapsulation by a host reference line

Virulence factors: Molecules employed by parasitoids to achieve their virulence tactic.

Virulence strategy: Set of means used by parasitoids to escape encapsulation. It includes the general effects on host encapsulation ability (local immunoevasion or overall immunosuppression), the underlying virulence tactics and the effects of each virulence factor on host targets.

Virulence tactic: Describes each of the mechanisms involved to achieve the virulence strategy, that is, the potential effects of the parasitoid on specific components of the encapsulation response, as cellular or humoral effectors.

FIGURE 6.2 (A) Geographical distribution of the encapsulation rate of *L. boulardi* populations in sympatric *D. melanogaster* populations (% represented by the black portion of the pie chart). (B) Geographic distribution of resistance in *D. melanogaster* populations. The resistance level (represented by the black portion of the pie chart) is estimated from the rate of encapsulation of the reference ISy line of *L. boulardi* in *D. melanogaster* populations. (C) Geographical distribution of the virulence in *L. boulardi* populations. The level of virulence (represented by the white portion of the pie chart) is estimated from the rate of encapsulation of various natural populations of *L. boulardi* by the reference resistant strain of *D. melanogaster*. *Note:* From Dupas et al. (2003).

virulence again being the most important factor (Kraaijeveld and Godfray–2001).

In order to investigate the genetic, physiological and molecular basis of the variation of virulence and resistance observed in natural populations of *L. boulardi*, the isofemale lines of *L. boulardi*, *D. melanogaster* and *D. yakuba* showing the most contrasting virulence and resistance abilities have been chosen. Each host–parasitoid combination results either in a very low (<10%) or a very high (>85%) percentage of encapsulated parasitoid eggs (Fig. 6.3). This matrix of interactions reflects the situation observed in natural populations. The parasitoid ISm line, which originates from Tunisia, represents the pattern observed in most places: it is highly virulent in *D. melanogaster*, whichever the host strain, but is completely unable to escape encapsulation in any *D. yakuba* strain (Dupas and Boscaro, 1999). The success of this strain is thus host-species specific. By contrast, the parasitoid ISy, which originates from Congo, can infect both *D. melanogaster* and *D. yakuba* but is host-genotype specific, which means that its success depends on the genotype of the host (susceptible *vs.* resistant; Dubuffet et al., 2007; Dupas et al., 2003). When observing the whole matrix of interactions, it appears that *L. boulardi* has specific interactions with its hosts, since the parasitoid success depends both on the host and parasitoid lines considered (Dubuffet et al., 2007).

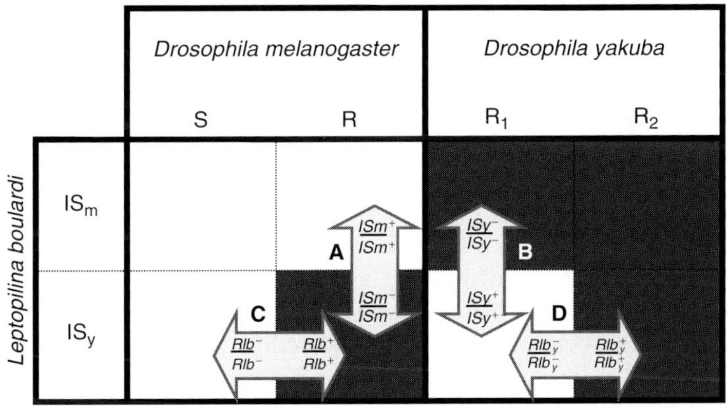

FIGURE 6.3 Genetic interactions between *L. boulardi* and its *Drosophila* hosts. Dark boxes: parasitoid failure (encapsulation); white boxes: parasitoid success. Variation in parasitoid virulence in *D. melanogaster* (A) and *D. yakuba* (B) is encoded by two distinct major biallelic loci (*ISm, ISy*). Variation in host resistance both in *D. melanogaster* (C) and *D. yakuba* (D) is encoded in each case by a single major biallelic locus (*Rlb*, Rlb_y). Success of the ISm line of *L. boulardi* is species-specific while the success of the ISy line is host-genotype specific. *Note:* Redrawn from Dubuffet et al. (2007).

6.3. HOST RESISTANCE: ORIGIN OF VARIATION

6.3.1. The actors of physiological resistance

An exhaustive description of the current knowledge regarding the molecular bases of immune defenses in *Drosophila* would largely exceed the purpose of this chapter (for a review, see Lemaitre and Hoffmann, 2007). However, data on the encapsulation process will be required to understand the next parts of this chapter fully. It has now been well described that large eukaryotic parasites, such as parasitoid eggs, that invade the hemocoel of insects generally provoke a series of immune responses mediated in large part by circulating blood cells (hemocytes) that form multilayer capsules around the foreign organism (Carton et al., 2008). In addition to some lepidopteran species, the model organism for studies on parasitoid encapsulation has been *D. melanogaster*. However, we have to keep in mind that if other species of the *melanogaster* subgroup such as *D. yakuba* and *D. simulans* use apparently rather similar immune components to those described in *D. melanogaster*, the death of the parasitoid is not associated with encapsulation in other species like *D. paramelanica* (Nappi, 1970; see Chapter 4 by Nappi et al.). Moreover, the specific hemocytes devoted to the formation of the capsule in the *melanogaster* group are not found in all *Drosophila* spp. (Eslin and Doury, 2006; see Chapter 7 by Eslin et al.). The mechanisms responsible for a parasitoid success or failure in *D. melanogaster* might thus strongly differ from those involved in the outcome of its interactions with other *Drosophila* spp., leading to a diversity in virulence strategies as well as resistance systems.

One of the first detectable events following parasitism in *D. melanogaster* larvae is the proliferation, release and/or differentiation of host hemocytes (Carton et al., 2008; Markus et al., 2009). In *Drosophila*, plasmatocytes and lamellocytes are the principal cells involved in cellular encapsulation. The proportion of lamellocytes, which are rarely observed in nonparasitized flies, is greatly enhanced in parasitized larvae (Lanot et al., 2001; Rizki and Rizki, 1992; Russo et al., 2001; Sorrentino et al., 2002). Six h following infection, a thin layer of melanin is observed on the surface of the parasitoid (Russo et al., 2001), which suggests that biochemical reactions associated with the production of melanin, for example, activation of the phenol oxidase (PO) cascade, are triggered very early following infection (Williams et al., 2005). They are associated with the production of cytotoxic radicals that are thought to be responsible for the parasitoid death. By 24 h after infection, the wasp egg is completely surrounded by plasmatocytes. By 40 h, lamellocytes are found attached around the egg and at 48 h after infection a fully formed melanotic capsule is visible in the host hemocoel (Williams et al., 2005). Lamellocytes also appear as sources for PO-mediated melanogenesis (Irving et al., 2005;

Nam et al., 2008), as is a third type of hemocyte, the crystal cell (Rizki and Rizki, 1985; Rizki et al., 1980). Besides physical damage such as rupture of the basal membrane, parasitism but also injection of female parasitoid venom can induce the proliferation of hemocytes and specifically of lamellocytes (Labrosse et al., 2005a) but at the moment, no "immune-inducing" component has been identified yet and the mechanisms that lead to the "recognition" of the invader are also largely unknown.

Regulation of the hemocyte number is controlled by different pathways including genes from the Ras-mitogen-activated protein kinase pathway, while the Jak/Stat and Jun kinase pathways strongly affect lamellocyte formation (Zettervall et al., 2004; Fig. 6.4). Other signals, such as those mediated by aop(ACT), Toll(10b) or Rac1, cause a simultaneous increase in lamellocytes and total hemocyte number. Adhesion and cell-shape changes are also an essential part of the encapsulation process. One family of proteins central to the processes involved in cell shape is the Rac GTPases. Once activated, Racs are involved in many cellular processes including: cytoskeletal organization, regulation of cellular adhesion, cellular polarity and transcriptional activation. Both *Drosophila* Rac1 and Rac2 genes are required for proper encapsulation of *L. boulardi* eggs (Williams et al., 2005, 2006). Rac2 is necessary for hemocyte spreading and cell–cell contact formation and melanization is disrupted in capsules recovered from Rac2 mutants (Williams et al., 2005). Rac1 is involved in the increase in hemocyte number as well as induction of lamellocyte formation.

In insects, PO is present as inactive prophenoloxidase (PPO) and cleaved into active PO by a serine protease (prophenoloxidase-activating enzyme; PPAE), that itself becomes activated through a sequential process involving other serine proteases (Cerenius et al., 2008; Jiang et al., 2003; Satoh et al., 1999). In *D. melanogaster*, the expression of several serine

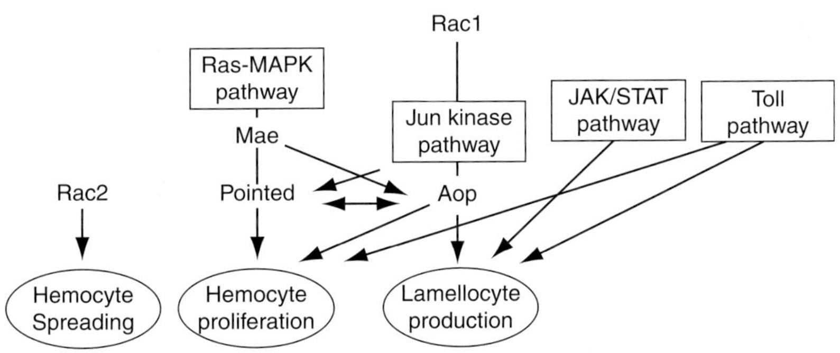

FIGURE 6.4 Part of the known pathways and genes involved in proliferation and spreading of hemocytes as well as production of lamellocytes in *D. melanogaster*. The mae gene is a candidate for the *Rlb* resistance gene to the ISy line of *L. boulardi*.

proteases is significantly increased within the few h following infection, which is concomitant with the melanin deposition (Schlenke et al., 2007; Wertheim et al., 2005). Serine proteases are themselves negatively regulated by serine protease inhibitors, the better described in *D. melanogaster* being Spn27A, which inhibits PPAE (De Gregorio et al., 2002). Also, introducing Spn27A into otherwise immune reactive *D. melanogaster* larvae reduces the frequency of melanotic encapsulation of eggs of *L. boulardi* (Nappi et al., 2005).

6.3.2. Variation and genetic determinism of *Drosophila* resistance to parasitoids

Naturally occurring resistance variation between populations, evidenced by the use of parasitoid reference lines, has been well described in *D. yakuba* against *L. boulardi* (Dubuffet et al., 2007) and in *D. melanogaster* against *L. boulardi* and *A. tabida* (Dupas et al., 2003; Kraaijeveld and van Alphen, 1995). Interestingly, *D. melanogaster* resistance levels toward the Congolese parasitoid line ISy of *L. boulardi*, are quite high while encapsulation rates toward sympatric parasitoid populations are in general very low (Dupas et al., 2003; Fig. 6.2A and B). This comes from the fact that virulence overcomes the effects of resistance in this model: encapsulation only occurs if the host is resistant and the parasitoid has a low virulence, a situation observed in Congo (Dupas et al., 2003; Fig. 6.2C). In this area, within-population variation of resistance has been described, encapsulation rates of the reference line ISy ranking between 12% and 90% (Carton and Boulétreau, 1985; Carton et al., 1992). In other areas, *L. boulardi* is highly virulent, which results in low encapsulation rates in sympatric associations. It is difficult to explain the high levels of resistance to ISy parasitoids found in these areas since it is totally inefficient toward *Leptopilina* parasitoids found in sympatry. They could be maintained as the result of a selection pressure coming from cooccurring parasitoid species. In the south of France, for instance, *L. boulardi* was shown to cooccur sometimes with *L. heterotoma* and other parasitoids (Fleury et al., 2004). Alternatively, resistance might be pleiotropic and its polymorphism maintained by completely different selection factors. By comparison, most populations of *D. melanogaster* encapsulate a significant proportion of *A. tabida* eggs in sympatric associations, due both to moderate virulence of the parasitoid as well as a significant proportion of resistant hosts (Kraaijeveld and Godfray, 2001). In *D. yakuba*, levels of encapsulation of *L. boulardi* in sympatric associations are not known, but are likely to be high, due to the low virulence of *L. boulardi* toward this host—the ISy parasitoid line is, to our knowledge, the only one described that can escape encapsulation in this species—and the high resistance levels to this line we found in populations (Dubuffet et al., 2007). So far, parasitoid success has only been reported in Congo (Dupas and Boscaro, 1999). This might be correlated with the fact that *D. yakuba* is intrinsically a much better encapsulator than *D. melanogaster*, probably because unparasitized larvae have on average more hemocytes (Carton and Kitano, 1981; Eslin and Prévost, 1998).

To analyze the genetics of resistance to parasitoid wasps in *D. melanogaster*, selected inbred resistant (R) and susceptible (S) lines were obtained from the same population (Brazzaville, Congo) using the ISy parasitoid line (Carton et al., 1992). Resistance to *A. tabida* WOV was analyzed using the same R strain that proved resistant also to *A. tabida* and Canton S as a susceptible strain (Benassi et al., 1998). Resistance of *D. yakuba* was analyzed using the isofemale lines 1880-D (R1 line, susceptible) and 1907 (R2 line, resistant) chosen from two populations in Tanzania using the ISy line of *L. boulardi* (Dubuffet et al., 2007).

Considering the high number of genes potentially involved in insect immune response to parasitoids (Irving et al., 2005; Zettervall et al., 2004), variation in resistance was expected to be multigenic. However, in both *D. melanogaster* and *D. yakuba* host species, resistance to parasitoids is always explained by a single diallelic locus inherited autosomally, with the resistant phenotype showing complete dominance over the susceptible one. In *D. melanogaster*, the loci were named *Rlb* (resistance to *L. boulardi*) and *Rat* (resistance to *A. tabida*; Benassi et al., 1998; Carton et al., 1992), and in *D. yakuba* the locus was named Rlb_y (resistance to *L. boulardi*; Dubuffet et al., 2007; Fig. 6.3). The use of isofemale lines might have favored the recovery of simple genetic systems, but a study dealing with genetic variation of resistance to *A. tabida* in *D. melanogaster* from different localities in Europe also concluded on a simple genetic basis of resistance (Orr and Irving, 1997).

These results raised the question whether the same *D. melanogaster* locus was involved in resistance to *A. tabida* and *L. boulardi*. Using recombination experiments, we showed that *Rlb* and *Rat* are 35 cM apart (Poirié et al., 2000). Besides, there is no correlation between the field capacity to encapsulate these two parasitoid species (Kraaijeveld and van Alphen, 1995) and a strain susceptible to *L. boulardi* was resistant to *A. tabida* (Vass et al., 1993). This suggested that resistance has parasitoid-specific components and that at least two separate genetic systems explain resistance to parasitic wasps in the same host species.

6.3.3. Physiological and molecular bases of *Drosophila* resistance to parasitoids

Occurrence of different genes responsible for resistance to *L. boulardi* and *A. tabida* in *D. melanogaster* was in agreement with selection experiments showing that lines selected for resistance to *L. boulardi* also increased in resistance to *A. tabida* while only a slight increase in resistance to *L. boulardi* was observed in lines selected against *A. tabida* (Fellowes et al., 1999). This led to consider that improved resistance had a nonspecific component more or less effective against both wasps and a specific component required for encapsulation of *L. boulardi*.

The nonspecific component might correspond to an increase in hemocyte number as observed in lines selected for increased resistance to *A. tabida* (Kraaijeveld et al., 2001). Accordingly, it has been shown that the hemocyte number can affect the resistance potential of *Drosophila* hosts against *A. tabida* (Eslin and Prévost, 1998). The *Rat* locus has not been cloned yet but it has been localized on chromosome 2R, near the centromere (Poirié et al., 2000) and may correspond to the major resistance locus characterized in a QTL mapping experiment (Orr and Irving, 1997).

The physiological basis of variation of resistance to *L. boulardi* in *D. melanogaster* and *D. yakuba* is still unknown. However, one difference between the R and S strains of *D. melanogaster* has been described by Russo et al. (2001): in larvae parasitized by ISy parasitoids, the number of hemocytes is about twofold higher in the R strain than in the S strain at 15 h postinfestation and higher in the S strain than in the R strain at 24 h postinfestation. It is then possible that the earlier ''proliferation response'' in the R strain plays a role in variation of resistance. Genetic experiments have been used to localize the locus *Rlb* on the right arm of chromosome 2, at a genetic location of ca. 2-86.7 (Poirié et al., 2000). Its localization was then restricted in a 300 kb region, in 55E2-E6; F3, using strains bearing deletions (Hita et al., 1999). Indeed, despite dominance of the Rlb^+ allele, F1 larvae bearing a deletion in front of the Rlb^+-containing region show a decreased encapsulation rate, probably because of transvection effects. The *Rlb*-containing region was then restricted to 100 kb by controlling the molecular limits of the deletions using *in situ* hybridization and Southern-blotting experiments. Finally, male recombination experiments were performed to localize *Rlb* to the right or to the left of a P-element inserted in this region. Results showed that *Rlb* was close to the P-element leading to characterization of two candidates, the *mae/edl* gene and CG15086 of unknown function (Hita et al., 1999). *mae* (modulator of the activity of Ets)/*edl* (Ets-domain lacking) is the more likely candidate for *Rlb*. It encodes a protein with an ETS-specific pointed domain (SAM domain) and acts as a signaling intermediate that directly links the RTK/RAS/MAPK signaling pathway to its downstream transcription factor targets (Baker et al., 2001). *mae/edl* mediates MAPK phosphorylation of the Ets transcription factors yan/aop and pointed P2, yan/aop being involved in cell choice between cell proliferation and differentiation following RTK signaling (Rogge et al., 1995). The fact that ectopic expression of *yan/aopACT*, a *yan/aop* constitutively active allele, stimulates both proliferation of hemocytes and formation of lamellocytes in *Drosophila* larvae (Zettervall et al., 2004) supports the possible involvement of *mae/edl* in resistance to *L. boulardi*. Differences between resistant and susceptible alleles, their expression or their regulation might explain differences in the timing of hemocyte proliferation in response to parasitism (Fig. 6.4).

It would be interesting now to determine whether the locus that determine the variation of resistance to *L. boulardi* in *D. yakuba* is homologous to the locus *Rlb*, or if completely different loci explain variation to the same parasitoid in the two different host species. Explaining the high level of resistance against *L. boulardi* in the field both *D. yakuba* and *D. melanogaster* will indeed require understanding the function of resistance genes as well as their degree of specificity.

6.4. PARASITOID VIRULENCE: ORIGIN OF VARIATION

Among the few well-studied *Drosophila* parasitoids, variation of virulence has largely been evidenced in some species, such as *L. boulardi* (Carton et al., 1989; Dupas and Boscaro, 1999) and *A. tabida* (Kraaijeveld and van Alphen, 1994) but not in other species like *A. citri* or *Ganaspis xanthopoda*. Occurrence and genetic analysis of such a variation in *L. heterotoma* was reported by Walker in 1959 but it has never been documented since then. Recent analyses of virulence of six *Leptopilina* spp. against three *Drosophila* host species (Dupas, unpublished data) also suggest that intraspecific variation in virulence can be easily observed in some but not all species of the same genus.

L. boulardi is undoubtedly the species whose variation of virulence has been best described, both for its occurrence in natural populations and its physiological and molecular mechanisms. The only description of intra-population variation in virulence concerns the host *D. simulans* (Carton et al., 1989), whereas interpopulation variation has been documented for *D. melanogaster*, *D. simulans* and *D. yakuba* (Dupas and Boscaro, 1999). The mechanisms underlying the interpopulation variations, which we present below, have been investigated in *D. melanogaster* and *D. yakuba* using the parasitoid lines ISm and ISy, which originate from different populations.

6.4.1. Genetic determinism of virulence variation

So far, *Leptopilina* spp. remain the only parasitoid genus for which the genetic determinism of virulence variation has been investigated. Both in *L. heterotoma* and *L. boulardi*, these analyses revealed that the success of parasitoids is more related to the genotype of their mothers than to their own genotype, since the success of hybrid eggs issued from crosses between virulent and avirulent lines remained the same as that of the maternal line (Dupas and Carton, 1999; Dupas et al., 1998; Walker, 1959). This suggested that variations in maternal secretions, like venoms, may determine the intraspecific variations of success of *Leptopilina* spp.

In *L. boulardi*, genetic crosses have been performed during two generations between the lines ISm and ISy, which have opposite virulence abilities on the species *D. melanogaster* and *D. yakuba* (Fig. 6.3). These crosses revealed that variations of virulence on each of these host species have a simple determinism, with a diallelic locus explaining these variations (Dupas and Carton, 1999; Dupas et al., 1998). In *D. melanogaster*, virulence and avirulence alleles are semidominant while in *D. yakuba* there is dominance of the avirulence phenotype.

Dupas and Carton (1999) mixed the lines ISy and ISm for 16 generations and tested the females obtained from this experimental population for their virulence abilities on *D. melanogaster* and *D. yakuba*. They found no correlation between these virulence abilities, which led them to conclude that the locus responsible for variation of virulence on *D. melanogaster*, called *ISm*, is distinct from the locus responsible for the variation of virulence on *D. yakuba*, called *ISy*. The absence of correlation in parasitoid virulence on the three species *D. melanogaster*, *D. yakuba and D. simulans* in natural populations tallies with distinct virulence genetic systems against each *Drosophila* spp. (Dupas and Boscaro, 1999).

6.4.2. Physiological determinism of virulence variation

As described elsewhere, parasitoids use various strategies to escape encapsulation (Pennacchio and Strand, 2006; Poirié et al., 2009). Some evade encapsulation due to surface characteristics that make them inaccessible to the host immune system, or due to a local decrease in efficiency of the immune response, which does not impair the overall host encapsulation response (local immunoevasion). Others modulate or suppress the whole host encapsulation response (systemic immunosuppression). Among *Drosophila* parasitoids, the first mechanism has been described in *A. tabida* (Prévost et al., 2005; see Chapter 9 by Prévost et al.) while immunosuppression has been reported for *A. citri*, *G. xanthopoda*, *L. heterotoma* and *L. victoriae* (Chiu et al., 2000; Morales et al., 2005; Prévost et al., 2005; Rizki et al., 1990). In *L. boulardi*, we have combined description of the virulence strategy used by successful parasitoids and investigation of the causes of failure of avirulent parasitoids to understand the physiological causes of virulence variation. Variations of parasitoid virulence can roughly originate from two main mechanisms: either they differ in their ability to evade locally the host immune system, or they differ in their ability to suppress the whole encapsulation response.

One way to distinguish between local immunoevasion and systemic immunosuppression strategies is to determine whether parasitism of a host larva by a virulent parasitoid can protect or not from encapsulation another foreign body that would be normally encapsulated in the same host. This foreign body can be the egg of a nonvirulent parasitoid or a drop of paraffin oil injected into the host larva. According to this criterion, virulent lines of *L. boulardi* have a systemic immunosuppression strategy on *D. yakuba* and *D. melanogaster*. In *D. yakuba*, a drop of paraffin oil is protected from encapsulation for 24 h postparasitization by the ISy line. However, this protection is only transient since the drop is fully encapsulated 48 h postparasitization (Dubuffet et al., 2008), at a time a parasitoid egg has reached the larval stage. The parasitoid larva might then use a local immunoevasion strategy that follows the initial systemic immunosuppression. Interestingly, the line ISm, avirulent on *D. yakuba*, does not affect its capacities to encapsulate the oil drop, which suggests that the variation of success of *L. boulardi* on *D. yakuba* is linked to a variation of the immunosuppressive abilities between the parasitoid lines (Dubuffet et al., 2008). On *D. melanogaster*, multiparasitism experiments have been performed using the lines ISm and ISy, respectively, virulent and avirulent on *D. melanogaster* (Labrosse et al., 2003). About 48 h postparasitization, ISy parasitoids are normally found encapsulated. However, in multiparasitized host larvae, larvae of the two parasitoid lines, easily distinguishable, were found free in the host hemolymph. This indicates that the ISm line can protect ISy parasitoids from encapsulation in *D. melanogaster*. In that case, ISm immunosuppression in *D. melanogaster* might be more durable than ISy immunosuppression in *D. yakuba*, lasting at least 48 h, or it might protect ISy eggs only until they hatch, where a local immunoevasion mechanism would then again take over the protection.

As in *L. heterotoma* and *L. victoriae*, the venom injected during oviposition by *L. boulardi* females was shown to be responsible for the suppression of the encapsulation response (Dubuffet et al., 2008; Labrosse et al., 2003; Morales et al., 2005; Rizki and Rizki, 1990). Both in *D. melanogaster* and *D. yakuba*, injection of venom from the virulent line (ISm for *D. melanogaster*, ISy for *D. yakuba*) can protect from encapsulation a foreign body that is usually encapsulated. In contrast, injection of venom from the avirulent line (ISy for *D. melanogaster*, ISm for *D. yakuba*) does not confer any protection. These observations led us to conclude that qualitative and/or quantitative variations in the venoms of the two parasitoid lines were responsible for the observed variations of virulence (Dubuffet et al., 2008; Labrosse et al., 2003). This assumption is strengthened by results from genetic analyses that suggested that the maternal secretions were responsible for the success/failure of the parasitoid progeny (see Section 6.4.1).

To investigate the functional basis of the variation of immunosuppressive effects in *L. boulardi* further, a first approach was to describe and quantify the physiological modifications that parasitism induces in *Drosophila* hosts. As for most parasitoids, studies focused on effects on the host hemocytes and the PO cascade since changes in the levels of production/effects of cytotoxic radicals are particularly difficult to evidence (see Chapter 4 by Nappi et al.).

Hemocytes were observed and counted after parasitization by the ISm and ISy lines, both in *D. yakuba* and *D. melanogaster* (Dubuffet et al., 2008; Russo et al., 2001). In all cases, the total number of hemocytes and specifically of plasmatocytes and lamellocytes increased significantly after parasitization in comparison to unparasitized controls. This suggests that *L. boulardi* does not prevent the production/release of hemocytes following recognition of the intruder. However, larvae parasitized by virulent parasitoids show a much lower increase in the number of some categories of hemocytes compared to larvae attacked by avirulent parasitoids: this is the case for the lamellocyte number in the *D. melanogaster*/ISm interaction and for the plasmatocyte number in the *D. yakuba*/ISy interaction. These interactions might simply elicit a weaker immune cellular response. Alternatively, immunosuppressive factors injected by ISm and ISy females might be responsible for these effects. The parasitoid wasps *L. heterotoma*, *A. citri* and *G. xanthopoda* are known to induce the atrophy of the *D. melanogaster* hematopoietic organ (Chiu et al., 2000; Prévost et al., 2005). Parasitism by *L. heterotoma* also leads to apoptosis of circulating plasmatocytes and hematopoietic precursors in the lymph gland, as well as to destruction of circulating lamellocytes (Chiu et al., 2000; Rizki and Rizki, 1984, 1990). However, none of these modifications are observed following parasitism of *D. melanogaster* with the ISy avirulent strain of *L. boulardi* (Chiu et al., 2000). The potential effects of *L. boulardi* virulent wasps on hemocytes or hematopoietic organs remain to be elucidated.

In addition to a variation of effects of ISy and ISm parasitoids on lamellocytes number in *D. melanogaster*, we observed a variation on the lamellocytes morphology: a significant proportion of lamellocytes (up to 50%) became bipolar following parasitization by the virulent line ISm, but not in ISy-parasitized larvae (Russo et al., 2001). Such change in lamellocyte morphology has also been reported in *L. heterotoma* and *L. victoriae*, and was suggested to correlate with a decreased ability to adhere and to form capsules (Morales et al., 2005; Rizki and Rizki, 1984). In both species, the factor responsible for these effects, called lamellolysin in *L. heterotoma*, is localized in the venom gland (Morales et al., 2005; Rizki and Rizki, 1991). Accordingly, injection of venom from the ISm line of *L. boulardi* mimicked parasitism effects on lamellocytes number and shape, while injection of venom from the ISy line had no effect, which indicates that variation in venoms is responsible for variations of effects on lamellocytes (Labrosse et al., 2005a). In *D. yakuba*, no effect on lamellocyte morphology or number has been observed, neither following parasitism by the ISm line nor in ISy-infected larvae (Dubuffet et al., 2008), which suggests that, considering these two host species only, effects on lamellocytes morphology are specific to *D. melanogaster*.

We recently started investigating the effects of *L. boulardi* venom on the humoral components of the encapsulation response. We showed that venom from the line ISy, virulent on *D. yakuba*, inhibits the activation of the proenzyme prophenoloxidase into PO in this species, in a dose-dependent manner (Colinet et al., 2009; Fig. 6.5A). In contrast, venom from the avirulent line ISm did not show such an inhibiting effect (Dubuffet, unpublished data). This suggests that variations of virulence in *D. yakuba* could be linked to a variation in the venom capability to inhibit the PO cascade of *D. yakuba*. The question remains if parasitoid success of ISm on *D. melanogaster* is also associated with an inhibition of the PO cascade.

Our physiological data, even incomplete, support the idea that some, if not all, mechanisms underlying the variation of virulence of *L. boulardi* on *D. melanogaster* and *D. yakuba* differ from each other: the variation of virulence in *D. melanogaster* is correlated to a variation of effects on the lamellocytes number and morphology while the variation of virulence in *D. yakuba* is correlated to variation of effects on plasmatocyte number and on the phenoloxidase cascade. The existence of mechanisms underlying the variation of virulence, different for each host species, is supported by the existence of the two distinct loci for virulence evidenced by Dupas and Carton (1999).

6.4.3. Parasitoid components at the origin of virulence variation

Both in *D. melanogaster* and *D. yakuba*, it appears that the variation of virulence observed between the ISy and ISm lines of *L. boulardi* is due to a variation of the immunosuppressive effects induced by the venom. Characterization of the virulence factors contained in these venoms, and particularly their quantitative and/or qualitative variations is thus crucial to determine the basis of virulence variation.

6.4.3.1. Variation in virus-like particles

In all figitid parasitoids studied to date, *L. heterotoma* (Rizki and Rizki, 1990, 1994), *L. victoriae* (Morales et al., 2005) and *L. boulardi* (Dupas et al., 1996; Labrosse et al., 2003), virus-like particles (VLPs) are observed in the venom of females. This characteristic is not unique to figitidae since other parasitoid families, including the well-studied braconidae and ichneumonidae, produce VLPs either in the venom apparatus or in the ovaries (Barratt et al., 1999; Reineke et al., 2006; Suzuki and Tanaka, 2006).

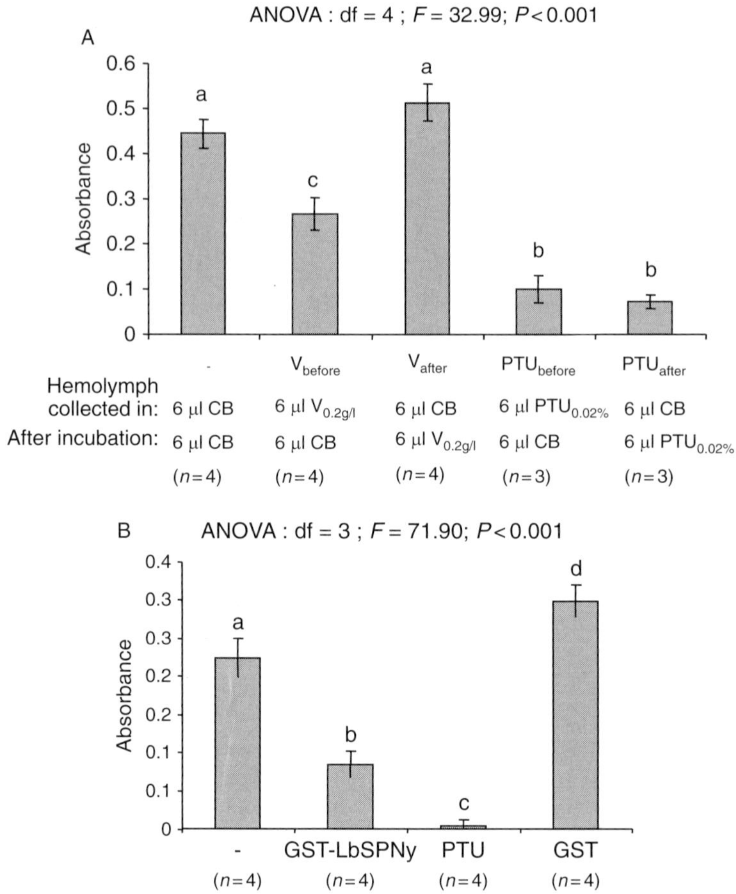

FIGURE 6.5 (A) Effect of ISy venom on PO activation in *D. yakuba*. PO activity was measured in hemolymph samples in which parasitoid venom (0.2 g/l) or PTU (a PO inhibitor) was present before or after a 45-min incubation at 37 °C. The proPO proenzyme is normally activated into active PO during this incubation period by the serine protease cascade. In samples collected in presence of ISy venom, PO activity is lower. This effect is not due to an inhibition of the PO enzyme itself, since it is not observed if the venom is added after the incubation period. In PTU controls, however, inhibition is observed whatever the moment where PTU is added. CB: cacodylate buffer; V 0.2 g/l: venom extract. PO activity is measured through the conversion of L-DOPA into dopachrome, which absorbance is measured at 490 nm. Mean values (standard error (SE)) are given for each category and numbers within brackets indicate numbers of replicate. (B) Effect of the recombinant serpin LbSPNy on PO activity. Hemolymph of *D. yakuba* larvae was collected in cacodylate buffer alone (–), in presence of the GST-LbSPNy fusion protein (0.3 g/l), in presence of a PO inhibitor (PTU) or in presence of the GST protein alone (0.3 g/l). Mean values (SE) are given for each category and numbers within brackets indicate numbers of replicates. *Note*: From Colinet et al. (2009).

The nature of these VLPs, which do not contain deoxyribonucleic acid (DNA) and can have very different aspects, remains to be elucidated. VLPs are injected into the host together with the eggs and have been described, for instance, to target hemocytes, inducing morphological changes and/or apoptosis (Rizki and Rizki, 1990; Suzuki and Tanaka, 2006). In *L. heterotoma*, VLPs have been shown to enter *Drosophila* host hemocytes (Chiu et al., 2006; Rizki and Rizki, 1994). They can be observed free in the cytoplasm of lamellocytes or in engulfed vesicles in plasmatocytes, suggesting that these last cells are able to phagocyte them (Rizki and Rizki, 1990, 1994). Besides, the so-called "lamellolysin" factor, injected by this parasitoid and responsible for changes in host lamellocyte morphology, has been demonstrated to be composed of VLPs (Rizki and Rizki, 1990, 1994). However, neither the nature of these VLPs nor the molecular nature of the factors responsible for these changes has been identified in this species.

In *L. boulardi*, VLPs have been detected in all lines studied including the ISm and ISy lines, but they strongly differ in the morphology and the number of the particles (Dupas et al., 1996; Labrosse et al., 2003; Fig. 6.6). The particles of the ISm line as well as of two other lines also virulent on *D. melanogaster*, are round shaped and contain several vesicles, while the particles in the ISy line, avirulent on *D. melanogaster*, are more elongated and contain fewer vesicles. F1 hybrid females resulting from the cross between the ISm and ISy lines produce VLPs of intermediate morphology with less elongated particles containing more vesicles than in the ISy line (Dupas et al., 1996). Interestingly, these hybrids exhibit half-immune suppressive ability toward *D. melanogaster*. These data suggest that the morphology of the VLPs might be somehow related with the parasitoid virulence level against *D. melanogaster*. However, the morphology of VLPs might not be related to virulence toward *D. yakuba* since these hybrids are completely avirulent on this host species (Dupas and Carton, 1999). The ISm type of *L. boulardi*, but not the ISy type, induces changes in the morphology of *D. melanogaster* lamellocytes. If ISm VLPs have the ability to enter lamellocytes as *L. heterotoma* VLPs, they might be responsible for these changes, either by themselves or by transporting the responsible factor(s) inside these hemocytes. Purification experiments and proteomic analysis will allow identifying the proteins that constitute and/or are transported by VLPs. This will give more insights into how the VLPs are formed what explain the observed intraspecific morphological differences, and what role they play on observed differences in virulence levels.

6.4.3.2. Variation in proteinic content of venoms

Biochemical approaches have provided recently valuable information on the nature and variation of the immune suppressive factors in *L. boulardi*. Native and sodium dodecyl sulfate polyacrylamide gel electrophoresis

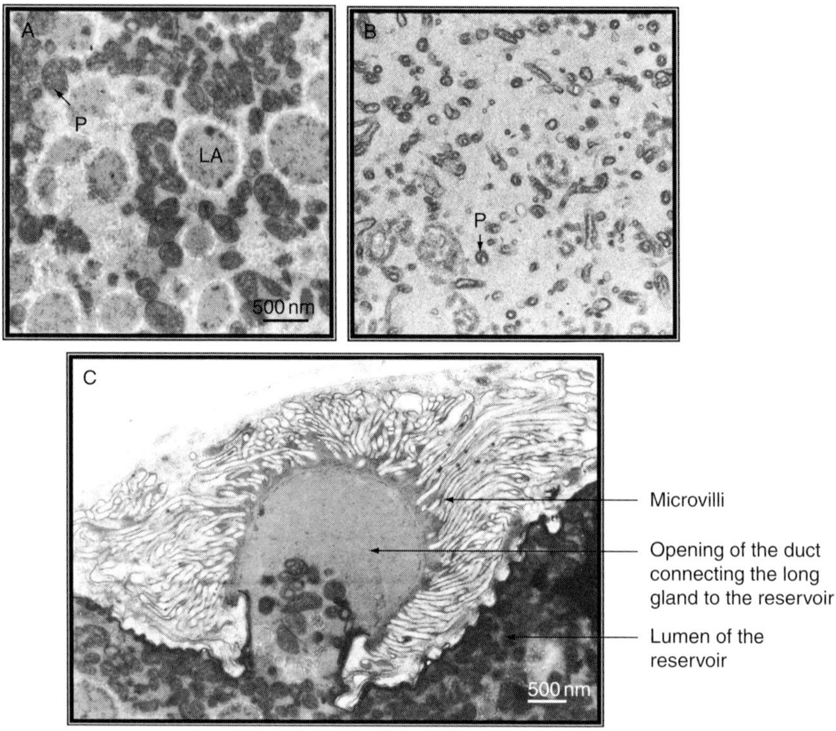

FIGURE 6.6 Results of transmission electron microscopy. (A) VLPs in the reservoir of a G431 virulent female. LA, large aggregates; P, particles. (B) VLPs in the reservoir of a G486 avirulent female. P, particles. (C) Transversal section of the apical part of the reservoir of a G431 virulent female. Numerous microvilli are observed. *Note:* From Labrosse et al. (2003).

(SDS-PAGE) of the protein content of the venom apparatus of the ISy and ISm *L. boulardi* revealed an impressive variation between these two lines (Colinet et al., 2009; Labrosse et al., 2005b; Fig. 6.7A and B). Nevertheless, all the lines virulent on *D. melanogaster* we tested harbored a proteinic profile more or less similar to the one observed for ISm parasitoids (Fig. 6.7C). This suggests that the intraspecific variation of virulence between the ISy line and the lines virulent on *D. melanogaster* is correlated with differences in venom gland protein profiles, resulting from qualitative and/or important quantitative differences in the protein content of these glands.

6.4.3.3. The LbGAP virulence factor and its variation

Among the major native proteinic bands in the venom of the line ISm and of all the tested lines virulent on *D. melanogaster*, the P1 and P4 bands have been the most studied. Each of these two bands, eluted from native PAGE,

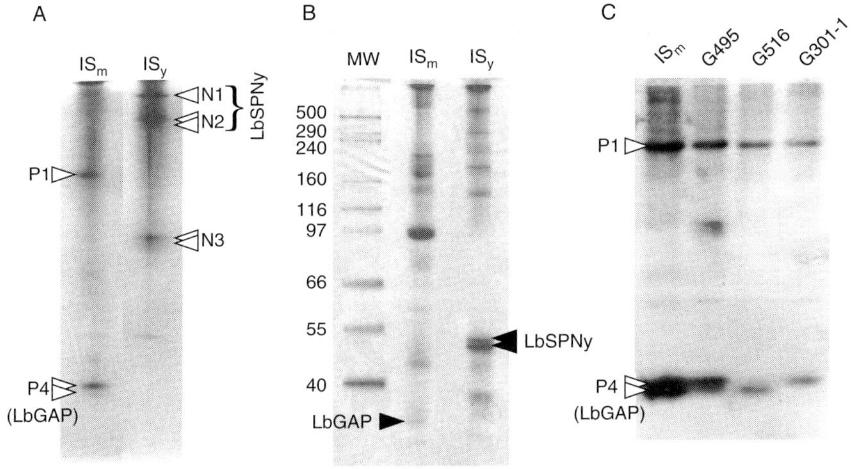

FIGURE 6.7 Comparison of the proteinic profiles of venom glands of *L. boulardi* from various localities. (A) Native-PAGE profiles of ISm and ISy strains (gel 8%). (B) SDS-PAGE profiles of ISm and ISy strains (gel 8%). (C) Native-PAGE profiles of *L. boulardi* strains virulent on *D. melanogaster* (gel 8%). Origin of strains: ISm: Nasr'allah, Tunisia; ISy: Brazzaville, Congo; G495: Lamto, Ivory Coast; G516:Toulouse, France; G301-1: Guadeloupe. White arrowheads: major native bands; black arrowheads: SDS bands containing identified virulence factors. MW: molecular weight in kDa. *Notes:* The profile of the ISy strain was published in Colinet et al. (2009); Fig. 6.7C is from Labrosse et al. (2005). (See Page 5 in Color Section at the back of the book.)

had a significant effect on the encapsulation rate of avirulent ISy eggs by *D. melanogaster* resistant larvae (Labrosse et al., 2005b). The strongest effect was nevertheless obtained with the band P4, for which injection had the same effect as that of whole venom gland extracts. Injection of this band also mimicked changes in the morphology of lamellocytes induced by parasitism (Labrosse et al., 2005a). These results led us to conclude that this band contains the major virulence factor of the line ISm, and that modification of lamellocytes is an essential part of the virulence strategy used by this line to escape encapsulation by *D. melanogaster*.

The protein band P4, eluted from native PAGE, was submitted to N-terminal sequencing allowing cloning the complete complementary DNA (cDNA). It encodes a RhoGAP (Rho GTPase-activating protein) domain-containing protein that was then renamed LbGAP (Labrosse et al., 2005b). Using Western blot experiments with a specific antibody against a recombinant LbGAP protein, it was confirmed that LbGAP is abundant in venom glands of ISm females, but it was not detected in the rest of the body (Labrosse et al., 2005b). Using immunofluorescence experiments, we showed that LbGAP enters plasmatocytes and lamellocytes and is directly involved in affecting the morphology of lamellocytes

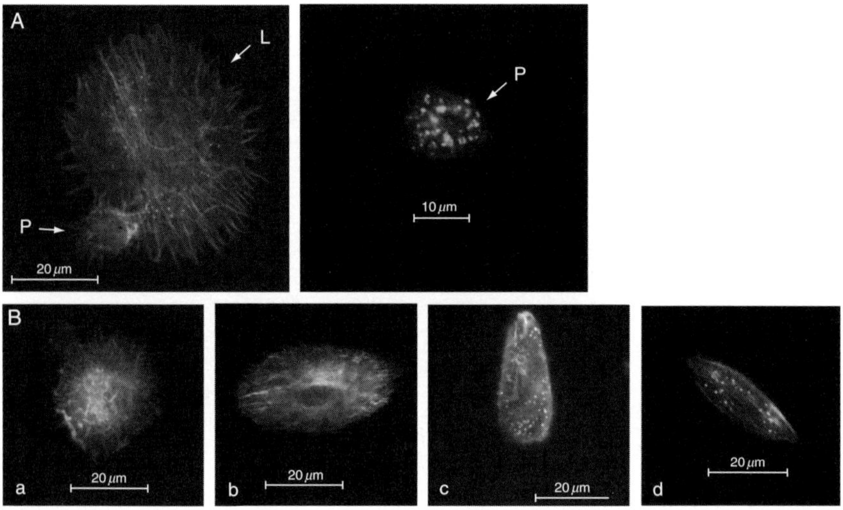

FIGURE 6.8 LbGAP enters *Drosophila* lamellocytes and plasmatocytes and affects lamellocyte morphology. (A) Example of LbGAP-containing hemocytes. L: Lamellocyte, P: plasmatocyte. (B) Classification of lamellocytes into four categories according to their morphological changes. a: Unmodified lamellocyte, b: slightly modified lamellocyte with 1–10 LbGAP spots, c: fairly modified lamellocyte with more than 30 LbGAP spots, d: strongly modified lamellocyte with 10–30 LbGAP spots. Hemocyte actin cytoskeleton was visualized using phalloidin (green). LbGAP was detected using a specific rabbit polyclonal antibody (red). *Note:* From Colinet et al. (2007). (See Page 5 in Color Section at the back of the book.)

(Fig. 6.8; Colinet et al., 2007). The quantity of LbGAP in a lamellocyte is indeed correlated with the degree of modification in the lamellocyte shape. Interestingly, LbGAP is observed as large spots in *Drosophila* hemocytes, which suggests that the protein is associated with larger structures.

The molecular bases of LbGAP effects have been further determined: using biochemical assays we showed that LbGAP has a RacGAP activity, and two-hybrid experiments allowed to characterize its targets in *D. melanogaster*. LbGAP specifically targets and inactivates the two Rac GTPases, Rac1 and Rac2 (Colinet et al., 2007). Rac GTPases are known to regulate cytoskeletal rearrangements necessary for cell-shape change and adhesion (Burridge and Wennerberg, 2004), which are an essential part of the insect cellular response against endoparasitoids. Moreover, both Rac1 and Rac2 were precisely reported to be required for successful encapsulation of *L. boulardi* eggs (Williams et al., 2005, 2006), thus explaining the physiological effects of LbGAP on host lamellocytes. These results were the first to describe the physiological effects of a parasitoid virulence factor together with its molecular function and its protein targets in the host.

By contrast to ISm, the ISy line of *L. boulardi* is encapsulated by resistant *D. melanogaster* flies and does not induce changes in lamellocyte morphology in *D. melanogaster* or in *D. yakuba* (Dubuffet et al., 2008; Russo et al., 2001). Further studies will be needed to determine whether this intraspecific variation of virulence results from qualitative differences in terms of functional activity of LbGAP or interaction with host targets or from a quantitative difference in its production. In agreement with the last hypothesis, LbGAP could not be detected in Western blots of ISy venom glands (Labrosse et al., 2005b). Besides, partial sequencing of the major bands in ISy venom did not reveal any peptide with similarities to Rho-GAP proteins (Colinet, unpublished data). This suggests that the LbGAP protein is not produced or is in a small amount in ISy venom but these data remain to be confirmed.

The characterization of LbGAP also allows us to address the question of host specificity of virulence. Indeed, the ISm line does not induce any modification of lamellocytes in *D. yakuba* (Dubuffet et al., 2008) and is totally avirulent on this host. Since there are no differences in Rac1 and Rac2 sequences between *D. melanogaster* and *D. yakuba*, the observed host specificity cannot be explained by a difference in the nature of the target of LbGAP. Another hypothesis would involve a difference in the capacity of LbGAP to enter lamellocytes between both *Drosophila* spp. The mode of entry of LbGAP in *D. melanogaster* lamellocytes is thus a central point to be elucidated. As reported above, VLPs can enter *Drosophila* hemocytes and we then suspect that LbGAP might be associated with VLPs thus facilitating its entry. This would explain why LbGAP is detected as "large spots" inside *D. melanogaster* hemocytes in immunofluorescence experiments. Moreover, we know that among the proteins characterized from samples of VLPs purified from the parasitoid *Venturia canescens* figures VLP2, a RhoGAP domain-containing protein such as LbGAP (Reineke et al., 2002). If VLPs act as "transporters of virulence factors," then the difference in host specificity of ISm females might come from a difference in VLPs ability to target and enter lamellocytes of *D. melanogaster* and *D. yakuba*. A detailed comparison of these hemocytes in the two species should address this question.

6.4.3.4. The serpin SPNy virulence factor and its variation

Analysis of the protein content of the venom apparatus of ISy females led to the analysis of two major bands, named N1 and N2, which were not observed in the venom of ISm females (Fig. 6.7A). Mass spectrometry led to identification of similar peptides from the two bands, suggesting they contain the same protein. The corresponding cDNA encodes a serpin-domain-containing protein, LbSPNy (Colinet et al., 2009). Using real time polymerase chain reaction (qRT-PCR) experiments, this factor was shown to be specifically overexpressed in ISy venom glands compared to the rest of the body (460-fold higher expression). Moreover, the recombinant LbSPNy protein reproduced the inhibition of the PO cascade observed with the venom (Fig. 6.5B; Colinet et al., 2009). LbSPNy is thus the first serpin demonstrated to be used as a virulence factor by a parasitoid wasp. Little is known on the *Drosophila* PO cascade as compared to Lepidopteran models. However, it seems that activation of the PO cascade, as well as melanization, occurs during the early parasitism period (Nappi and Christensen, 2005; Russo et al., 1996; Wertheim et al., 2005). These events seem to be important for the encapsulation to take place, since the injection of natural or synthetic serine protease inhibitors inhibits the encapsulation response (Ling and Yu, 2005; Nappi et al., 2005). Serpins act as suicide-substrate inhibitors, which means they become inactive once they inhibited their serine protease target (Law et al., 2006). The expression of many serine proteases is increased in the first 24 h following parasitization (Wertheim, 2005), which suggests that their production might at some point overcome the number of serpin molecules. The PO cascade could then be triggered, and the encapsulation could subsequently take place. Inhibition of the activation of the PO cascade by the serpin LbSPNy we described could thus be responsible for the transient immunosuppression observed in *D. yakuba* parasitized by ISy parasitoids. Our work opens the way to identification of the serine protease(s) targeted by LbSPNy, which will provide information on the regulatory pathways of *Drosophila* PO activation. An open area of research is now to determine how important is the use of serpins as virulence factors among parasitoids since they are known to be used by other parasites, such as nematodes, to evade the host immune responses (Knox, 2007; Zang and Maizels, 2001).

Interestingly, the venom of ISm *L. boulardi* females does not seem to contain any abundant protein potentially corresponding to LbSPNy (electrophoresis experiments and partial sequencing of major proteins, data not shown). Moreover, preliminary data suggest that ISm venom does not inhibit the PO cascade in *D. yakuba*. This supports the essential role of LbSPNy in targeting this cascade and might be one of the reasons why ISm females are not virulent on *D. yakuba* hosts. Further studies will be needed to explain the bases and evolutionary origin of these intraspecific differences.

6.5. DISCUSSION

6.5.1. On intra- and interspecific variability of virulence strategies in the *Leptopilina* genus

It remains a challenge to determine what makes a parasitoid "virulent" or "avirulent" against a given host, even when focusing on the immune aspects of the interaction. At first, the virulence strategies that allow a successful parasitoid to escape the host immune defenses have to be known so that the aspect in which the avirulent parasitoid fails can be evidenced. In *Drosophila* parasitoids, the strategies that have been described are diverse and can differ even at the genus level. This corresponds to a difference in the virulence ''tactics'' used by these parasitoids. Most of our work aimed at elucidating *L. boulardi* virulence strategy, with the final objective of characterizing the processes underlying variations of virulence toward *D. melanogaster* and *D. yakuba*. However, it is also of interest to question the occurrence of variations in the means used to escape encapsulation within the *Leptopilina* genus.

One of our major results was to evidence the key role of LbGAP in the high virulence of the Tunisian ISm line of *L. boulardi*, and its ability to induce modification in lamellocyte shape (Colinet et al., 2007; Labrosse et al., 2005a,b). Such a strategy is probably the most common in *L. boulardi*. Indeed, the Guadeloupean line G301-1 also modifies the lamellocytes of *D. melanogaster* (Poirié, unpublished data). Besides, proteinic patterns of venoms of this line as well as lines from the south of France and Ivory Coast, all highly virulent on *D. melanogaster*, are roughly similar, and all include the bands that correspond to LbGAP (Labrosse et al., 2005b; Fig. 6.7C). *L. heterotoma* and *L. victoriae* have also been reported to suppress the ability of the host to encapsulate a foreign body (Morales et al., 2005; Rizki and Rizki, 1990; Schlenke et al., 2007) and to induce modifications in *D. melanogaster* lamellocytes. Considering these results, it would be tempting to conclude that the virulence strategy used toward the host *D. melanogaster* is largely conserved within the *Leptopilina* genus.

Data obtained with two *L. boulardi* lines, however, appear to question this conclusion: the Congolese line ISy and the Californian line Lb17 are consistently able to achieve successful parasitization of *D. melanogaster* larvae without inducing any modification of lamellocytes. Schlenke et al. (2007) performed microarrays to compare the transcriptional response of *D. melanogaster* larvae infected by Lb17 *L. boulardi* females and *L. heterotoma*. Based on the results that showed few changes in the transcription level of immune genes in hosts infested by *L. heterotoma* but upregulation or downregulation of several of these genes in Lb17-infected larvae, they concluded that *L. boulardi* and *L. heterotoma* have totally different virulence strategies. *L. heterotoma* would escape encapsulation by *D. melanogaster* through a ''near complete failure of attacked flies to mount an immune transcriptional response,'' while *L. boulardi* would escape encapsulation by attaching to the host tissues, a feature previously reported in other strains of *L. boulardi* (Rizki and Rizki, 1990). Whether this ''egg-sticking'' strategy explains the success of the line Lb17 remains to be determined. However, in the lines ISm, 301.1 and ISy, where we observed sometimes such egg attachment, we did not find any correlation between attachment and parasitoid success, neither in *D. melanogaster* nor in *D. yakuba* (Dubuffet, unpublished data).

The Congolese line ISy of *L. boulardi* protects its eggs from encapsulation in the ''susceptible'' genotype of *D. melanogaster* but not in the ''resistant'' genotype (Dubuffet et al., 2007). Susceptible hosts are nevertheless immunocompetent, since they can encapsulate *A. tabida* (Poirié et al., 2000). The parasitoid success is not correlated with a modification of lamellocytes, which is consistent with the absence of immunodetection of the virulence factor LbGAP in the venom of the ISy line (Russo and Labrosse, 2005a). This result, together with a venom proteinic profile completely different from that of the other lines, suggests that the ISy line relies on an alternative virulence strategy to escape encapsulation in the susceptible larvae of *D. melanogaster*. Interestingly, this parasitoid line can also infest the ''susceptible genotype'' of *D. yakuba*. In this species, it inhibits the PO cascade activation due to the serpin LbSPNy and delays the proliferation of plasmatocytes (Colinet et al., 2009; Dubuffet et al., 2008). Future investigations will determine whether the virulence strategy used by ISy females on *D. melanogaster* is similar to the one described for *D. yakuba*.

The existence of lines such as ISy and Lb17 raises also an important question: is the modification of lamellocytes a conserved feature of the virulence strategy used by *Leptopilina* wasps (that was lost by both lines) or an example of convergence of effects? Such a convergence of effects is indeed commonly observed in host–parasitoid interactions. Roughly similar effects can be induced by various parasitoids on hosts that are as different as Lepidopteran caterpillars or *Drosophila* larvae, and due to completely different virulence factors. For example, disruption of actin cytoskeleton of hemocytes is induced by completely unrelated proteins such as those encoded by the polydnavirus gene CrV1 of *Cotesia rubecula* (braconid) or the polydnavirus gene VHv1.1 of *Campoletis sonorensis* (Ichneumonid), or by the factor LbGAP of *L. boulardi* (cynipid; Glatz et al., 2004; Labrosse et al., 2005b). Inhibition of PO activation is also caused by factors as various as a serine protease homolog in the braconid *C. rubecula*, a smapin in the braconid *Microplitis demolitor*, or a serpin in *L. boulardi* (Asgari et al., 2003; Beck and Strand, 2007; Colinet et al., 2009). The modification of the shape of lamellocytes observed after parasitism by *L. heterotoma*, *L. victoriae* or the lines ISm and G301-1 of *L. boulardi* could similarly result from totally different virulence factors that converge in their effects. Accordingly, preliminary data strongly suggest that RhoGAP proteins are not involved in *L. heterotoma* virulence against *D. melanogaster* (Colinet, unpublished data). To compare properly the virulence strategies used by parasitoids, we thus think that it is actually necessary to distinguish three levels within the term ''virulence strategies'': (1) the general strategy of the parasitoid, assessed through the effects on host encapsulation ability (systemic immunosuppression or local immunoevasion); (2) the ''tactic(s)'' used to achieve this general strategy, that is, the immune components targeted by the parasitoid; and (3) the virulence factors used to achieve each of these tactics. Only the characterization of these virulence factors, and their resulting effects on specific components of the host immune system and on the whole encapsulation ability, will allow the full comprehension of the diversity of virulence strategies used by parasitoids.

Altogether, available data for different *L. boulardi* lines suggest that alternative virulence strategies exist in *L. boulardi*, the ISm/G301.1 strategy ''resembling'' more that of *L. heterotoma*. Surprisingly, within-species variability in the means to escape encapsulation is a question which has never been explored so far. Parasitoids species are usually considered as ''invariants,'' and comparisons between the virulence strategies used by different parasitoid species always rely on comparisons between single laboratory lines, as in the study performed by Schlenke (2007). In this case, comparison of the species *L. boulardi* and *L. heterotoma* could have resulted in quite different conclusions if other *L. boulardi* lines such as ISm were used in addition to the line Lb17.

6.5.2. On the variation of outcome in host–parasitoid interactions

The virulence strategy of parasitoids comprises multiple tactics, each achieved by one or many virulence factors. These tactics are used on diverse components of the host immune system, and in many models it appears that various tactics are employed at different periods of the parasitoid development (Dubuffet et al., 2008; Glatz et al., 2004; Schmidt et al., 2001). Similarly, encapsulation is a complex immune reaction that involves the coordination between recognition molecules, signaling pathways and immune effectors (Carton et al., 2008; Govind, 2008). Variations in the outcome of any host–parasitoid interactions can potentially originate from variations of any of the components of the parasitoid virulence strategy or host resistance.

Linking the molecular bases that underlie the variations of resistance and virulence in a host–parasitoid interaction is a thrilling objective in the field of evolutionary biology, since it aims to determine which genes in the host and in the parasitoid populations are potentially involved in coevolutionary processes. The achievement of this objective requires three important points: first, a genetic variation for resistance and/or virulence has to exist in the model. Second, it is necessary to have elucidated both the cellular and molecular processes leading to encapsulation (in unsuccessful infections) and the nature and function of effector virulence factors preventing encapsulation (in successful infections).

Third, tools have to be available to study these variations. When all these requirements are fulfilled, it is possible to determine what makes the difference between an avirulent and a virulent parasitoid, and/or between a resistant and a susceptible host and to assess whether these traits are under coevolution or not. Parasitoids of *Drosophila* are a model of choice to solve this puzzle, since extensive variations in the outcome of their interactions with *Drosophila* hosts are regularly reported in natural populations (Dubuffet et al., 2007; Dupas et al., 2003; Kraaijeveld and Godfray, 1999). The use of isofemale lines allows study of each factor that originate these variations, that is, the genetic variations of resistance and virulence. Moreover, *Drosophila* is the insect model for which the encapsulation response is the most studied, and the existence of genetic markers throughout the genome allows determination of which genes underlie the variation of resistance (Hita et al., 2006). We recently also characterized the virulence strategy used by the parasitoid *L. boulardi* on the hosts *D. yakuba* and *D. melanogaster*, and developed a method based on the comparison of physiological effects and virulence factors between avirulent and virulent lines to study the mechanisms underlying the variations of virulence (Colinet et al., 2009; Dubuffet et al., 2008; Labrosse et al., 2003, 2005a,b).

Studies performed on the parasitoids *L. boulardi* and *A. tabida* revealed that their success depends on both host and parasitoid genotypes (Dubuffet et al., 2007; Kraaijeveld and Godfray, 2001). However, the geographic variation in host–parasitoid outcomes is more explained by the variations in parasitoid virulence than by the variations in host resistance (see Section 6.2; Dupas et al. 2003; Kraaijeveld and Godfray, 2001). From other models, we know that environmental factors can also influence the host immune responses and the infective abilities of parasitoids (Blumberg, 1997; Calatayud et al., 2002; Delpuech et al., 1996; Fytrou et al., 2006; Karimzadeh and Wright, 2008). In order to investigate deeply the factors that influence the outcome of host–parasitoid interactions in the field, it is now necessary to determine whether the effects of environmental factors overcome those of host and parasitoid genetic factors, have on the contrary minor effects, or if all these factors interact altogether.

6.5.3. On the ways to reconcile the genetic and molecular data

Because host immune response and parasitoid virulence strategy are multifactorial, their variation was expected to be multigenic. However, genetic crosses or quantitative trait locus (QTL) analyses performed between resistant and susceptible host lines, or between virulent and avirulent parasitoid lines, always concluded in a simple genetic determinism (Benassi et al., 1998; Carton et al., 2005; Dubuffet et al., 2007; Orr and Irving, 1997). In other invertebrate–parasite systems, QTL analyses revealed that most of the genetic variation for resistance is generally explained by few loci (2.47 on average), and in about 20% of the cases studied, resistance is explained by a single locus (Wilfert and Schmid-Hempel, 2008). However, when different parasites or host isolates were used, different QTLs were generally found (Wilfert and Schmid-Hempel, 2008). Whether this also applies in *Drosophila*–parasitoids systems has to be determined. For example, it would be interesting to assess whether the locus *Rlb* also explains genetic variation of resistance to the semivirulent strain Tasagil of *L. boulardi* (Kraaijeveld and van Alphen, 1995) or if it would be recovered from genetic analyses of resistance to the ISy line in other *D. melanogaster* strains. It would also be interesting to determine whether the gene corresponding to *Rlb* is also responsible for the genetic variation of resistance to the same ISy parasitoid line in *D. yakuba*.

Variation of virulence of *L. boulardi* toward *Drosophila* hosts was also found to be determined by single loci, the nature of which remains to be determined. Data from the field and from laboratory crosses both evidenced that the *ISm* and *ISy* loci, responsible for virulence against *D. melanogaster* and *D. yakuba* respectively, are distinct (Dupas and Boscaro, 1999; Dupas and Carton, 1999). This is in agreement with the variation of effects of ISm and ISy female venoms on the two host species, the ISm line containing the ISm^{+} allele but not the ISy^{+} and vice versa. A parasitoid which is "strong" on a host species is neither especially "strong" nor "weak" on the other, which would be the case if virulence alleles had positive pleiotropic effects on different hosts, or if virulence on each host species was allelic. Interactions between *L. boulardi* and these two host species (if not all *Drosophila* spp.) thus has to be considered independently (see Table 6.3).

There is an interesting challenge in linking each of the two loci *ISm* and *ISy* with variations in virulence factors contained in the venom of *L. boulardi* females. On *D. melanogaster*, we suspect that variation of virulence between the lines ISy and ISm might be linked to the presence or the quantity of the LbGAP protein in the venom (Labrosse et al., 2005b). It would be interesting to focus now on parasitoid lines that have intermediate levels of virulence, like the Turkish strain Tasagil, to see whether it is correlated with intermediate amounts of LbGAP. It also would be interesting to determine whether intrapopulation variations of virulence are linked with the variation of this factor. On *D. yakuba*, future investigations will determine whether qualitative or quantitative variations in serpins like LbSPNy could originate the variation of virulence between ISm and ISy lines (Colinet et al., 2009).

The loci *ISm* and *ISy* might thus encode for qualitative or quantitative variations of the factors LbGAP and LbSPNy, respectively, acting then as major loci for virulence. However, the presence/quantity of several venom proteins potentially involved in virulence, other than LbGAP and LbSPNy, differ between the lines ISm and ISy (Labrosse et al., 2005a, unpublished data). *ISm* and *ISy* loci might thus contain clusters of

TABLE 6.3 Genetic, immunologic and molecular determinants of virulence variation in *L. boulardi*

	Variation of virulence in *D. melanogaster*	Variation of virulence in *D. yakuba*
Parasitoid/host lines used to study the underlying mechanism	ISy ("avirulent") and ISm ("virulent")/ host reference line "R"	ISy ("virulent") and ISm ("avirulent")/ host reference line "R1"
Genetic determinism of virulence	Variation of encapsulation of the parasitoid eggs determined by the mother genotype One major biallelic locus, called *ISm*. ISm^+ allele, associated with virulence, is semidominant over ISm^-	Variation of encapsulation of the parasitoid eggs determined by the mother genotype One major biallelic locus, called *ISy*. ISy^+ allele, associated with virulence, is recessive to ISy^-
Variation of effects on the host encapsulation ability (virulence factor which variation might be responsible for the variation of virulence)	Variation in immunosuppressive effects of parasitoids (venom, LbGAP-containing proteinic band P4 in particular)	Variation in immunosuppressive effects (venom)
Variation of effects on the cellular immune response (virulence factor which variation might be responsible for the variation of	Less important proliferation of lamellocytes in ISm parasitized hosts Alteration of lamellocyte shape in hosts parasitized by	Less important plasmatocyte proliferation in ISy parasitized hosts No effect on lamellocyte shape

(*continued*)

TABLE 6.3 *(continued)*

	Variation of virulence in *D. melanogaster*	Variation of virulence in *D. yakuba*
parasitoid success)	virulent ISm parasitoids (venom, LbGAP contained in proteinic band P4, in particular)	
Variation of effects on the humoral immune response (virulence factor which variation might be responsible for the variation of parasitoid success)	?	Inhibition of the PO cascade activation and subsequent melanization by ISy parasitoids (venom, SPNy in particular). No effect of ISm parasitoid venom

virulence genes encoding these proteins, and/or correspond to a single gene with pleiotropic effects (e.g., it could encode a transcription factor responsible for increased expression of different genes in venom tissues). Alternatively, variations in the presence/quantity of proteins other than LbGAP and SPNy could originate from other virulence loci, each determining the virulence of *L. boulardi* on a specific host species. Differences in minor proteinic bands have been noticed between the venoms of lines having similar virulence properties on *D. yakuba* and *D. melanogaster* but not on *D. simulans* (Labrosse, unpublished data). Further investigation of the proteins contained in these bands, along with the characterization of the physiological mechanisms underlying variation of virulence of *L. boulardi* on *D. simulans* will generate interesting data for the comprehension of the diversity of virulence factors contained in parasitoid venoms.

6.5.4. On intraspecific variation of virulence and host specificity in parasitoids

Thompson hypothesized that host or parasitoid populations from various geographical areas should differ in their traits involved in the interaction with the interacting species, due to a geographical mosaic of selection

(Thompson, 2005). Such mosaic of selection is likely to occur in many parasitoid species, especially if the range of species available is highly variable depending on the localities, as for *L. boulardi* (Dupas et al. 1999; 2003; see Chapter 11 by Dupas et al.). The availability of each host species as well as their respective levels of resistance may then shape the evolution of the virulence strategies in each parasitoid population. They can eventually become quite different, with the involvement of different virulence factors. Such variation in the nature or quantity of virulence factors resulting from this geographical mosaic of selection could then lead to strong variations of virulence, as observed in *L. boulardi*.

Most parasitoids rely on factors contained in their venom glands and/or calyx fluids to escape encapsulation by their hosts. Many of these factors, injected during oviposition, are proteins produced by the wasps themselves (Moreau and Guillot, 2005), but others are viruses (polydnaviruses, PDVs) encoding for virulence factors which are expressed due to the host machinery during the parasitoid development (Bezier et al., 2009; Glatz et al., 2004; Renault et al., 2005). Future investigations will determine whether variations in parasitoid success are common, and whether they can be correlated with qualitative and/or quantitative variations in venom or calyx fluid secretions. To our knowledge, genetic variation of virulence has been documented in only three species in addition to parasitoids of *Drosophila*: *Cotesia sesamiae*, *Aphidius ervi* and *Lysiphlebus fabarum* (Henter, 1995; Ngi-Song et al., 1998; Vorburger et al., 2009). From these, only the braconid *C. sesamiae* was studied for the molecular basis underlying the variations of virulence. Injection of virulent wasp calyx fluid in hosts infected by the avirulent wasp allows development of the avirulent parasitoid (Mochiah et al., 2002). The virulent and avirulent lines differ in the presence of few proteinic bands in calyx fluid analyses (Gitau et al., 2006) and show qualitative and quantitative differences at the level of the CrV1 PDV gene, known to induce inactivation of host hemocytes (Gitau et al., 2007). CrV1 variants between virulent and avirulent parasitoids strains are also submitted to positive Darwinian selection (Dupas et al., 2008), which suggests that diversity is selected in this PDV gene, maybe in relation with changes in the host range.

The ability of parasitoids to parasitize a new host species successfully can rely either on the *de novo* production or overproduction of molecules that complete the repertoire of virulence factors already present or to subtle changes in the present virulence molecules. Such changes could allow the virulence factors to "match" with their targets in new host species, which might also present some subtle differences. These mechanism could explain the diversity of some gene families in PDVs as well as the positive selection pressures observed on some genes of these families, like CrV1 (Dupas et al., 2008; Espagne et al., 2004; Serbielle et al., 2008). Of course, the targets of these parasitoid virulence factors have to be identified, as well as their own variation in order to determine whether virulence factors can diversify as a result of their coevolution with host targets.

Altogether, the opening of the "black box" containing the mechanisms underlying the variations of outcomes in host–parasitoid interactions results in the opening of an exciting area of research. It gives insights about the role of virulence factors contained in venoms or other secretions in the evolution of parasitoid host ranges, and raises also questions about the molecular basis of the specificity of these virulence factors. Hopefully, future studies on these challenging questions will include more parasitoid models and will provide interesting data about the overall evolution and diversification of parasitoids.

ACKNOWLEDGMENTS

We acknowledge Christine Coustau and Steve Sait for their comments of the manuscript.

REFERENCES

Asgari, S., Zhang, G., Zareie, R., Schmidt, O., 2003. A serine proteinase homolog venom protein from an endoparasitoid wasp inhibits melanization of the host hemolymph. Insect Biochem. Mol. Biol. 33, 1017–1024.

Baker, D.A., Mille-Baker, B., Wainwright, S.M., Ish-Horowicz, D., Dibb, N.J., 2001. Mae mediates MAP kinase phosphorylation of Ets transcription factors in *Drosophila*. Nature 411, 330–334.

Barratt, B.I.P., Evans, A.A., Stoltz, D.B., Vinson, S.B., Easingwood, R., 1999. Virus-like particles in the ovaries of *Microctonus aethiopoides* Loan (Hymenoptera: Braconidae), a parasitoid of adult weevils (Coleoptera: Curculionidae). J. Invert. Pathol. 73, 182–188.

Beck, M.H., Strand, M.R., 2007. A novel polydnavirus protein inhibits the insect prophenoloxidase activation pathway. Proc. Natl. Acad. Sci. USA 104, 19267–19272.

Benassi, V., Frey, F., Carton, Y., 1998. A new specific gene for wasp cellular immune resistance in *Drosophila*. Heredity 80, 347–352.

Bensadia, F., Boudreault, S., Guay, J.-F., Michaud, D., Cloutier, C., 2006. Aphid clonal resistance to a parasitoid fails under heat stress. J. Insect Physiol. 52, 146–157.

Bezier, A., Annaheim, M., Herbiniere, J., Wetterwald, C., Gyapay, G., Bernard-Samain, S., et al., 2009. Polydnaviruses of braconid wasps derive from an ancestral nudivirus. Science 323, 926–930.

Blumberg, D., 1997. Parasitoid encapsulation as a defense mechanism in the Coccoidea (Homoptera) and its importance in biological control. Biol. Control. 8, 225–236.

Blumberg, D., Van Driesche, R.G., 2001. Encapsulation rates of three encyrtid parasitoids by three mealybug species (Homoptera: Pseudococcidae) found commonly as pests in commercial greenhouses. Biol. Control. 22, 191–199.

Brodeur, J., Vet, L.E.M., 1995. Relationships between parasitoid host range and host defence: a comparative study of egg encapsulation in two related parasitoid species. Physiol. Entomol. 20, 7–12.

Calatayud, P.A., Polania, M.A., Seligmann, C.D., Bellotti, A.C., 2002. Influence of water-stressed cassava on *Phenacoccus herreni* and three associated parasitoids. Entomol. Exp. Appl. 102, 163–175.

Carius, H.J., Little, T.J., Ebert, D., 2001. Genetic variation in a host–parasite association: potential for coevolution and frequency-dependant selection. Evolution 55, 1136–1145.
Carton, Y., Boulétreau, M., 1985. Encapsulation ability of *Drosophila melanogaster*: a genetic analysis. Dev. Comp. Immunol. 9, 211–219.
Carton, Y., Kitano, H., 1981. Evolutionary relationships to parasitism by seven species of the *Drosophila melanogaster* subgroup. Biol. J. Linn. Soc. 16, 227–241.
Carton, Y., Capy, P., Nappi, A.J., 1989. Genetic variability of host–parasite relationship traits: utilisation of isofemale lines on a *Drosophila* parasitic wasp. Genet. Select. Evol. 21, 437–446.
Carton, Y., Frey, F., Nappi, A.J., 1992. Genetic determinism of the cellular immune reaction in *Drosophila melanogaster*. Heredity 69, 393–399.
Carton, Y., Nappi, A., Poirié, M., 2005. Genetics of anti-parasite resistance in invertebrates. Dev. Comp. Immunol. 29, 9–32.
Carton, Y., Poirié, M., Nappi, A.J., 2008. Insect immune resistance to parasitoids. Insect Sci. 15, 67–87.
Cerenius, L., Lee, B.L., Soderhall, K., 2008. The proPO-system: pros and cons for its role in invertebrate immunity. Trends Immunol. 29, 263–271.
Chiu, H., Morales, J., Govind, S., 2006. Identification and immuno-electron microscopy localization of p40, a protein component of immunosuppressive virus-like particles from *Leptopilina heterotoma*, a virulent parasitoid wasp of *Drosophila*. J. Gen. Virol. 87, 461–470.
Chiu, H., Sorrentino, R.P., Govind, S., 2000. Suppression of the *Drosophila* cellular immune response by *Ganaspis xanthopoda*. Adv. Exp. Med. Biol. 484, 161–167.
Colinet, D., Dubuffet, A., Cazes, D., Moreau, S., Drezen, J.M., Poirié, M., 2009. A serpin from the parasitoid wasp *Leptopilina boulardi* targets the *Drosophila* phenoloxidase cascade. Dev. Comp. Immunol. 33, 681–689.
Colinet, D., Schmitz, A., Depoix, D., Crochard, D., Poirié, M., 2007. Convergent use of RhoGAP toxins by eukaryotic parasites and bacterial pathogens. PLoS Pathog. 3, e203.
David, J.R., Gibert, P., Legout, H., Petavy, G., Capy, P., Moreteau, B., 2005. Isofemale lines in *Drosophila*: an empirical approach to quantitative trait analysis in natural populations. Heredity 94, 3–12.
De Gregorio, E., Han, S.-J., Lee, W.-J., Baek, M.-J., Osaki, T., Kawabata, S.-I., et al., 2002. An immune-responsive serpin regulates the melanization cascade in *Drosophila*. Dev. Cell. 3, 581–592.
Delpuech, J.M., Frey, F., Carton, Y., 1996. Action of insecticides on the cellular immune reaction of *Drosophila melanogaster* against the parasitoid *Leptopilina boulardi*. Environ. Toxicol. Chem. 15, 2267–2271.
Dubuffet, A., Doury, G., Labrousse, C., Drezen, J.M., Carton, Y., Poirié, M., 2008. Variation of success of *Leptopilina boulardi* in *Drosophila yakuba*: the mechanisms explored. Dev. Comp. Immunol. 32, 597–602.
Dubuffet, A., Dupas, S., Frey, F., Drezen, J.-M., Poirié, M., Carton, Y., 2007. Genetic interactions between the parasitoid wasp *Leptopilina boulardi* and its *Drosophila* hosts. Heredity 98, 21–27.
Dupas, S., Boscaro, M., 1999. Geographic variation and evolution of immunosuppressive genes in a *Drosophila* parasitoid. Ecography 22, 284–291.
Dupas, S., Carton, Y., 1999. Two non-linked genes for specific virulence of *Leptopilina boulardi* against *Drosophila melanogaster* and *D. yakuba*. Evol. Ecol. 13, 211–220.
Dupas, S., Brehélin, M., Frey, F., Carton, Y., 1996. Immune suppressive virus-like particles in a *Drosophila* parasitoid: significance of their intraspecific morphological variations. Parasitology 113, 207–212.
Dupas, S., Carton, Y., Poirié, M., 2003. Genetic dimension of the coevolution of virulence-resistance in *Drosophila*-parasitoid wasp relationships. Heredity 90, 83–89.

Dupas, S., Frey, F., Carton, Y., 1998. A single parasitoid segregating factor controls immune suppression in *Drosophila*. J. Heredity 89, 306–311.

Dupas, S., Gitau, C.W., Branca, A., Le Ru, B.P., Silvain, J.F., 2008. Evolution of a polydnavirus gene in relation to parasitoid-host species immune resistance. J. Heredity 99, 491–499.

Eslin, P., Doury, G., 2006. The fly *Drosophila subobscura*: a natural case of innate immunity deficiency. Dev. Comp. Immunol. 30, 977–983.

Eslin, P., Prévost, G., 1998. Hemocyte load and immune resistance to *Asobara tabida* are correlated in species of the *Drosophila melanogaster* subgroup. J. Insect Physiol. 44, 807–816.

Espagne, E., Dupuy, C., Huguet, E., Cattolico, L., Provost, B., Martins, N., et al., 2004. Genome sequence of a polydnavirus: insights into symbiotic virus evolution. Science 306, 286–289.

Fellowes, M.D., Kraaijeveld, A.R., Godfray, H.C., 1998. Trade-off associated with selection for increased ability to resist parasitoid attack in *Drosophila melanogaster*. Proc. R. Soc. Lond. B 265, 1553–1558.

Fellowes, M.D.E., Kraaijeveld, A.R., Godfray, H.C.J., 1999. Cross-resistance following artificial selection for increased defense against parasitoids in *Drosophila melanogaster*. Evolution 53, 966–972.

Fleury, F., Ris, N., Allemand, R., Fouillet, P., Carton, Y., Boulétreau, M., 2004. Ecological and genetic interactions in *Drosophila*-parasitoids communities: a case study with *D. melanogaster*, *D. simulans* and their common *Leptopilina* parasitoids in south-eastern France. Genetica 120, 181–194.

Fytrou, A., Schofield, P.G., Kraaijeveld, A.R., Hubbard, S.F., 2006. *Wolbachia* infection suppresses both host defence and parasitoid counter-defence. Proc. R. Soc. Lond. B 273, 791–796.

Gitau, C.W., Dupas, S., Ngi-Song, A.J., Mbugi, J.P., Schulthess, F., 2006. Calyx fluid proteins of two *Cotesia sesamiae* (Cameron) (Hymenoptera: Braconidae) biotypes in Kenya: implications to biological control of the stem borer *Busseola fusca* (Fuller) (Lepidoptera: Noctuidae). Ann. Soc. Entomol. France 42, 433–441.

Gitau, C.W., Gundersen-Rindal, D., Pedroni, M., Mbugi, J.P., Dupas, S., 2007. Differential expression of the CrV1 haemocyte inactivation-associated polydnavirus gene in the African maize stem borer *Busseola fusca* (Fuller) parasitized by two biotypes of the endoparasitoid *Cotesia sesamiae* (Cameron). J. Insect Physiol. 53, 676–684.

Glatz, R.V., Asgari, S., Schmidt, O., 2004. Evolution of polydnaviruses as insect immune suppressors. Trends Microbiol. 12, 545–554.

Govind, S., 2008. Innate immunity in *Drosophila*: pathogens and pathways. Insect Sci. 15, 29–43.

Haine, E.R., 2008. Symbiont-mediated protection. Proc. R. Soc. Lond. B 275, 353–361.

Henter, H.J., 1995. The potential for coevolution in a host–parasitoid system. II. Genetic variation within a population of wasps in the ability to parasitize an aphid population. Evolution 49, 439–445.

Hita, M., Espagne, E., Huguet, E., Pascual, L., Poirié, M., 2006. Mapping candidate genes for *Drosophila melanogaster* resistance to the parasitoid wasp *Leptopilina boulardi*. Genet. Res. 88, 81–91.

Hita, M.T., Poirié, M., Leblanc, N., Lemeunier, F., Lutcher, F., Frey, F., et al., 1999. Genetic localization of a *Drosophila melanogaster* resistance gene to a parasitoid wasp and physical mapping of the region. Genome Res. 9, 471–481.

Irving, P., Ubeda, J.-M., Doucet, D., Troxler, L., Lagueux, M., Zachary, D., et al., 2005. New insights into *Drosophila* larval haemocyte functions through genome-wide analysis. Cell. Microbiol. 7, 335–350.

Jiang, H., Wang, Y., Yu, X.-Q., Zhu, Y., Kanost, M.R., 2003. Prophenoloxidase-activating proteinase-3 (PAP-3) from *Manduca sexta* hemolymph: a clip-domain serine proteinase regulated by serpin-1J and serine proteinase homologs. Insect Biochem. Mol. Biol. 33, 1049–1060.

Kanost, M.R., Jiang, H., Yu, X.-Q., 2004. Innate immune responses of a lepidopteran insect: *Manduca sexta*. Immunol. Rev. 198, 97–105.

Karimzadeh, J., Wright, D.J., 2008. Bottom-up cascading effects in a tritrophic system: interactions between plant quality and host–parasitoid immune responses. Ecol. Entomol. 33, 45–52.

Kraaijeveld, A.R., 1999. Kleptoparasitism as an exploitation for paradoxical oviposition decisions of the parasitoid *Asobara tabida*. J. Evol. Biol. 12, 123–133.

Kraaijeveld, A.R., Godfray, H.C.J., 1997. Trade-off between parasitoid resistance and larval competitive ability in *Drosophila melanogaster*. Nature 389, 278–280.

Kraaijeveld, A.R., Godfray, H.C.J., 1999. Geographic patterns in the evolution of resistance and virulence in *Drosophila* and its parasitoids. Am. Nat. 153, S61–S74.

Kraaijeveld, A.R., Godfray, H.C.J., 2001. Is there local adaptation in *Drosophila*–parasitoid interactions? Evol. Ecol. 3, 107–116.

Kraaijeveld, A.R., van Alphen, J.J.M., 1994. Geographical variation in resistance of the parasitoid *A. tabida* against encapsulation by *D. melanogaster* larvae: the mechanism explored. Physiol. Entomol. 19, 9–14.

Kraaijeveld, A.R., van Alphen, J.J.M., 1995. Geographical variation in encapsulation ability of *Drosophila melanogaster* larvae and evidence for parasitoid-specific components. Evol. Ecol. 9, 10–17.

Kraaijeveld, A.R., Limentani, E.C., Godfray, H.C.J., 2001. Basis of the trade-off between parasitoid resistance and larval competitive ability in *Drosophila melanogaster*. Proc. R. Soc. Lond. B 268, 259–261.

Kraaijeveld, A.R., van Alphen, J.J.M., Godfray, H.C.J., 1998. The coevolution of host resistance and parasitoid virulence. Parasitology 116, S29–S45.

Labrosse, C., Carton, Y., Dubuffet, A., Drezen, J.M., Poirié, M., 2003. Active suppression of *D. melanogaster* immune response by long gland products of the parasitic wasp *Leptopilina boulardi*. J. Insect Physiol. 49, 513–522.

Labrosse, C., Eslin, P., Doury, G., Drezen, J.-M., Poirié, M., 2005a. Haemocyte changes in *D. melanogaster* in response to long gland components of the parasitoid wasp *Leptopilina boulardi*: a Rho-GAP protein as an important factor. J. Insect Physiol. 51, 161–170.

Labrosse, C., Stasiak, K., Lesobre, J., Grangeia, A., Huguet, E., Drezen, J.-M., et al., 2005b. A RhoGAP protein as a main immune suppressive factor in the *Leptopilina boulardi* (Hymenoptera, Figitidae)–*Drosophila melanogaster* interaction. Insect Biochem. Mol. Biol. 35, 93–103.

Lambrechts, L., Halbert, J., Durand, P., Gouagna, L.C., Koella, J.C., 2005. Host genotype by parasite genotype interactions underlying the resistance of anopheline mosquitoes to *Plasmodium falciparum*. Malaria J. 4(3), 16–23.

Lanot, R., Zachary, D., Holder, F., Meister, M., 2001. Postembryonic hematopoiesis in *Drosophila*. Dev. Biol. 230, 243–257.

Law, R.H.P., Zhang, Q., McGowan, S., Buckle, A.M., Silverman, G.A., Wong, W., et al., 2006. An overview of the serpin superfamily. Genome Biol. 7(5), 216.

Lazzaro, B.P., Little, T.J., 2009. Immunity in a variable world. Philosophical Trans. R. Soc. B 364, 15–26.

Ling, E., Yu, X.-Q., 2005. Prophenoloxidase binds to the surface of hemocytes and is involved in hemocyte melanization in *Manduca sexta*. Insect Biochem. Mol. Biol. 35, 1356–1366.

Markus, R., Laurinyecz, B., Kurucz, E., Honti, V., Bajusz, I., Sipos, B., et al., 2009. Sessile hemocytes as a hematopoietic compartment in *Drosophila melanogaster*. Proc. Natl. Acad. Sci. USA 106(12), 4805–4809.

Mochiah, M.B., Ngi-Song, A.J., Overholt, W.A., Stouthamer, R., 2002. Effects of calyx fluid from two populations of *Cotesia sesamiae* on *Busseola fusca* (Lepidoptera: Noctuidae). Insect Sci. Appl. 21, 369–374.

Mohamed, S.A., Overholt, W.A., Wharton, R.A., Lux, S.A., Eltoum, E.M., 2003. Host specificity of *Psyttalia cosyrae* (Hymenoptera: Braconidae) and the effect of different host species on parasitoid fitness. Biol. Cont. 28, 155–163.
Morales, J., Chiu, H., Oo, T., Plaza, R., Hoskins, S., Govind, S., 2005. Biogenesis, structure, and immune-suppressive effects of virus-like particles of a *Drosophila* parasitoid, *Leptopilina victoriae*. J. Insect Physiol. 51, 181–195.
Moreau, S.J.M., Guillot, S., 2005. Advances and prospects on biosynthesis, structures and functions of venom proteins from parasitic wasps. Insect Biochem. Mol. Biol. 35, 1209–1223.
Nam, H.J., Jang, I.H., Asano, T., Lee, W.J., 2008. Involvement of pro-phenoloxidase 3 in lamellocyte-mediated spontaneous melanization in *Drosophila*. Mol. Cells. 26(6), 606–610.
Nappi, A.J., 1970. Defense reactions of *Drosophila euronotus* larvae against the hymenopterous parasite *Pseudeucoila bochei*. J. Invert. Pathol. 16, 408–418.
Nappi, A.J., Frey, F., Carton, Y., 2005. *Drosophila* serpin 27A is a likely target for immune suppression of the blood cell-mediated melanotic encapsulation response. J. Insect Physiol. 51, 197–205.
Ngi-Song, A.J., Overholt, W.A., Stouthamer, R., 1998. Suitability of *Busseola fusca* and *Sesamia calamistis* (Lepidoptera: Noctuidae) for the development of two populations of *Cotesia sesamiae* (Hymenoptera: Braconidae) in Kenya. Biol. Cont. 12, 208–214.
Ojala, K., Julkunen-Tiitto, R., Lindstrom, L., Mappes, J., 2005. Diet affects the immune defense and life-history traits of an Artiid moth *Parasemia plantaginis*. Evol. Ecol. Res. 7, 1153–1170.
Oliver, K.M., Russell, J.A., Moran, N.A., Hunter, M.S., 2003. Facultative bacterial symbionts in aphids confer resistance to parasitic wasps. Proc. Natl. Acad. Sci. USA 100, 1803–1807.
Orr, H.A., Irving, S., 1997. The genetics of adaptation – the genetic basis of resistance to wasp parasitism in *Drosophila melanogaster*. Evolution 51, 1877–1885.
Pennacchio, F., Strand, M.R., 2006. Evolution of developmental strategies in parasitic hymenoptera. Ann. Rev. Entomol. 51, 233–258.
Poirié, M., Carton, Y., Dubuffet, A., 2009. Virulence strategies in parasitoid Hymenoptera as an example of adaptive diversity. C. R. Biol. 332, 311–320.
Poirié, M., Frey, F., Hita, M., Huguet, E., Lemeunier, F., Periquet, G., Carton, Y., 2000. *Drosophila* resistance genes to parasitoids: chromosomal location and linkage analysis. Proc. R. Soc. Lond. B 267, 1417–1421.
Prévost, G., Eslin, P., Doury, G., Moreau, S.J.M., Guillot, S., 2005. *Asobara*, braconid parasitoids of *Drosophila* larvae: unusual strategies to avoid encapsulation without VLPs. J. Insect Physiol. 51, 171–179.
Reineke, A., Asgari, S., Schmidt, O., 2006. Evolutionary Origin of *Venturia canescens* virus-like particles. Arch. Insect Biochem. Physiol. 61, 123–133.
Reineke, A., Asgari, S., Ma, G., Beck, M., Schmidt, O., 2002. Sequence analysis and expression of a viral-like particle protein, VLP2, from the parasitoid wasp *Venturia canescens*. Insect Mol. Biol. 11, 233–239.
Renault, S., Stasiak, K., Federici, B., Bigot, Y., 2005. Commensal and mutualistic relationships of reoviruses with their parasitoid wasp hosts. J. Insect Physiol. 51, 137–148.
Rizki, R.M., Rizki, T.M., 1984. Selective destruction of a host blood cell type by a parasitoid wasp. Proc. Natl. Acad. Sci. USA 81, 6154–6158.
Rizki, R.M., Rizki, T.M., 1990. Parasitoid virus-like particles destroy *Drosophila* cellular immunity. Proc. Natl. Acad. Sci. USA 87, 8388–8392.
Rizki, R.M., Rizki, T.M., 1991. Effects of lamellolysin from a parasitoid wasp on *Drosophila* blood cells *in vitro*. J. Exp. Zool. 257, 236–244.
Rizki, T., Rizki, R., Grell, E., 1980. A mutant affecting the crystal cells in *Drosophila melanogaster*. Wilhelm Roux's Arch. 188, 91–99.
Rizki, T.M., Rizki, R.M., 1985. Paracristalline inclusions of *Drosophila melanogaster* hemocytes have prophenoloxidases. Genetics 110, 598.

Rizki, T.M., Rizki, R.M., 1992. Lamellocyte differentiation in *Drosophila* larvae parasitized by *Leptopilina*. Dev. Comp. Immunol. 16, 103–110.

Rizki, T.M., Rizki, R.M., 1994. Parasitoid-induced cellular immune deficiency in *Drosophila*. Ann. N. Y. Acad. Sci. 712, 178–194.

Rizki, T.M., Rizki, R.M., Carton, Y., 1990. *Leptopilina heterotoma* and *L. boulardi*: strategies to avoid cellular defense responses of *Drosophila melanogaster*. Exp. Parasitol. 70, 466–475.

Rogge, R., Green, P.J., Urano, J., Horn-Saban, S., Mlodzik, M., Shilo, B.Z., Hartenstein, V., Banerjee, U., 1995. The role of yan in mediating the choice between cell division and differentiation. Development 121, 3947–3958.

Russo, J., Brehélin, M., Carton, Y., 2001. Haemocyte changes in resistant and susceptible strains of *D. melanogaster* caused by virulent and avirulent strains of the parasitic wasp *Leptopilina boulardi*. J. Insect Physiol. 47, 167–172.

Sagarra, L.A., Peterkin, D.D., Vincent, C., Stewart, R.K., 2000. Immune response of the hibiscus mealybug, *Maconellicoccus hirsutus* Green (Homoptera: Pseudococcidae), to oviposition of the parasitoid *Anagyrus kamali* Moursi (Hymenoptera: Encyrtidae). J. Insect Physiol. 46, 647–653.

Satoh, D., Horii, A., Ochiai, M., Ashida, M., 1999. Prophenoloxidase-activating enzyme of the silkworm, *Bombyx mori*. Purification, characterization, and cDNA cloning. J. Biol. Chem. 274, 7441–7453.

Schlenke, T.A., Morales, J., Govind, S., Clark, A.G., 2007. Contrasting infection strategies in generalist and specialist wasp parasitoids of *Drosophila melanogaster*. PLoS Pathog. 3, e158.

Schmidt, O., Theopold, U., Strand, M., 2001. Innate immunity and its evasion and suppression by hymenopteran endoparasitoids. Bioessays 23, 344–351.

Serbielle, C., Chowdhury, S., Pichon, S., Dupas, S., Lesobre, J., Purisima, E.O., et al., 2008. Viral cystatin evolution and three-dimensional structure modelling: a case of directional selection acting on a viral protein involved in a host–parasitoid interaction. BMC Biol. 6, 38.

Sorrentino, R.P., Carton, Y., Govind, S., 2002. Cellular immune response to parasite infection in the *Drosophila* lymph gland is developmentally regulated. Dev. Biol. 243, 65–80.

Suzuki, M., Tanaka, T., 2006. Virus-like particles in venom of *Meteorus pulchricornis* induce host hemocyte apoptosis. J. Insect Physiol. 52, 602–613.

Thompson, J.N., 2005. The geographic mosaic of coevolution. University of Chicago Press, Chicago, IL.

Vass, E., Nappi, A.J., Carton, Y., 1993. Comparative study of immune competence and host susceptibility in *Drosophila melanogaster* parasitized by *Leptopilina boulardi* and *Asobara tabida*. J. Parasitol. 79, 106–112.

Vorburger, C., Sandrock, C., Gouskov, A., Castañeda, L.E., Ferrari, J., 2009. Genotypic variation and the role of defensive endosymbionts in all-parthenogenetic host–parasitoid interaction. Evolution 63(6), 1439–1450.

Walker, I., 1959. Die Abwehrreaktion des Wirtes *Drosophila melanogaster* gegen die zoophage Cynipidae *Pseudocoila bochei* Weld. Rev. Suisse Zool. 66, 569–632.

Wertheim, B., Kraaijeveld, A.R., Schuster, E., Blanc, E., Hopkins, M., Pletcher, S.D., Strand, M.R., Partridge, L., Godfray, H.C.J., 2005. Genome-wide gene expression in response to parasitoid attack in *Drosophila*. Genome Biol. 6, R94.

Wilfert, L., Schmid-Hempel, P., 2008. The genetic architecture of susceptibility to parasites. BMC Evol. Biol. 8, 187.

Williams, M.J., Ando, I., Hultmark, D., 2005. *Drosophila melanogaster* Rac2 is necessary for a proper cellular immune response. Genes Cells 10, 813–823.

Williams, M.J., Wiklund, M.-L., Wikman, S., Hultmark, D., 2006. Rac1 signalling in the *Drosophila* larval cellular immune response. J. Cell Sci. 119, 2015–2025.

Zettervall, C.-J., Anderl, I., Williams, M.J., Palmer, R., Kurucz, E., Andó, I., 2004. A directed screen for genes involved in *Drosophila* blood cell activation. Proc. Natl. Acad. Sci. USA 101, 14192–14197.

CHAPTER 7

Immune Resistance of *Drosophila* Hosts Against *Asobara* Parasitoids: Cellular Aspects

Patrice Eslin, Geneviève Prévost, Sébastien Havard, and **Géraldine Doury**

Contents

Laboratoire de Biologie des Entomophages, EA 3900 BioPI, Université de Picardie-Jules Verne, 33 rue Saint Leu, 80039 Amiens cedex, France

Advances in Parasitology, Volume 70
ISSN 0065-308X, DOI: 10.1016/S0065-308X(09)70007-7

Abstract The immunity of *Drosophila* relies on a variety of defenses cooperating to fight parasites and pathogens. The encapsulation reaction is the main hemocytic response neutralizing large parasites like endophagous parasitoids. The diversity of the mechanisms of immunoevasion evolved by *Asobara* parasitoids, together with the wide spectrum of *Drosophila* host species they can parasitize, make them ideal models to study and unravel the physiological and cellular aspects of host immunity. This chapter summarizes what could be learnt on the cellular features of the encapsulation process in various *Drosophila* spp., and also on the major role played by *Drosophila* hosts hemocytes subpopulations, both in a quantitative and qualitative manner, regarding the issue of the immune *Asobara*–*Drosophila* interactions.

7.1. INTRODUCTION

Like all living organisms, *Drosophila melanogaster* has evolved a powerful and sophisticated network of molecules and cells that cooperate to fight pathogens and parasites, including parasitoids. This complex of cooperating entities forms what is called the innate immune defense system, conferring broad protection against a variety of aggressors. Composed of both humoral and cellular components, the innate immunity represents the only line of defense for fruit flies. Due to the strength of genetic and molecular approaches, the very species *D. melanogaster* has become a major model organism to study the humoral canvas of the innate immune response. It has also recently emerged as a powerful model organism to explore cellular and evolutionary aspects of immune defenses.

Cynipids from the *Leptopilina* genus and Braconids from the *Asobara* genus have been the most studied larval endoparasitoids of *Drosophila*. Over the last 20 years, extensive genetic and physiological knowledge has been accumulated on the interactions between hosts of the *melanogaster* subgroup and their larval parasitoids. Molecular data regarding immune processes and resistance in *D. melanogaster* are now available (Colinet et al., 2007; Dubuffet et al., 2007; Labrosse et al., 2005a,b; see also Chapter 6 by Dubuffet et al.) and the virulence strategies of these parasitoids are beginning to be well characterized (Prévost et al., 2005; see also Chapter 8 by Moreau et al. and Chapter 9 by Prévost et al.). The genetics of host resistance has also been undertaken, and *Drosophila* physiological defenses against endoparasitoids have been shown to be under genetic control, with some strains being unable to encapsulate the parasitoid eggs (Carton and Nappi, 2001; Dubuffet et al., 2007; Dupas et al., 2003; Hita et al., 1999). In particular, the *D. melanogaster* gene *Rat* conferring resistance to the Braconid *Asobara tabida* has been localized (Poirié et al., 2000) and the autosomal inheritance of the encapsulation capacity has been analyzed (Benassi et al., 1998).

In the *Asobara* genus, recent work has highlighted the existence of an unexpected diversity of mechanisms to avoid *Drosophila* hosts immune defenses. Interestingly, these mechanisms do not seem to be associated with the presence of symbiotic agents such as polydnavirus or virus-like particles unlike what is known in many other Braconids (Prévost et al., 2005). The singularity of the mechanisms developed by *Asobara* parasitoids to avoid host defenses makes them particularly interesting models for a physiological approach of the immune interactions in host–parasitoid systems.

7.2. THE IMMUNE SYSTEM IN *D. MELANOGASTER*

Despite their lack of vertebrate-like immunoglobulin and histoincompatibility system, fruit flies are able to discriminate between self and nonself and to elicit an efficient response against ''aggressors.'' Hemolymph is the circulatory fluid filling the interior hemocoel of the body cavity. It is the site of humoral reactions and contains both circulating and sessile hemocytes. Regulation of antimicrobial peptide production during humoral responses is well characterized in *D. melanogaster* larvae. Conversely, the recognition, signaling and development of hemocytes in the different cellular compartments of the *Drosophila* immune system are only partially understood. Therefore, study of the molecular mechanisms underlying cellular immune responses and the origin and differentiation of hemocytes have been subjects of growing interest over the last decade.

7.2.1. Humoral and cellular responses

In *D. melanogaster*, humoral responses mainly involve the synthesis of various antimicrobial peptides and the rapid induction of enzymatic cascades. Infection is followed by the massive expression of novel peptides. Antimicrobial peptides undergo rapid synthesis in the fat body and are then secreted into the hemolymph where they target rather specific types of microintruders. Defensin is produced in response to infection by Gram-positive bacteria whereas attacin, cecropin, diptericin and drosocin all target Gram-negative bacteria, and drosomycin or metchikowin are directed against fungi (Hultmark, 2003). The genes encoding antimicrobial peptide genes are regulated by a balance between two signaling pathways. Gram-positive bacteria and fungi activate the Toll pathway while the Imd pathway is mainly triggered by Gram-negative bacteria (Lemaitre and Hoffmann, 2007).

Proteolytic cascades enable local reactions to occur such as hemolymph clotting, opsonization of intruders and melanization. Melanization reactions can be rapidly activated at the site of injury, subsequently to a wound or the intrusion of a large parasite in the hemocoel. The deposition of melanin on damaged tissues and around parasites contributes to wound clotting and to neutralization of endoparasitoid eggs, together with the production of toxic intermediates including reactive oxygen species (ROS; Nappi and Christensen, 2005; Nappi and Vass, 2001).

In *Drosophila*, cellular immune reactions rely on the activity of blood immunocompetent cells, the so-called hemocytes that are freely circulating through the body cavities, or are sessile, associated with various tissues and organs (Lavine and Strand, 2002). Cellular responses include two very distinct processes: endocytosis and encapsulation. Individual immunocytes typically phagocytose small pathogens due to the formation of pseudopodes that bind the foreign targets and allow their capture through engulfment in a phagosome. Encapsulation of aggressors such as metazoan parasites is a prime cellular defense mechanism occurring in most insects. Indeed when the intruders are too large, their elimination cannot rely on endocytosis processes. Instead, manifold specialized immunocytes accumulate and form a melanized capsule of overlapping layers of hemocytes around the foreign body.

7.2.2. Hemolymph cellular components in *Drosophila* larvae

Drosophila larvae contain several thousand hemocytes, which can be divided into three main cell types on the basis of their structural and functional features: plasmatocytes, lamellocytes and crystal cells (Brehélin, 1982; Lanot et al., 2001; Ribeiro and Brehélin, 2006; Fig. 7.1).

Hematopoiesis in *Drosophila* occurs in two phases during development. In the embryo, a first population of hemocyte precursors is specified at the blastoderm stage from the head mesoderm (Bataille et al., 2005; Lebestky et al., 2000). Toward the end of embryogenesis, the precursors of a specialized hematopoietic tissue, the lymph gland, form in the lateral mesoderm from where they migrate dorsally. A second population of hemocytes can then be released from the hematopoietic organ in the hemolymph of the larva (Lanot et al., 2001; Rizki and Rizki, 1984; Sorrentino et al., 2002). The lymph gland is composed of four to six lobes located along the dorsal vessel and representing the major source of hemocytes in *Drosophila* larvae. The posterior lobes contain mainly undifferentiated precursor cells, called prohemocytes, whereas the anterior lobes comprise large amounts of differentiated hemocytes

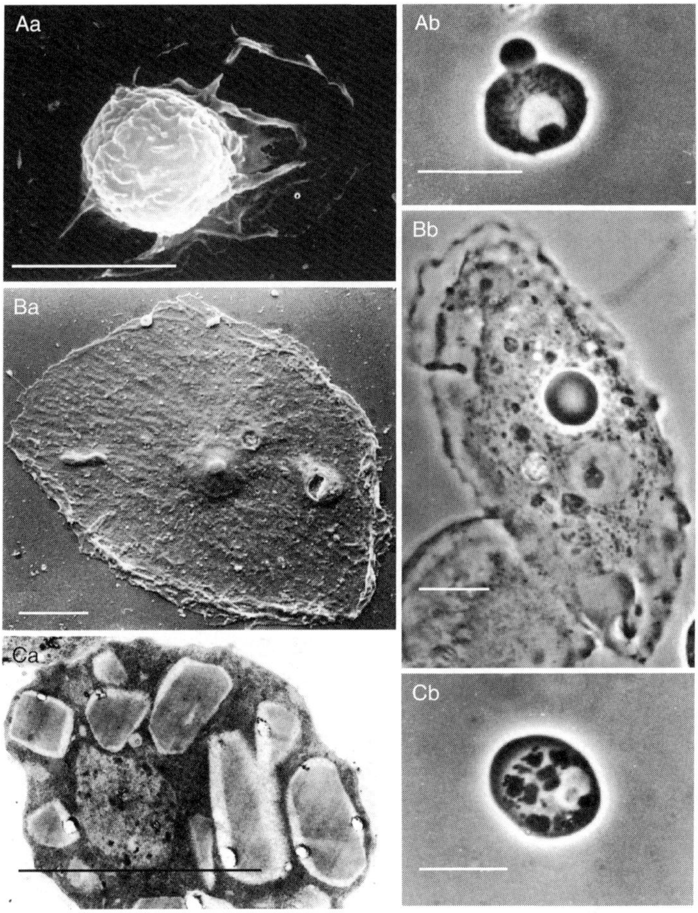

FIGURE 7.1 Scanning (Aa and Ba) or transmission electron microscopy (Ca) and phase contrast microscopy (Ab, Bb and Cb) of circulating hemocytes in third-instar larvae of *D. melanogaster*. (A) *Plasmatocyte*; (B) *Lamellocyte*; (C) *Crystal Cell*; Bar: 10 μm.

(Meister, 2004). Both embryonic and larval hemocyte precursors give rise to plasmatocytes and crystal cells (Evans et al., 2003; Meister and Lagueux, 2003). In larvae, the hematopoietic organ also contains precursors of a third type of immunocytes, the lamellocytes. Plasmatocytes represent the main and most abundant cell type circulating in the hemolymph of larvae, but crystal cells and lamellocytes are also present (Lavine and Strand, 2002). Lamellocytes differentiate in large amounts only in response to a specific immune challenge, for example, parasitization by Hymenopteran endoparasitoids (Lanot et al., 2001; Meister, 2004).

The understanding of *Drosophila* hematopoiesis derives mainly from the genetic and molecular analyses of mutant phenotypes in embryos. The transcriptional regulation of larval hematopoiesis has been studied by several research teams during the last decade. However, the hematopoietic origin of lamellocytes has been controversial. They were proposed to develop directly from plasmatocytes, although this remained purely hypothetical (Rizki and Rizki, 1980b). The induction of lamellocyte differentiation and the formation of melanized capsules in *D. melanogaster* involve a signaling cascade. In this process, lamellocytes have been found to derive from precursor cells within the anterior lobes of the lymph glands, from where they may be released (Crozatier et al., 2004; Lanot et al., 2001). Present hypothesis is that the detection and/or recognition of the parasitoid would lead plasmatocytes to send a signal to the posterior signaling center (PSC), a region in the anterior lobes of the hematopoietic organ which, in return, would induce the neighbor cells to differentiate as lamellocytes. Recently, evidence has been given that sessile subepidermal hemocytes are a major source of larval hemocytes and represent the main site producing lamellocytes precursors (Markus et al., 2009).

However, the genetic and molecular control of the hematopoiesis during *Drosophila* larval development remains little understood compared to what is known on embryogenesis (Crozatier and Meister, 2007). Several genes are candidates for controlling the lamellocytes specification (Krzemien et al., 2007). What is well established is that lamellocyte differentiation is induced either by particles too large to be phagocytosed or by any disruption in basal membranes (Markus et al., 2005; personal observations). Signaling and induction mechanisms involved in this process remain so far unidentified.

7.2.3. Circulating immunocytes: Characteristics and functions

The roles and functions of hemocytes are controversial. In the larval stages of *D. melanogaster* referred to as the model species, three main types of hemocytes have been described (Brehélin, 1982; Lanot et al., 2001; Ribeiro and Brehélin, 2006). The plasmatocytes are small rounded cells (Fig. 7.1Aa and Ab) and represent the most abundant hemocyte type circulating in the hemolymph. They count for about 90% of the total hemocyte count (THC) in unparasitized *D. melanogaster* larvae (Brehélin 1982; Eslin and Prévost, 1996). They are involved in phagocytosis and share morphological and functional properties with the mammalian monocyte/macrophage lineage. Plasmatocytes are also involved in the initial and terminal steps of encapsulation (Russo et al., 1996). Crystal cells (Fig. 7.1Ca and Cb), the least abundant hemocyte type (2–5% of THC; Eslin, 1998), contain elements—substrate and enzymes—of the phenoloxidase cascade. They are involved in melanization of foreign bodies, in clotting and in the production of cytotoxic radicals (Carton et al., 2008; Evans et al., 2003; Meister, 2004; Meister and Lagueux, 2003; Nappi and

Streams, 1969; Nappi and Vass, 1993; Nappi et al., 1995; Rizki and Rizki, 1959; Russo et al., 1996). The third hemocyte type is the lamellocyte, consisting of large, round flattened cells (Fig. 7.1Ba and Bb) that represent at the most 6% of the THC in nonimmune-challenged *Drosophila* larvae (Eslin, 1998; Lanot et al., 2001; Russo et al., 1996). The single known function of lamellocytes is to participate in the encapsulation reaction against intruders that exceed the size limit for phagocytosis by plasmatocytes. Indeed, they present all the cellular features allowing them to be mutually adhesive and form multilayered sheaths around foreign targets. The proportion of adhesive cells involved in the encapsulation of large foreign bodies can reach up to 50% of the THC in larvae parasitized by a parasitoid wasp (Eslin and Prévost, 1996, 1998; Russo et al., 2001).

The three major hemocyte types play a role in the formation of capsules in *Drosophila* larvae. The edification of a capsule starts by the surrounding of the parasitoid egg by a layer of plasmatocytes, followed by the accumulation of several layers of lamellocytes, and finally melanization of the hence constructed capsule, which leads to neutralization of the egg (Russo et al., 1996).

7.3. ENCAPSULATION: A STORY BASED ON QUANTITIES

The hallmark of *Drosophila* larvae defense against endoparasitoids developing in their body cavity is the construction of melanized multicellular layers around the parasitic eggs leading to what is called their "encapsulation" (Salt, 1963). In *Drosophila*, the process of encapsulation is the main physiological defense against endoparasitoids described to date. The parasite is enclosed within the capsule, which undergoes a progressive blackening due to melanization (Carton and Nappi, 1997). Phenoloxidase is a key enzyme leading to the production of melanin. Present as an inactive zymogen (prophenoloxidase) in the hemolymph, its activation by a cascade of serine proteases catalyzes the oxidization of tyrosine and other phenols into reactive compounds that crosslink proteins and polymerize into melanin. Melanization ensures the efficient killing of endoparasitoids, possibly by asphyxiation during their precocious development (Fisher, 1963) and by the local production of cytotoxic intermediates or by-products of the melanization cascade, and cytotoxic radicals such as ROS (superoxide anions, hydrogen peroxide), nitric oxide, quinones or semiquinones (Nappi et al., 1995, 2000). The dichotomy often made between "cellular response" and "humoral response" poorly applies to the encapsulation process during which the cooperation between the different components of immunity should be emphasized (see Chapter 4 by Nappi et al.).

7.3.1. Importance of the hemocyte load

Salt (1963) suggested that the concentration of the hemocytic cells circulating in the hemolymph of host insects could be an important factor conditioning their ability to mount a cellular encapsulation response against macroparasites. Later, Götz (1986) reported that among several insect orders he tested, larvae carrying a greater hemocyte load were also more likely to form cellular capsules around foreign targets than those with fewer hemocytes. More recently, the ability to encapsulate synthetic beads was shown to be positively and independently correlated with hemocyte load in the house cricket (Ryder and Siva-Jothy, 2001).

In *Drosophila*, several studies proved that one main factor influencing the success of the immune response to endophagous parasitoids is the number and maturity of the hemocytes circulating at the time of parasitization. This factor appears to be of particular importance in parasitoid species that have evolved a "passive" strategy to avoid encapsulation (Eslin and Prévost, 1998, 2000).

A. tabida (Braconidae) is a larval parasitoid attacking various *Drosophila* spp. in Europe and North America. It is the only described species among *Drosophila* larval parasitoids avoiding encapsulation without exerting any depressive effect on the host's hemocyte population. Instead, *A. tabida* eggs get embedded within the host tissues in the hemocoel, therefore preventing the attack by the host's hemocytes (Eslin et al., 1996; Kraaijeveld and van Alphen, 1994; Monconduit and Prévost, 1994). Success of *A. tabida* development strongly relates on its geographical origin as well as on the *Drosophila* host species (Kraaijeveld and van Alphen, 1994; see Chapter 10 by Kraaijeveld and Godfray). Most of *A. tabida* geographical strains successfully develop in a large proportion of *D. melanogaster* larvae, while *D. simulans*, a sibling of *D. melanogaster*, usually encapsulates the parasite (Eslin and Prévost, 1998). Hemocyte counts performed on both host species demonstrated that *D. simulans* larvae carried three to five times more hemocytic cells in their hemolymph compared to *D. melanogaster* larvae raised in the same conditions (Fig. 7.2). The same striking differences in hemocyte counts were observed between the reactive *D. simulans* hosts building a capsule around the parasitic egg in 24 h and the unreactive hosts unable to encapsulate the parasitoid (Fig. 7.3).

Our interest in the cellular composition of hemolymph as a key factor in host–parasitoid interactions arose from these original studies. These experiments also provided information on the hemocyte loads in the young *Drosophila* larval stages, which are naturally subject to parasitization by larval endoparasitoids. Comparison of the hemocyte counts between newly hatched first-instar larvae "reactive" and "unreactive" to *A. tabida* eggs suggested that a minimum threshold of the concentration in immunocytes may constitute a prerequisite for the encapsulation

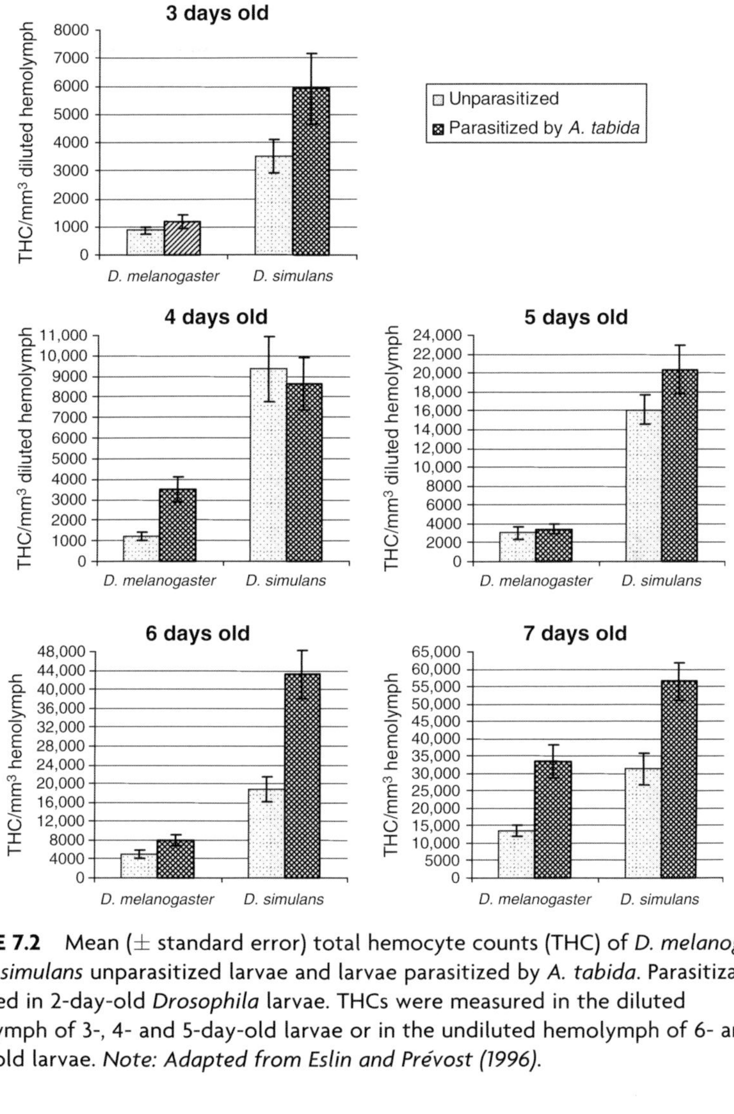

FIGURE 7.2 Mean (± standard error) total hemocyte counts (THC) of *D. melanogaster* and *D. simulans* unparasitized larvae and larvae parasitized by *A. tabida*. Parasitization occurred in 2-day-old *Drosophila* larvae. THCs were measured in the diluted hemolymph of 3-, 4- and 5-day-old larvae or in the undiluted hemolymph of 6- and 7-day-old larvae. *Note: Adapted from Eslin and Prévost (1996).*

reaction to be achieved and successfully arrest parasitoid development within less than 24 h after parasitization (Fig. 7.4). Since the putative threshold of the concentration in immunocytes is likely to vary with age and physiological development, it is conceivable that it could be reached sooner by a large proportion of the *D. simulans* hosts, while only a few *D. melanogaster* larvae would build, within the same time period, a hemocyte pool sufficient to counteract the development of the parasitoid by a proper cellular defense. A first hypothesis would be that *Drosophila* larvae

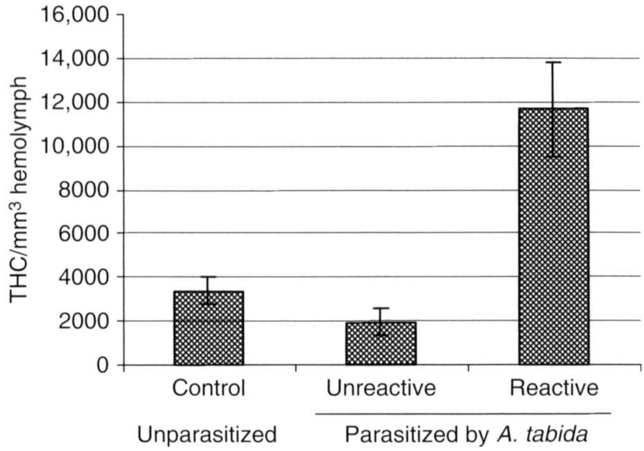

FIGURE 7.3 Mean (± standard error) total hemocyte counts (THC) of 3-day-old *D. simulans* unparasitized larvae and 3-day-old larvae parasitized by *A. tabida*. Parasitization occurred in 2-day-old *Drosophila* larvae. THCs were measured in the diluted hemolymph. The "reactive" *D. simulans* parasitized larvae encapsulated *A. tabida* eggs within 24 h of parasitization, while the "unreactive" larvae did not. *Note: Adapted-from Eslin and Prévost (1996).*

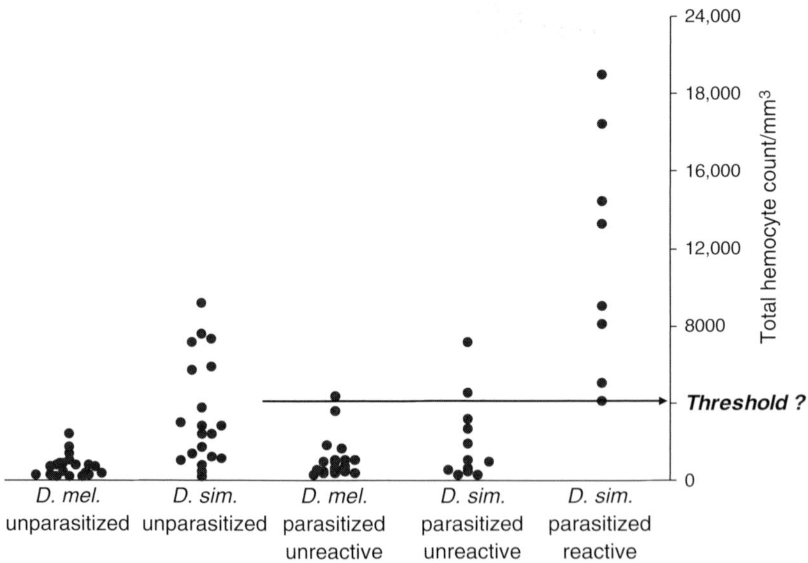

FIGURE 7.4 Individual values of the total hemocyte count (THC) measured in the diluted hemolymph of 3-day-old *D. melanogaster* (*D. mel.*) and *D. simulans* (*D. sim.*), unparasitized larvae and larvae parasitized by *A. tabida*. Parasitization occurred in 2-day-old larvae. The "reactive" *D. simulans* parasitized larvae encapsulated *A. tabida* eggs within 24 h of parasitization, while the "unreactive" larvae did not. The 3-day-old *D. melanogaster* hosts were all "unreactive." The arrow points at an approximate value for a THC threshold, under which encapsulation of *A. tabida* eggs would not occur. Dilution resulted from adding 0.2 μl Phosphate Buffered Saline to each individual hemolymph sample; factors of dilution and final concentrations of hemocytes in hemolymph could not be estimated. *Note: From Eslin and Prévost (1996).*

may vary regarding to the number of circulating hemocytes, and that *D. simulans* larvae carrying a heavy load of hemocytic cells before parasitization may build cellular capsules faster, especially around certain foreign targets such as *A. tabida* eggs.

Therefore, the success of the *D. melanogaster* and *D. simulans* larvae defense reaction against *A. tabida* eggs was considered to relate, at least partially, to the number of hemocytes circulating in the hemolymph within a few hours of parasitization (Eslin and Prévost, 1996, 1998). Regarding the hosts, the variations of parasitic success for the species *A. tabida* thus appear to be associated with both interspecific and intraspecific differences in their hemocyte loads.

The original study was then extended to most of the species of the *melanogaster* subgroup. Larvae from six *Drosophila* spp. of this subgroup were compared with each other for both their hemocytes concentration in the hemolymph and their capacity to encapsulate the parasitoid *A. tabida*. Results showed a high correlation between the concentration of circulating hemocytes in the parasitized host larvae and their aptitude to form a hemocytic capsule around *A. tabida* eggs (Fig. 7.5). This correlation showed that associating "immune" resistance to *A. tabida* with high hemocyte load (Eslin and Prévost, 1996) was not fortuitous and could be extended to other *Drosophila* spp. of the same subgroup.

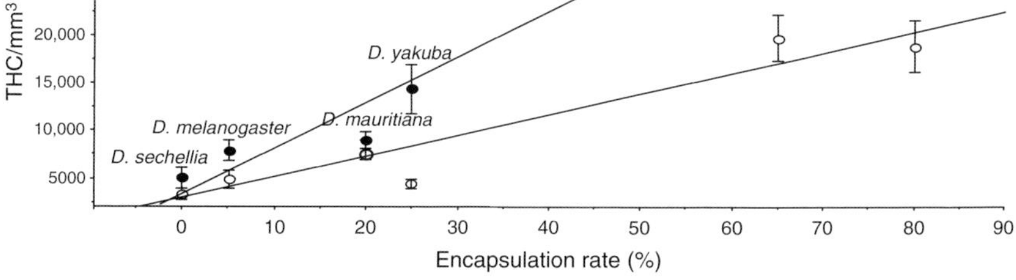

FIGURE 7.5 Regression between the encapsulation rate (%) and the total hemocyte count (THC) recorded in unparasitized larvae (o) and larvae parasitized by *A. tabida* (•). Mean values (± standard error) are given for six *Drosophila* spp. of the *melanogaster* subgroup: *D. sechellia* (*D. sech.*), *D. melanogaster* (*D. mel.*), *D. mauritiana* (*D. maur.*), *D. yakuba* (*D. yak.*), *D. teissieri* (*D. teis.*) and *D. simulans* (*D. sim.*). Correlation is highly significant ($r = 0.99$; $P < 0.001$). *Note: From Eslin and Prévost (1998).*

Two conditions seem to be required for the encapsulation of the parasitic eggs to succeed. One is the occurrence of a primary hemocytic response, which may relate to the recognition of the parasite by the host defense system and the subsequent signalization leading to the amplification of the hemocyte population in the hemolymph. The other condition is the recruitment, in the parasitized hosts, of a hemocyte load sufficient enough for the cellular capsule to be completed before the *A. tabida* egg becomes protected by embedment within the host tissues. Since the concentration of hemocytes in the parasitized hosts is partially related to the concentration of hemocytes before parasitization, *Drosophila* spp. carrying a high hemocyte load would be more likely to resist *A. tabida*.

Strains of *D. melanogaster* selected for higher resistance to *A. tabida* were also found to carry approximately twice as many circulating hemocytes as the control strains (Kraaijeveld et al., 2001a; see Chapter 10 by Kraaijeveld and Godfray). This confirms the importance of hemocyte loads in the completion of the encapsulation process in the *D. melanogaster–A. tabida* system.

Unlike other species of the *Asobara* genus (Mabiala et al., unpublished data; Moreau et al., 2003) or many other parasitoids (e.g., *Leptopilina* spp.) that have been well described for their regulation effect on host immunity, *A. tabida* does not provoke any depressive effect on the hemocyte population of parasitized hosts. Instead, a characteristic feature of *A. tabida* eggs is that they possess a ''sticky'' chorion allowing them to get attached to a variety of host tissues in the hemocoel (digestive tube, tracheal cells, fat body; Kraaijeveld et al., 1994; Monconduit and Prévost, 1994). The outcome of the interaction between *A. tabida* and *Drosophila* hosts strongly depends on the number and maturity reached by the host's circulating hemocytes within a few hours following parasitization (Eslin and Prévost, 1998, 2000). It seems obvious that the importance of the ''host's hemocyte load'' for the issue of the host–parasitoid interaction is enhanced by this particular mean *A. tabida* has to avoid encapsulation, that is, attaching to the host tissues, hence becoming unreachable by the host hemocytes.

Within the *melanogaster* subgroup, *D. simulans* is known for being very immunoreactive against several endophagous parasitoids (Carton and Kitano, 1981). Among species tested within this subgroup, not only *D. simulans* was the one with the greatest amount of circulating hemocytes but also happened to be the most resistant toward *A. tabida* (Fig. 7.5). Moreau et al. (2005) investigated whether other species of *Asobara* parasitoids could circumvent the immune reaction of *D. simulans* larvae. The species *A. citri* was chosen because of the successful immunodepression it exerts on *D. melanogaster* larvae, leading to the near absence of encapsulation reactions in these hosts (Moreau et al., 2003). Observations under a stereomicroscope showed that the hematopoietic organs of *D. melanogaster* and *D. simulans* larvae parasitized by *A. citri* exhibited the same pattern of disruption. In the parasitized larvae of both *Drosophila* spp., the anterior lobes of the hematopoietic organs were strongly reduced in size (Moreau et al., 2005), which seemed to be due to a necrosis process (Prévost et al., 2005). Despite severe disruption of the hematopoietic organs, *D. simulans* parasitized larvae still managed to encapsulate 45% of the *A. citri* eggs while only 30% of the parasitoids successfully completed their development. These percentages are significantly different from the ones obtained in *D. melanogaster* hosts which encapsulated almost none of *A. citri* eggs, subsequently leading to 84% successful parasitism. These results demonstrate that targeted disruption of the host hematopoietic organ by the parasitoid is not always sufficient to prevent encapsulation. Instead, the high hemocyte load of *D. simulans* larvae at the time of parasitization probably plays a significant role in their resistance toward *A. citri*.

Interestingly, *D. simulans* larvae also resist *Leptopilina boulardi* better than *D. melanogaster* larvae (Carton and Kitano, 1981), although the strategy of host immune depression developed by this figitid parasitoid is very different from *A. citri* (Labrosse et al., 2003). *D. melanogaster* and *D. simulans* are two cosmopolitan species often living in sympatry and both are submitted to a variety of natural enemies (Carton et al., 1986; Dupas et al., 2004). Though performed on a single strain for each species, these studies strongly suggest that *D. simulans* has evolved a more potent cellular immune response than *D. melanogaster* to resist the guild of parasitoids potentially present in their habitat.

7.3.2. The burst of lamellocytes

Changes in the hemocytes populations have long been known to represent a key feature indicating whether the cellular immune system has been activated or not in *Drosophila* host larvae (Brehélin, 1982; Carton and Kitano, 1979; Nappi, 1981; Nappi and Streams, 1969). Although lamellocytes can hardly be observed in the hemolymph of healthy, nonparasitized *D. melanogaster* larvae, they are not totally absent from the population of circulating hemocytes. A few lamellocytes can systematically be observed in the hemolymph of almost every larva from the *Drosophila* spp. belonging to the *melanogaster* subgroup. The proportion of lamellocytes that can be recorded in "nonimmunely challenged" larvae of the *melanogaster* subgroup usually ranges from 0.1% to 6.0% of the total number of circulating hemocytes (Eslin and Doury, 2006; Eslin and Prévost, 1998).

Activation of the immune system by a large foreign body systematically induces important changes in *Drosophila* hemocytes populations, and particularly a significant increase in the total number of hemocytes. This can mostly be explained by the increase in the number of lamellocytes, the capsule-forming hemocytes, with manifold lamellocytes being quickly released in the hemolymph. It is now well established that this burst of circulating lamellocytes in the hemolymph of *Drosophila* larvae is characteristic of their defenses being stimulated, the abundance of the lamellocytes reflecting an overall activation of cellular immune mechanisms (Eslin and Prévost, 1996, 1998; Labrosse et al., 2005a; Rizki and Rizki, 1994).

Though difficult to measure, the implication of physical wounding induced by the female wasp ovipositor at the time of parasitization cannot be ruled out. Effects of physical injuries on the hemocyte loads of *D. melanogaster* and *D. yakuba* larvae have been previously reported. The effects of perforating the larval cuticle, injecting Ringer or iron saccharate solutions all contributed to induce significant elevations in the total number of circulating hemocytes (Brehélin, 1982). Similarly, injection of Ringer (Labrosse et al., 2005a) or paraffin oil droplets (Eslin and Doury, 2006) in *D. melanogaster* larvae systematically led to a significant increase in the total number of hemocytes, that is, of the number of both plasmatocytes and lamellocytes. The release of plasmatocytes recorded after a trauma could be linked to healing process and/or homeostasis. Rizki and Rizki (1992) studied the effects of a variety of physical injuries on the number of circulating lamellocytes. Using large metal needles to inflict a wound, or inject either saline buffer or glass beads always led to high lamellocyte loads, unlike microneedle-inflicted wounds. The process of microinjection using a glass needle certainly represents much more important trauma than that of a female wasp laying an egg. Indeed, injection of foreign material or infliction of wounds with large needles is supposed to damage the basement membrane (Rizki and Rizki, 1992), a sufficient condition to trigger immunostimulation (Markus et al., 2005; Schmidt et al., 2001; Strand and Pech, 1995). Consequently, tissues with damaged basement membranes that represent nonself will trigger lamellocyte differentiation, leading to the encapsulation of the damaged tissues and formation of melanotic masses (Rizki and Rizki, 1992; Strand and Pech, 1995).

Whatever the *Asobara* spp. studied so far, the increase of lamellocytes is a recurrent effect recorded after parasitization of reactive host larvae of the *melanogaster* subgroup. More particularly, hemocyte counts of *D. simulans* larvae parasitized by *A. citri* revealed that greater numbers of circulating lamellocytes could be recorded in the hemolymph of reactive *D. simulans* larvae (i.e., that had successfully encapsulated *A. citri*), compared to the unreactive ones (i.e., that had failed in encapsulating the parasitoid; Moreau et al., 2005). However, the concentrations of circulating lamellocytes measured in *D. simulans* hosts reactive to *A. citri* were much smaller (approximately 20 times) than those previously reported in *D. simulans* hosts reactive to *A. tabida* (Eslin and Prévost, 1998; Fig. 7.6).

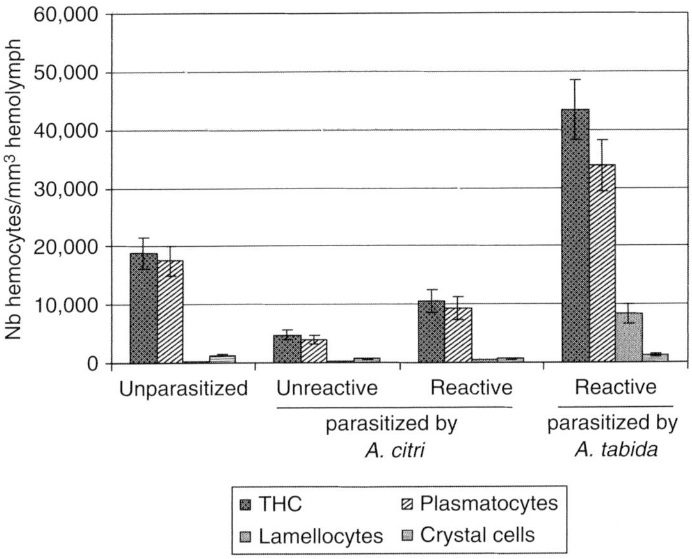

FIGURE 7.6 Mean (± standard error) total hemocyte counts (THC) and mean (± standard error) numbers of plasmatocytes, lamellocytes and crystal cells in the hemolymph of 6-day-old *D. simulans*, unparasitized larvae and larvae parasitized by *A. citri* or *A. tabida*. Parasitization occurred in 2-day-old *Drosophila* larvae. Parasitized larvae were designated as "reactive" or "unreactive" larvae depending on their ability or inability to encapsulate the parasitoid egg, respectively. *Note: Adapted from Eslin and Prévost (1998) and Moreau* et al. *(2005).*

This suggests that even a small increase in the number of circulating lamellocytes would be sufficient to allow the formation of a functional capsule around the parasitoid.

Drosophila larvae susceptible to *A. tabida* seem to be characterized by a low hemocyte load, which may or may not be associated with a weak ability to amplify the number of circulating lamellocytes after parasitization, in comparison with other species such as *D. simulans, D. teissieri or D. yakuba*, for which almost all individuals are able to recruit massive amounts of lamellocytes (Fig. 7.7). These two characteristics of the hemocyte population in susceptible hosts suggest the existence of different types of *Drosophila* immune reactions toward *A. tabida*:

1. Some *Drosophila* larvae may not react to the presence of the parasitoid egg in their hemocoel, as their hemolymph is devoid of lamellocytes after parasitization. The majority of *D. sechellia* and *D. mauritiana* larvae fall into this type, which leads to failure of encapsulating *A. tabida* eggs.

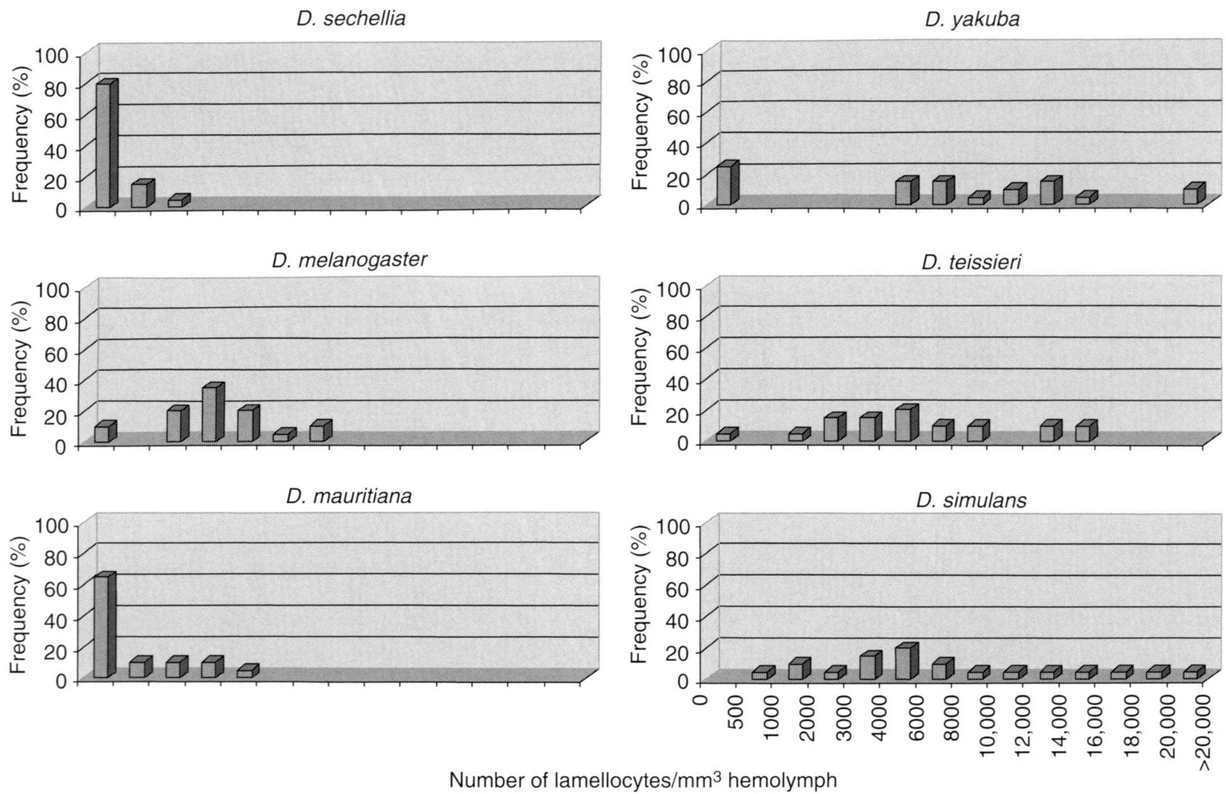

FIGURE 7.7 Distribution of the individual values of the number of lamellocytes per mm^3 of hemolymph measured in *Drosophila* larvae parasitized by *A. tabida*. The number of lamellocytes was recorded in 6-day-old *Drosophila* larvae from six species of the *melanogaster* subgroup, 4 days after parasitization. *Note: From Eslin and Prévost (1998).*

2. Other hosts, such as most of the *D. melanogaster* larvae and a small proportion of the *D. sechellia* and *D. mauritiana* larvae, may react to the presence of the parasitoid egg, as their numbers of circulating lamellocytes increase after parasitization. Nevertheless, these hosts may be unable to mount a successful encapsulation reaction because their hemocyte load, although being increased after parasitization, still remains lower than the threshold required (Eslin and Prévost, 1996).

7.3.3. A race against time

The parasitoid *A. tabida* has evolved a particular strategy to avoid the attack by the host's hemocytes. Embedment of parasitoid eggs within the host tissues provides them with an efficient protection against the attack by the host hemocytes (Kraaijeveld et al., 1994; Monconduit and Prévost, 1994).

The success of *A. tabida* protection thus depends on the speed of capsule completion, which was shown to be correlated with the concentration of the host's hemocytes present in the hemolymph after parasitization. Therefore, the outcome of the encapsulation reaction toward *A. tabida* eggs seems to depend on the issue of a "physiological race" occurring between the encapsulation reaction of the *Drosophila* host larvae and the mechanism of *A. tabida* avoidance of encapsulation. *D. simulans*, which is faster mounting a hemocyte capsule than *D. melanogaster*, therefore will be more willing to encapsulate the parasitoid eggs (Fig. 7.8). The "race" between the host's defense reaction and the parasitoid's protection is one of the key factors determining the issue of *A. tabida* parasitic success in all the *Drosophila* spp. tested within the *melanogaster* subgroup (Eslin and Prévost, 2000).

In addition, the observation of the capsules built by *D. simulans* larvae around the eggs of *A. citri* (Fig. 7.9) strongly suggested that the immune reaction was activated very promptly after parasitization. The capsules were of unusual small size and obviously formed around parasitic eggs at an early developmental stage. Therefore, the outcome of the encapsulation reaction may depend on a balance between the host potential immune reaction toward the parasitoid, and *A. citri* potential to disrupt the cellular processes of encapsulation. This "time race" appears very similar to what we reported on the *A. tabida–Drosophila* interaction.

Though a physiological race occurring among a host insect and a parasitoid is not a new concept (Thompson, 1982), this is the first time that such a concept can be addressed in the physiological context of immune reactions.

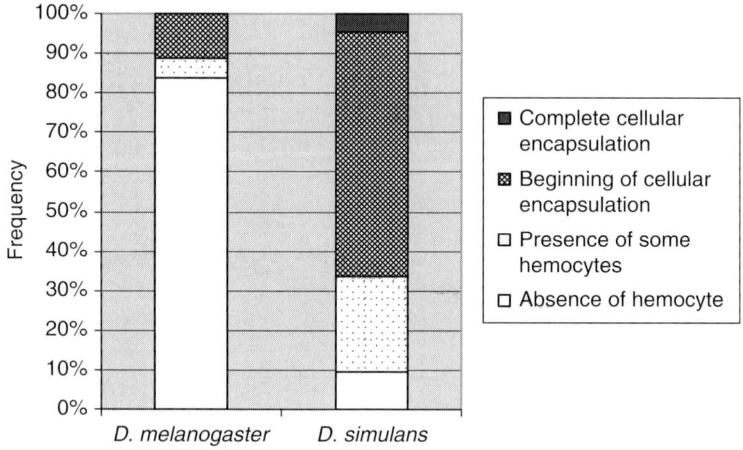

FIGURE 7.8 Degrees of encapsulation of *A. tabida* eggs measured in *D. melanogaster* and *D. simulans* larvae parasitized at 3 days old. Bars represent percentages of host larvae, 24 h after parasitization, whose level of encapsulation corresponds to one of the four categories defined hereafter: "absence of hemocyte" (total lack of hemocytes at the surface of the parasitoid egg), "presence of some hemocytes" (low number of hemocytes at the surface of the foreign body), "beginning of cellular encapsulation" (cellular capsule partially covering the surface of the foreign body), and "complete cellular encapsulation" (cellular capsule totally covering the surface of the foreign body). *Note: Adapted from Eslin and Prévost (2000).*

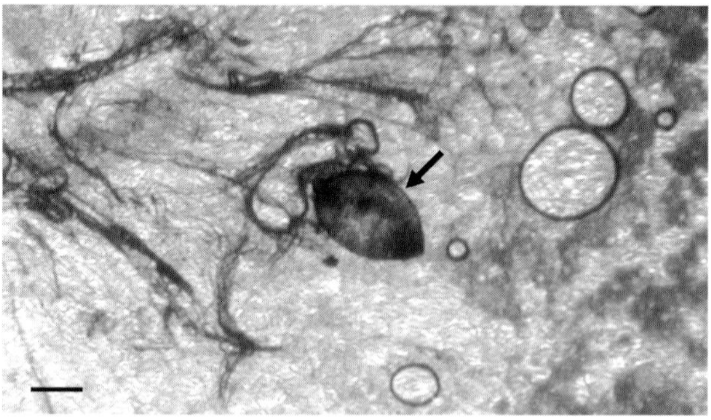

FIGURE 7.9 Egg of *A. citri* found encapsulated in *D. simulans* after total development of the fly, showing that the melanized layer surrounding the egg was very thin and obviously formed around the parasitoid egg at an early developmental stage. Scale bar, 100 μm. *Note: From Moreau* et al. *(2005).*

7.4. BUT DOES QUALITY MATTER? THE CASE OF THE *OBSCURA* GROUP

Almost all organisms face attacks by parasites and have evolved some defense mechanisms that allow survival. As for many developmental events, the formation of melanotic capsules has mostly been studied in the reference species *D. melanogaster* and very little information is available concerning the immune response to parasitization in other species of the *Drosophila* genus.

7.4.1. Deficiency in encapsulation ability associated with the absence of lamellocytes

The species *D. subobscura* is a drosophilid of the *obscura* group found in fermenting substrates, whose larvae have the potential to be natural hosts for several endoparasitoid species, particularly *A. tabida* and *Leptopilina heterotoma* (Vet and van Alphen, 1985). Eslin and Doury (2006) have reported that *D. subobscura* does not mount any obvious cellular immune reaction against these two species of parasitoids, nor to oil drops injected as foreign bodies. Also lamellocytes, the hemocyte lineage specialized in mounting the encapsulation reaction in *D. melanogaster*, could not be found in *D. subobscura*, neither after injection of foreign bodies nor parasitization. Therefore, the species *D. subobscura* has been considered as a case of "immunity deficiency," especially regarding voluminous parasites like endoparasitoid eggs.

Havard et al. (2009) demonstrated that none of the nine newly tested species of the *obscura* group possessed lamellocytes or underwent the stereotypical *D. melanogaster* melanotic encapsulation response. The species *D. azteca, D. bifasciata, D. guanche, D. miranda, D. persimilis* and *D. pseudoobscura*, just like *D. subobscura* (Eslin and Doury, 2006), were all incapable of encapsulating a large foreign body despite the fact their hemocyte loads were high compared to those of *D. melanogaster* larvae, with mostly circulating plasmatocytes. Instead, the incapacity to encapsulate was consistently associated with the lack of lamellocytes. These nonencapsulating species devoid of lamellocyte appeared unable, in immunity terms, to neutralize a large foreign body or a parasitoid egg. Therefore, they can also be regarded as immunity deficient against macroparasites. In conclusion, the cellular immunity deficiency originally found in *D. subobscura* extends to other species of the *obscura* group as well.

The cellular and molecular bases of this immunity deficiency in encapsulation have not been unraveled yet. Triggering of the lamellocyte program could be regulated at different levels. The molecular mechanisms of immunocytes differentiation are well studied in the reference model

D. melanogaster (Crozatier and Meister, 2007; Krzemien et al., 2007). Compared to *D. melanogaster*, flies of the *obscura* group represent an ideal microevolution paradigm to study the evolution of "integrator" genes since more and more genomes are being sequenced and annotated. These *Drosophila* spp. could become instrumental in understanding the molecular pathways that are activated when encapsulation is triggered.

7.4.2. Atypical encapsulating hemocytes

Cellular immunodeficiency is not a phylogenetic trait that can be generalized to the whole *obscura* group (Table 7.1). Despite the fact that lamellocytes have long been considered the major capsule-forming hemocyte type in *Drosophila* (Brehélin and Duvic, 1999), three species, *D. affinis*, *D. tolteca* and *D. obscura*, were able to initiate encapsulation in the absence of lamellocytes. All three species of *Drosophila* present new hemocyte cell types that have been named "atypical hemocytes" (Fig. 7.10) because they phenotypically differ from the typical lamellocytes described in *D. melanogaster*. The significant increase of atypical hemocytes in immuno-challenged larvae of these three species strongly suggested that these particular cells allowed for some encapsulation and were thus involved in the mounting of capsules (Havard et al., 2009).

However, it should be noted that the encapsulation reaction remained quantitatively and qualitatively restricted. More than 80% of the *D. melanogaster* larvae with oil droplets encapsulated these foreign bodies (Eslin and Doury, 2006), whereas less than 20% managed to do so in *D. affinis* and only a small minority (<4%) in *D. tolteca* and *D. obscura*. In addition, half of the capsules were only partial. A possible explanation is that the concentration of atypical hemocytes in larvae of these species was low (Havard et al., 2009) compared with the high concentration of lamellocytes that can be recorded in *D. melanogaster* larvae (Eslin and Prévost, 1998). Also, their shape and/or adherence properties could be less suitable to the formation of multicellular layers.

In *D. melanogaster* larvae, the capsule is initiated by plasmatocytes which build a precapsule, and the multilayer of lamellocytes is added later on (Russo et al., 1996). Conversely to what was believed so far, lamellocytes would not be strictly necessary to the process of capsule edification, but could just act like encapsulation improvers.

7.5. DISCUSSION AND CONCLUDING REMARKS

Parasitism by *Asobara* parasitoids in larvae from any *Drosophila* spp. of the *melanogaster* group always triggers a stimulation of the immune system, which is typically followed by a significant increase in the total number of

TABLE 7.1 Ability to initiate an encapsulation reaction (+, able; −, unable) toward a 10-nl paraffin oil drop or an egg of *A. tabida* as a foreign body, associated with the presence (+) or absence (−) of lamellocytes and atypical hemocytes in *Drosophila* spp. of the *obscura* group

	Ability to initiate encapsulation toward oil drop	Ability to initiate encapsulation toward *A. tabida* egg	Presence of lamellocytes	Presence of atypical hemocytes
affinis subgroup				
D. affinis	+	+	−	+
D. azteca	−	−	−	−
D. tolteca	+	−	−	+
pseudoobscura subgroup				
D. persimilis	−	−	−	−
D. pseudoobscura	−	−	−	−
D. miranda	−	−	−	−
obscura subgroup				
D. bifasciata	−	−	−	−
D. obscura	+	+	−	+
subobscura subgroup				
D. guanche	−	−	−	−
D. subobscura				
strains: Amiens	−	−	−	−
Gotheron	−	−	−	−
Madeira	−	−	−	−
Montgenèvre	−	−	−	−
Moscow	−	−	−	−
Nyon	−	−	−	−

Note: Adapted from Eslin and Doury (2006) and Havard et al. (2009).
Ten species and six strains of *D. subobscura* have been tested.

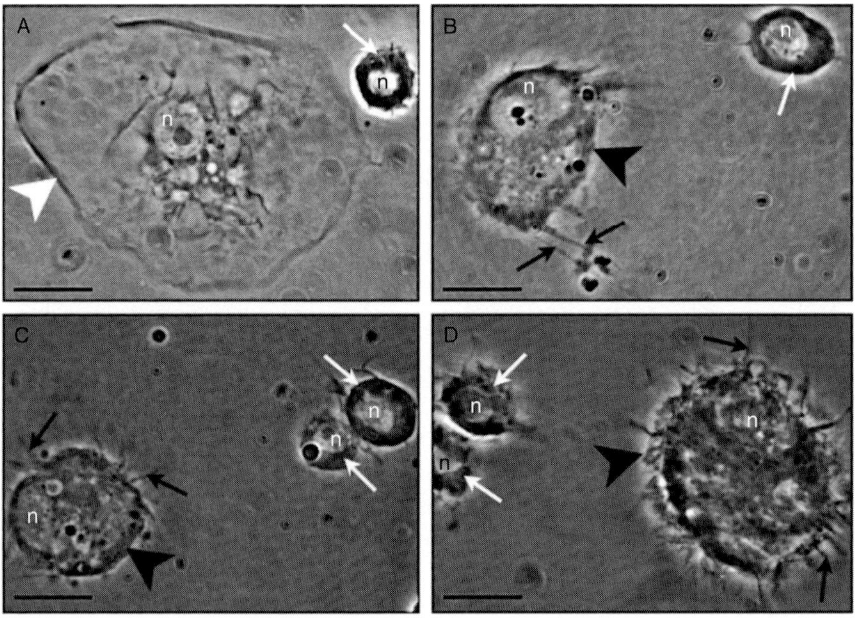

FIGURE 7.10 Phase contrast microscopy of circulating hemocytes in third-instar larvae at 72 h postinjection of a 10 nl oil droplet. (A) *Drosophila melanogaster*; (B) *Drosophila tolteca*; (C) *Drosophila obscura*; (D) *Drosophila affinis*. n: nucleus; white arrowhead: lamellocyte; white arrow: plasmatocyte; black arrowhead: atypical hemocyte; black arrow: digitations. Scale bar, 10 μm. *Note: From Havard* et al. *(2009).*

lamellocytes. When an increase in the total number of hemocytes is also recorded, it can largely be explained by this increase in the number of circulating lamellocytes.

7.5.1. Hemocyte load: An investment?

Whether a cost is related to the hemocyte load is not known and is difficult to assess. Kraaijeveld et al. (2001b) noted an association between the development of the hematopoietic organ and the head muscles during larval stages. The study of fate maps of *Drosophila* embryos revealed that the hemopoietic organ and the head musculature derive from the same region of the embryo. Possible preallocation of more tissue to the blood-forming organ would leave less tissue available for muscles enabling food intake. One possible explanation for a trade-off between hemocyte numbers and feeding rate is that investing more general resources in the immune system goes at the expense of being able to feed quickly. As hypothesized by Dupas et al. (2004), the production of hemocytes could be under balanced selection depending on the infection risk. Although the hemocyte load of the host insect can be considered less parasitoid-specific than other factors involved in the host–parasitoid interaction, the hemolymph concentration in hemocytes will depend on more general patterns of abundance of a large spectrum of parasite species. Insect hemocyte load is likely to be influenced by various environmental factors, and the presence or absence of different parasites or pathogens is probably an important factor if hemocyte concentration plays a role in other immune responses. However, cells that circulate in insect hemolymph are also involved in various physiological functions, such as metabolic transport (Sass et al., 1994; Wigglesworth, 1972) and enzyme synthesis (Crossley, 1979; Dvornik, 1992; Rizki and Rizki, 1980a). These cells also contribute to other mechanisms, such as the formation of the basement membrane (Wigglesworth, 1973) and wound healing (Brehélin, 1982). Therefore, it is possible that specific average numbers of hemocytes may be more or less stabilized by various physiological constraints. As a consequence, hemocytes of hosts would be unable (or physiologically limited) to respond quickly to a sudden environmental change, such as that of an invading parasitoid.

The implication in the host defense reaction of a quantitative and more or less unspecific character, such as the hemocyte load, is of primary importance. However, it should be noted that a physiological factor of little specificity as that of the hemocyte load cannot explain on its own the whole potential resistance of a host in a host–parasitoid system. It is also known that resistance specific to one parasitoid species may be different from the resistance that was observed in the correlation with hemocyte load, and this appears to be due to the contribution of other components of the immune reaction (Carton et al., 2008).

7.5.2. Cellular immunity: Are all *Drosophila* spp. equally armed?

The species *D. subobscura* lacks lamellocytes and is unable to form cellular capsules. Within the *obscura* group, several other tested species have been found to be phenotypically similar to *D. subobscura*, the observed phenotype being larvae failing to encapsulate. These phylogenetically related species of the genus *Drosophila* appear to be devoid of encapsulation immune response against foreign bodies, possibly caused by the lack of certain hemocytes. Indeed all these species that fail to encapsulate do not produce any lamellocytes. Flies of the *obscura* group appear immunocompromised for encapsulation reactions, and consequently immune deficient toward parasitoids. Ecological and evolutionary factors are likely to be involved in this incapacity of *Drosophila* spp. from the *obscura* group to mount an immune cellular response against parasites. European and American populations of *D. subobscura* can be exposed to larval parasitoids (Kraaijeveld and Van Alphen, 1994; Schlenke et al., 2007).

Although field data on the exposure of other species from the *obscura* group to parasitoids are drastically lacking, it is quite puzzling as it opposes the general assumption that all organisms, in natural conditions, are potentially able to fight parasitization.

Given that *D. subobscura* had been shown to be deficient in encapsulation, the fact that closely related species share the same deficiency is not surprising. However, the finding that three species that have weak encapsulation ability also possess a population of previously undescribed ''atypical'' hemocytes specifically induced upon parasitization is interesting, especially because this novel type of hemocyte seems to have encapsulation ability. The presence of lamellocytes in species of the *melanogaster* subgroup and the lack of lamellocytes in the *obscura* group begs the question of what the ancestral state was. Did the *melanogaster* lineage gain lamellocytes or did the *obscura* lineage lose lamellocytes? Is the *melanogaster* subgroup truly a reference model or should it be considered as an exception? Fitness tradeoffs might explain the ''immunity deficiency'' seen in the *obscura* group flies. However, species of the *obscura* group possess high numbers of hemocytes, and also have significantly greater numbers of plasmatocytes than the reference species *D. melanogaster*. Perhaps the flies of the *obscura* group trade off immunity against parasitoids (via encapsulation) in order to be more immune resistant against microorganisms (via phagocytosis). These questions are very interesting from an evolutionary point of view and of particular significance in the coevolutionary interactions between hosts and parasitoids.

ACKNOWLEDGMENT

This work was partially supported by the GDR-CNRS 2153.

REFERENCES

Bataille, L., Auge, B., Ferjoux, G., Haenlin, M., Waltzer, L., 2005. Resolving embryonic blood cell fate choice in *Drosophila*: interplay of GCM and RUNX factors. Development 132, 4635–4644.

Benassi, V., Frey, F., Carton, Y., 1998. A new specific gene for wasp cellular immune resistance in *Drosophila*. Heredity 80, 47–52.

Brehélin, M., 1982. Comparative study of structure and function of blood cells from two *Drosophila* species. Cell Tissue Res. 221, 607–615.

Brehélin, M., Duvic, B., 1999. Cellular defence reactions and their depression in insects. J. Soc. Biol. 193, 325–328.

Carton, Y., Kitano, H., 1979. Changes in the haemocyte population of *Drosophila* larvae after single and multiple parasitization by *Cothonaspis* (Parasitic Cynipidae). J. Invert. Pathol. 34, 88–89.

Carton, Y., Kitano, H., 1981. Evolutionary relationships to parasitism by seven species of the *Drosophila melanogaster* subgroup. Biol. J. Linn. Soc. 16, 227–241.

Carton, Y., Nappi, A., 1997. *Drosophila* cellular immunity against parasitoids. Parasitol. Today 13, 218–226.

Carton, Y., Nappi, A., 2001. Immunogenetics aspects of the cellular immune response of *Drosophila* against parasitoids. Immunogenetics 5, 157–164.

Carton, Y., Boulétreau, M., van Alphen, J.J.M., van Lenteren, J.C., 1986. The *Drosophila* parasitic wasps. In: Ashburner, Thomson, (Eds.), The Genetics and Biology of *Drosophila*. Vol. 3C. Academic Press, New York/London, pp. 347–394.

Carton, Y., Poirié, M., Nappi, A.J., 2008. Insect immune resistance to parasitoids. Insect Sci. 15, 67–77.

Colinet, D., Schmitz, A., Depoix, D., Crochard, D., Poirié, M., 2007. Convergent use of RhoGAP toxins by eukaryotic parasites and bacterial pathogens. PLoS Pathogens 3, e203.

Crossley, A.C., 1979. Biochemical and ultrastructural aspects of synthesis, storage, and secretion in hemocytes. In: Gupta, (Ed.), Insect Hemocytes: Development, Forms, Functions and Techniques. Cambridge University Press, Cambridge, pp. 423–474.

Crozatier, M., Ubeda, J.M., Vincent, A., Meister, M., 2004. Cellular immune response to parasitization in *Drosophila* requires the EBF orthologue Collier. PLoS Biol. 2, 1107–1113.

Crozatier, M., Meister, M., 2007. *Drosophila* haematopoiesis. Cell Microbiol. 9, 1117–1126.

Dubuffet, A., Dupas, S., Frey, F., Drezen, J.M., Poirié, M., Carton, Y., 2007. Genetic interactions between the parasitoid wasp *Leptopilina boulardi* and its *Drosophila* hosts. Heredity 98, 21–27.

Dupas, S., Carton, Y., Poirié, M., 2003. The genetic dimension of the coevolution of virulence-resistance in *Drosophila*-parasitoid wasp relationships. Heredity 90, 84–89.

Dupas, S., Morand, S., Eslin, P., 2004. Evolution of hemocyte concentration in the *melanogaster* subgroup species. Comp. Rend. Acad. Sci. 327, 139–147.

Dvornik, V.Y., 1992. Peroxidase and catalase in hemocytes of blackfly, *Wilhelmia salopiensis* (Edw.), during preimaginal development. Proc. XIX Int. Congr. Entomol. Abstract no. 41, IIIP.

Eslin, P., 1998. Les Mécanismes de l'interaction Virulence-Résistance Immunitaire Entre *Asobara tabida* et ses Hôtes Drosophile: Etude Cellulaire et Analyse Populationnelle [thesis]. University of Paris VI, France.

Eslin, P., Doury, G., 2006. The fly *Drosophila subobscura*: A natural case of innate immunity deficiency. Dev. Comp. Immunol. 30, 977–983.

Eslin, P., Prévost, G., 1996. Variation in *Drosophila* concentration of haemocytes associated with different ability to encapsulate *Asobara tabida* larval parasitoid. J. Insect Physiol. 42, 549–555.

Eslin, P., Prévost, G., 1998. Hemocyte load and immune resistance to *Asobara tabida* are correlated in species of *Drosophila melanogaster* subgroup. J. Insect Physiol. 44, 807–816.

Eslin, P., Prévost, G., 2000. Racing against host's immunity defenses: a likely strategy for passive evasion of encapsulation in *Asobara tabida* parasitoids. J. Insect Physiol. 46, 1161–1167.

Eslin, P., Giordanengo, P., Fourdrain, Y., Prévost, G., 1996. Avoidance of encapsulation in the absence of VLP by a braconid parasitoid of *Drosophila* larvae: an ultrastructural study. Can. J. Zool. 74, 2193–2198.

Evans, C.J., Hartenstein, V., Banerjee, U., 2003. Thicker than blood: conserved mechanisms in *Drosophila* and vertebrate hematopoiesis. Dev. Cell 5, 673–690.

Fisher, R.C., 1963. Oxygen requirements and the physiological suppression of supernumary insect parasitoids. J. Exp. Biol. 38, 605–658.

Götz, P., 1986. Mechanisms of encapsulation in Dipteran hosts. Symp. Zool. Soc. Lond. 56, 1–19.

Havard, S., Eslin, P., Prévost, G., Doury, G. (2009). Encapsulation ability: are all *Drosophila* species equally armed? An investigation in the *obscura* group. Can. J. Zool. 635–641.

Hita, M., Poirié, M., Leblanc, N., Lemeunier, F., Lutcher, F., Frey, F., et al., 1999. Genetic localization of a *Drosophila melanogaster* resistance gene to a parasitoid wasp and physical mapping of the region. Genome Res. 9, 471–481.

Hultmark, D., 2003. *Drosophila* immunity: paths and patterns. Curr. Opin. Immunol. 15, 12–19.
Kraaijeveld, A.R., van Alphen, J.M., 1994. Geographical variation in resistance of the parasitoid *Asobara tabida* against encapsulation by *Drosophila melanogaster* larvae: the mechanism explored. Physiol. Entomol. 19, 9–14.
Kraaijeveld, A.R., Hutcheson, K.A., Limentani, E.C., Godfray, H.C., 2001a. Costs of counterdefenses to host resistance in a parasitoid of *Drosophila*. Evolution 55, 1815–1821.
Kraaijeveld, A.R., Limentani, E.C., Godfray, H.C., 2001b. Basis of the trade-off between parasitoid resistance and larval competitive ability in *Drosophila melanogaster*. Proc. R. Soc. Lond. B 268, 259–261.
Krzemien, J., Dubois, L., Makki, R., Meister, M., Vincent, A., Crozatier, M., 2007. Control of blood cell homeostasis in *Drosophila* larvae by the posterior signalling centre. Nature 446, 325–328.
Labrosse, C., Carton, Y., Dubuffet, A., Drezen, J.M., Poirié, M., 2003. Active suppression of *D. melanogaster* immune response by the parasitic wasp *Leptopilina boulardi*. J. Insect Physiol. 49, 513–522.
Labrosse, C., Eslin, P., Doury, G., Drezen, J.M., Poirié, M., 2005a. Haemocyte changes in *D. melanogaster* in response to long gland components of the parasitoid wasp *Leptopilina boulardi*: a Rho-GAP protein as an important factor. J. Insect Physiol. 51, 161–170.
Labrosse, C., Stasiak, K., Lesobre, J., Grangeia, A., Huguet, E., Drezen, J.M., Poirié, M., 2005b. A RhoGAP protein as a main immune suppressive factor in the *Leptopilina boulardi* (Hymenoptera, Figitidae)-*Drosophila melanogaster* interaction. Insect Biochem. Mol. Biol. 35, 93–103.
Lanot, R., Zachary, D., Holder, F., Meister, M., 2001. Postembryonic hematopoiesis in *Drosophila*. Dev. Biol. 230, 243–257.
Lavine, M.D., Strand, M.R., 2002. Insect hemocytes and their role in immunity. Insect Biochem. Mol. Biol. 32, 1295–1309.
Lebestky, T., Chang, T., Hartenstein, V., Banerjee, U., 2000. Specification of *Drosophila* hematopoietic lineage by conserved transcription factors. Science 288, 146–149.
Lemaitre, B., Hoffmann, J., 2007. The host defense of *Drosophila melanogaster*. Ann. Rev. Immunol. 25, 697–743.
Markus, R., Kurucz, E., Rus, F., Ando, I., 2005. Sterile wounding is a minimal and sufficient trigger for a cellular immune response in *Drosophila melanogaster*. Immunol. Lett. 101, 108–111.
Markus, R., Laurinyecz, B., Kurucz, E., Honti, V., Bajusz, I., Sipos, B., et al., 2009. Sessile hemocytes as a hematopoietic compartment in *Drosophila melanogaster*. Proc. Natl Acad. Sci. USA 106, 4805–4809.
Meister, M., 2004. Blood cells of *Drosophila*: cell lineages and role in host defence. Curr. Opin. Immunol. 16, 10–15.
Meister, M., Lagueux, M., 2003. *Drosophila* blood cells. Cell. Microbiol. 5, 573–580.
Monconduit, H., Prévost, G., 1994. Avoidance of encapsulation by *Asobara tabida*, a larval parasitoid of *Drosophib* species. Norw. J. Agri. Sci. 16, 301–309.
Moreau, S.J.M., Eslin, P., Giordanengo, P., Doury, G., 2003. Comparative study of the strategies evolved by two parasitoids of the *Asobara* genus to avoid the immune response of the host, *Drosophila melanogaster*. Dev. Comp. Immunol. 27, 273–282.
Moreau, S.J.M., Guillot, S., Populaire, C., Doury, G., Prévost, G., Eslin, P., 2005. Conversely to its sibling *Drosophila melanogaster, D. simulans* overcomes the immunosuppressive effects of the parasitoid *Asobara citri*. Dev. Comp. Immunol. 29, 205–209.
Nappi, A.J., 1981. Cellular immune response of *Drosophila melanogaster* against *Asobara tabida*. Parasitology 83, 319–324.
Nappi, A.J., Christensen, B.M., 2005. Melanogenesis and associated cytotoxic reactions: Applications to insect innate immunity. Insect Biochem. Mol. Biol. 35, 443–459.
Nappi, A.J., Streams, F.A., 1969. Haemocytic reactions of *Drosophila melanogaster* to the parasites *Pseudocoila mellipes* and *P. hochei*. J. Insect Physiol. 15, 1551–1556.
Nappi, A.J., Vass, E., 1993. Melanogenesis and the generation of cytotoxic molecules during insect cellular immune reactions. Pigment Cell Res. 6, 117–126.

Nappi, A.J., Vass, E., 2001. Cytotoxic reactions associated with insect immunity. In: Berck, Sugumaran, Cooper, (Eds.), Phylogenetic Perspectives on the Vertebrate Immune System. Kluwer/Plenum, New York, pp. 329–348.
Nappi, A.J., Vass, E., Frey, F., Carton, Y., 1995. Superoxide anion generation in *Drosophila* during melanotic encapsulation of parasites. Eur. J. Cell Biol. 68, 450–456.
Nappi, A.J., Vass, E., Frey, F., Carton, Y., 2000. Nitric oxide involvement in *Drosophila* immunity. Nitric Oxide 4, 423–430.
Poirié, M., Frey, F., Hita, M., Huguet, E., Lemeunier, F., Periquet, G., et al., 2000. *Drosophila* resistance genes to parasitoids: chromosomal location and linkage analysis. Proc. R. Soc. Lond. B 267, 1417–1421.
Prévost, G., Eslin, P., Doury, G., Moreau, S.J.M., Guillot, S., 2005. *Asobara*, braconid parasitoids of *Drosophila* larvae: unusual strategies to avoid encapsulation without VLPs. J. Insect Physiol. 51, 171–179.
Ribeiro, C., Brehélin, M., 2006. Insect haemocytes: what type of cell is that? J. Insect Physiol. 52, 417–429.
Rizki, R.M., Rizki, T.M., 1980a. The direction of evolution in the *Drosophila melanogaster* species subgroup based on functional analysis of the crystal cells. J. Exp. Zool. 212, 323–328.
Rizki, R.M., Rizki, T.M., 1984. Selective destruction of a host blood cell type by a parasitoid wasp. Proc. Natl. Acad. Sci. USA 81, 6154–6158.
Rizki, T.M., Rizki, R.M., 1959. Functional significance of crystal cells in the larva of *Drosophila melanogaster*. J. Biophys. Biochem. Cytol. 5, 235–240.
Rizki, T.M., Rizki, R.M., 1980b. Properties of the larval hemocytes of *Drosophila melanogaster*. Experientia 36, 1223–1226.
Rizki, T.M., Rizki, R.M., 1992. Lamellocyte differentiation in *Drosophila* larvae parasitized by *Leptopilina*. Dev. Comp. Immunol. 16, 103–110.
Rizki, T.M., Rizki, R.M., 1994. Parasitoid-induced cellular immune deficiency in *Drosophila*. Ann. N.Y. Acad. Sci. 712, 178–194.
Russo, J., Brehélin, M., Carton, Y., 2001. Haemocyte changes in resistant and susceptible strains of *D. melanogaster* caused by virulent and avirulent strains of the parasitic wasp *Leptopilina boulardi*. J. Insect Physiol. 47, 167–172.
Russo, J., Dupas, S., Frey, F., Carton, Y., Brehélin, M., 1996. Insect immunity: early events in encapsulation process of parasitoid (*Leptopilina boulardi*) eggs in reactive and non reactive strains of *Drosophila*. Parasitology 112, 135–142.
Ryder, J.J., Siva-Jothy, M.T., 2001. Quantitative genetics of immune function and body size in the house cricket, *Acheta domesticus*. J. Evol. Biol. 14, 646–653.
Salt, G., 1963. The defense reactions of insects to metazoan parasites. Parasitology 53, 320–327.
Sass, M., Kiss, A., Locke, M., 1994. Integument and hemocyte peptides. J. Insect Physiol. 40, 407–421.
Schlenke, T.A., Morales, J., Govind, S., Clark, A., 2007. Contrasting infection strategies in generalist and specialist wasp parasitoids of *Drosophila melanogaster*. PLoS Pathog. 3, 1486–1501.
Schmidt, O., Theopold, U., Strand, M., 2001. Innate immunity and its evasion and suppression by hymenopteran endoparasitoids. Bioessays 23, 344–351.
Sorrentino, R.P., Carton, Y., Govind, S., 2002. Cellular immune response to parasite infection in the *Drosophila* lymph gland is developmentally regulated. Dev. Biol. 243, 65–80.
Strand, M.R., Pech, L.L., 1995. Immunological basis for compatibility in parasitoid-host relationships. Ann. Rev. Entomol. 40, 31–56.
Thompson, J.N., 1982. In: Interaction and Coevolution. Wiley, New York.
Wigglesworth, V.B., 1972. In: The Principles of Insect Physiology. 7th edn. Chapman & Hall, London.
Wigglesworth, V.B., 1973. Haemocytes and basement membrane formation in *Rhodnius*. J. Insect Physiol. 19, 831–844.

CHAPTER 8

Components of *Asobara* Venoms and their Effects on Hosts

Sébastien J.M. Moreau,* **Sophie Vinchon,**[†]
Anas Cherqui,[†] and **Geneviève Prévost**[†]

Contents

Abstract Hymenoptera of the *Asobara* genus are endophagous parasitoids of *Drosophila* larvae. In these apocrita insects whose venom gland is associated with the female reproductive tract, the wasp venom is

* UMR 6035 CNRS, Institut de Recherche sur la Biologie de l'Insecte, Faculté des Sciences et Techniques, Université François-Rabelais, Parc Grandmont, 37200 Tours, France
† Laboratoire de Biologie des Entomophages, EA 3900 BioPI, Université de Picardie-Jules Verne, 33 rue Saint Leu, 80039 Amiens cedex, France

Advances in Parasitology, Volume 70
ISSN 0065-308X, DOI: 10.1016/S0065-308X(09)70008-9

injected into the host along with the parasitoid egg during oviposition. We conducted a comparative study of the venom apparatuses from three *Asobara* spp.: the European *Asobara tabida*, the Asiatic *A. japonica* and the African *A. citri*. Light and electron microscopy of venom glands, together with the biochemical analysis of their contents, revealed important differences between *Asobara* spp. In addition, the physiological effects of female wasp's venom injected into *Drosophila* larvae differed greatly between the tested *Asobara* spp.

8.1. INTRODUCTION

Soliman (1941) and Bender (1943) were pioneers in describing the morphology of the venom apparatus (VA) of parasitoid Hymenoptera. Subsequent work by King and Ratcliffe (1969), van Marle (1977), and Edson and others (Blass and Ruthmann, 1989; Edson and Vinson, 1979; Edson et al., 1982; van Marle and Piek, 1986) strongly contributed to the understanding of the anatomical and functional organization of venom-producing organs in parasitoid wasps. These last decades, parasitoids' venoms received a growing interest attested by the works of numerous investigators in the fields of insect physiology, biochemistry and molecular biology (for reviews, see Asgari, 2006, 2007; Moreau and Guillot, 2005). It is now established that venoms play a key role in the virulence of parasitoids toward their insect hosts, notably in species of parasitic wasps devoid of polydnaviruses.

Parasitoids from the genus *Asobara* (Hymenoptera: Braconidae) have evolved various strategies to overcome the immune defenses of *Drosophila* larvae in the absence of such endosymbiotic viruses (Prévost et al., 2005). Since venoms play an active role in the virulence of many braconid parasitoids, study of the constituents of *Asobara* secretions should help understand how they contribute to the parasitoid's success.

In this chapter, we will focus on the anatomy and functional organization of the VA in three *Asobara* spp.: *Asobara tabida* (from Europe), *A. japonica* (from Japan) and *A. citri* (from Africa). Comparison between these three species illustrates the striking richness and diversity observed in hymenopteran parasitoids, even in closely related species, regarding the anatomy and physiology of apparatuses adapted to the production of virulence factors. We will first compare the morphology of the VA within the *Asobara* genus, with respects to other known parasitoids, and then describe our present knowledge on the physiological effects and composition of each species' venom. Prospects raised from studying venoms of the *Asobara* genus will then be discussed.

8.2. ANATOMY OF THE VENOM APPARATUS WITHIN THE *ASOBARA* GENUS

8.2.1. Comparative approach of the morphology of venom apparatuses

VA within the posterior end of the abdomen of female parasitoids are organs of ectodermal origin, classically composed of one or more secreting glands and a reservoir connected by a thin duct to the base of the ovipositor (Edson and Vinson, 1979).

Venom glands are sometimes designated as "acid glands," due to the acid pH of venomous secretions in parasitoid Hymenoptera. In contrast, the "alkaline gland" of these insects corresponds to the Dufour's gland, which is specialized in the synthesis of marking pheromones (Barrera et al., 1994; Guillot et al., 1974; Howard and Baker, 2003; Robertson, 1968; Ueno and Tanaka, 1996; Vinson and Guillot, 1972). Both glands are associated with the reproductive system of parasitoid females. The venom glands permanently fill the content of the reservoir of the VA. The shape of these glands varies from spherical to more elongated with filament-like structures. However, the morphology and organization of parasitoid VAs are highly variable between species. In some cases, the VA even lacks the reservoir and is then composed of a unique secreting gland (Edson and Vinson, 1979).

In *Asobara* parasitoids, the number of glands varies considerably from one species to another and even within a particular species. In *A. citri*, three glands generally surround the reservoir (Fig. 8.1), with an impaired gland always smaller than the other two. In *A. tabida*, the number of glands ranges from 3 to 10 with a median distribution centered on six glands per VA in the French A1 strain and seven to eight glands in the Dutch WOPV strain (Fig. 8.2). In *A. japonica*, the VA often possesses more than 10 glands (Fig. 8.3). These venom-secreting glands are similarly pear shaped in the *Asobara* genus and measure approximately 85–95 μm in mean diameter and 120–300 μm in length. The glands unload their secretions via thin ducts that converge to one small, afferent collecting duct just before reaching the upper part of the reservoir. In some other Braconids, the secreting glands can be independently linked to the reservoir and the venomous secretions are discharged either in the lower, middle or upper part of the reservoir (Edson and Vinson, 1979).

The reservoir serves both to store and deliver venom secretions to the ovipositor. In *Asobara* parasitoids, the reservoir functions like a pump, due to the presence of an internal helicoidal chitin layer associated with numerous external muscular fibers (Fig. 8.4). These muscular fibers form a thick wall around the reservoir. Muscle contractions compress the chitin helix and reduce the volume of the reservoir's central cavity. Venom is

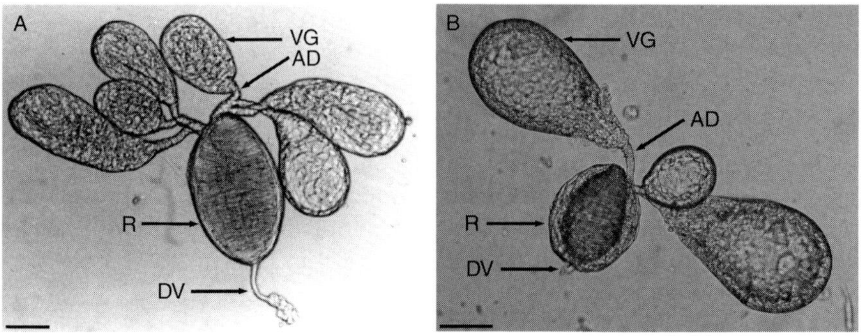

FIGURE 8.1 Venom apparatuses (VAs) from the braconid wasps *A. tabida* (A) and *A. citri* (B). Each VA consists of a central reservoir (R) collecting secretions from venom glands (VG) through an afferent duct (AD). At the basis of the reservoir an efferent duct, the ductus venatus (DV), connects the VA to the ovipositor. This morphological organization is typical of the VA of many Braconids. Scale bars represent 50 μm.

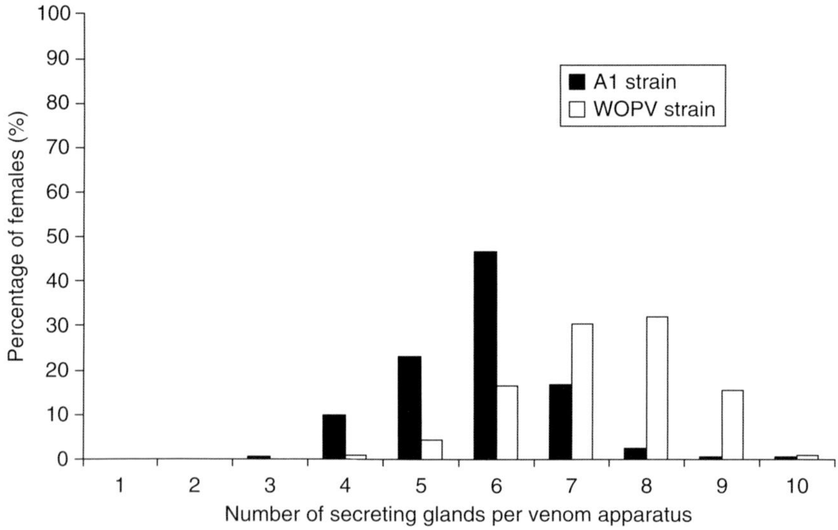

FIGURE 8.2 Intraspecific variations in the number of secreting glands per venom apparatus (VA) in *A. tabida* braconid wasps. VA from 208 females from the A1 strain and 90 females from the WOPV strain of *A. tabida* were dissected under a stereomicroscope in phosphate buffer saline (PBS) 0.15 M (pH 7.2) and the number of glands was immediately counted.

then ejected from the reservoir to the ovipositor. Conversely, a relaxing muscle causes the chitin helix to extend, which in turn induces the straining of the reservoir lumen and the influx of venom from the secreting glands to the reservoir. This function is similar to that of the reservoirs of well-studied

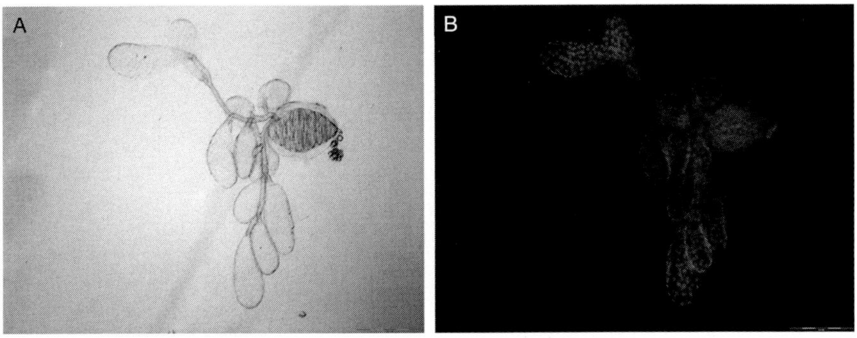

FIGURE 8.3 Venom apparatus (VA) of *A. japonica.* A VA was dissected in phosphate buffer saline (PBS) and then incubated for 15 min in a solution containing 0.1% (v/v) Triton X100 and 0.001% (w/v) Hoechst 33258 in PBS. The sample was transferred to a glass slide, mounted in PBS and covered with a glass coverslip. Images were obtained under an Olympus BX51 microscope, using a bright-field (A) or an epifluorescence microscope system equipped with a 330–385 nm excitation filter (B). The Hoechst 33258 staining of DNA reveals the positions of the nuclei. Note that the large nuclei of secreting cells are organized in a single layer within each gland. Scale bars represent 200 μm. (See Page 6 in Color Section at the back of the book.)

Braconids such as *Bracon hebetor* and *Aphidius ervi* (Beard, 1978; Edson and Vinson, 1979; van Marle, 1977). In other species (e.g., *Chelonus* spp. near *curvimaculatus, Nasonia vitripennis, Cardiochiles nigriceps, Eupelmus orientalis, Anisopteromalus calandrae, Pteromalus cerealellae, Euplectrus* spp. near *plathypenae*), the reservoirs are only composed of a thin epithelium surrounded by a few muscles (Doury et al., 1997; Edson and Vinson, 1979; Howard and Baker, 2003; King and Ratcliffe, 1969; Nakamatsu and Tanaka, 2003; Robertson, 1968; van Marle, 1977). In such cases, strong abdominal contractions would ensure the ejection of venom to the ovipositor (van Marle, 1977), while the reservoir would be filled passively.

Newly emerged *Asobara* females have fully formed glands and a small spherical reservoir that progressively becomes elongated as it is filled by venomous secretions within the first 3 days after ecdysis. The reservoir reaches its maximum volume between days 3 and 5 after ecdysis. Its average length is approximately 300 μm (Ø about 110 μm) in *A. tabida* and *A. japonica,* and only 140 μm (Ø about 60 μm) in *A. citri*. There is no correlation between the width and length of the reservoir and the number of glands surrounding the reservoir. In *A. tabida,* but not in *A. citri,* the parameters of width and length of the reservoir are strongly associated with the size of the female wasp, the bigger females having the bigger reservoirs. Due to the continued secretion activity of the glands, the content of the reservoir is restored even after a significant number of successive ovipositions. The reservoir content thus does not significantly vary quantitatively or qualitatively with the female wasp's age and

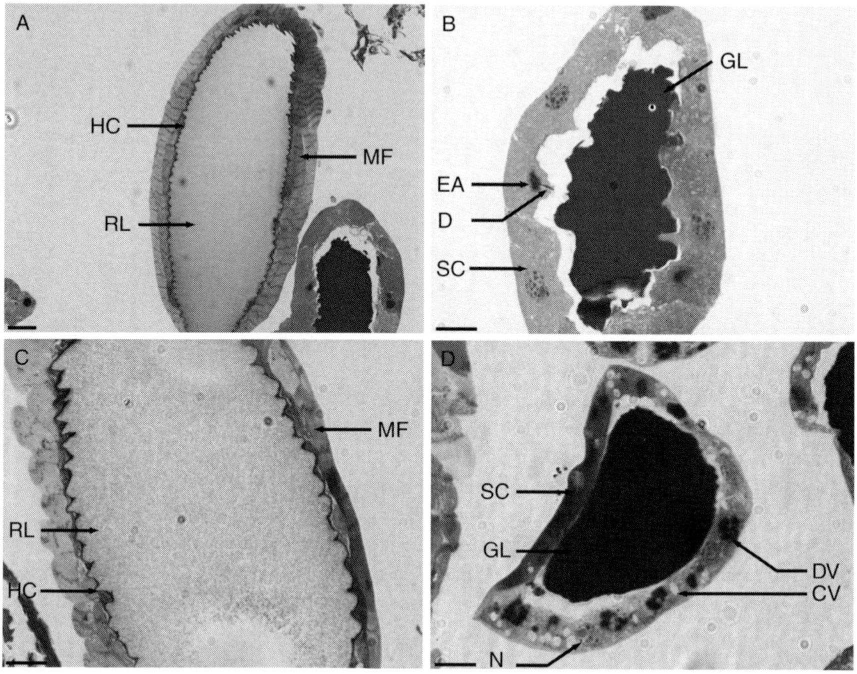

FIGURE 8.4 Semi-thin sections of the venom apparatus (VA) of *A. japonica* and *A. tabida*. Semi-thin sections of reservoirs and glands of VA of *A. japonica* (A, B) and *A. tabida* (C, D), respectively, are shown after toluidine blue staining. Samples were dissected in phosphate buffer saline (PBS), then fixed in 1% (v/v) glutaraldehyde and 4% (w/v) paraformaldehyde in PBS. VAs were rinsed in PBS added with 0.2% (w/v) sodium chloride and postfixed in a 2% (w/v) osmium tetroxide solution in 0.1 M phosphate buffer. The sample was dehydrated in ethanol and finally embedded overnight in Epon. Semi-thin sections (0.5 μm) were stained with 0.5% toluidine blue in 1% sodium borate, mounted and observed under an Olympus BX51 microscope. CV, clear vacuole; D, ductule; DV, dense vacuole; EA, end apparatus; GL, gland's lumen; HC, internal helix of chitin; MF, muscular fiber; N, nucleus; RL, reservoir's lumen; SC, secretory cell. Scale bars represent 20 μm. (See Page 6 in Color Section at the back of the book.)

experience (personal observations). In detergent-free conditions of extractions (i.e., by gently squeezing VA in phosphate buffered saline (PBS), approximately 1 μg of soluble venom proteins per VA can be recovered from a single 3–5 days old *A. tabida* female.

Reservoirs in *Asobara* spp. do not exhibit a secondary secreting ability like the reservoirs of the ectoparasitoid wasps *B. hebetor* (van Marle, 1977) and *N. vitripennis* (King and Ratcliffe, 1969), the endoparasitoid *Diadromus collaris* (Li et al., 2006) and the solitary sphecid wasp *Liris niger* (Gnatzy and Volknandt, 2000).

8.2.2. Ultrastructural study of venom glands in *A. japonica* and *A. tabida*

In *A. japonica*, each gland contains one layer of secreting cells (Fig. 8.5) filled with an extended, rough, endoplasmic reticulum, transparent and electron-dense vesicles, and numerous Golgi apparatuses that reveal their highly secretory activity. In each secretory cell, vacuoles containing venom secretions are directed to a microvillar surface that goes on extra-cellular spaces called secreting organelles or ductules (Fig. 8.5). Each

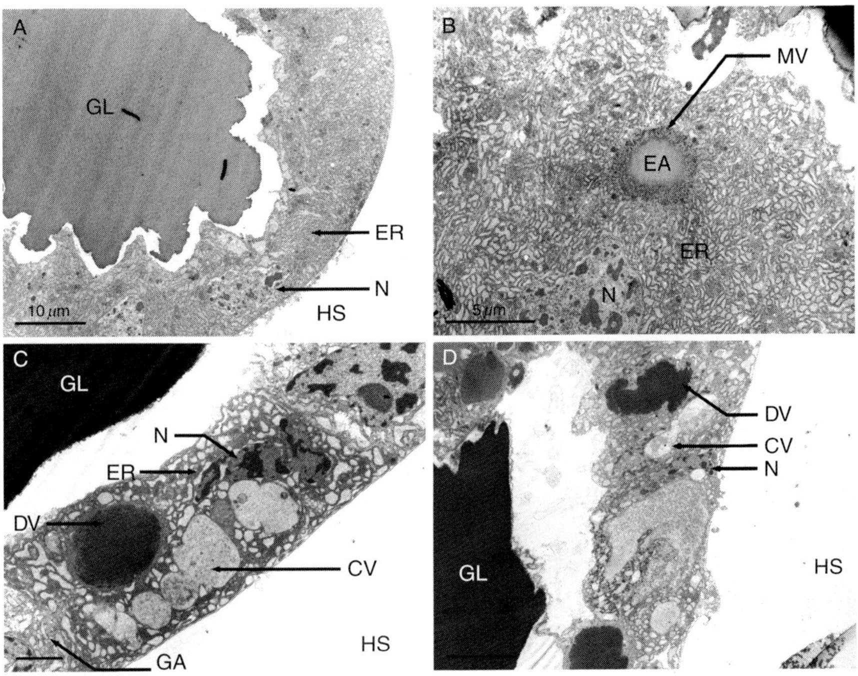

FIGURE 8.5 Structural and ultrastructural views of the venom glands in *A. japonica* and *A. tabida*. Ultrathin sections (80–90 nm) of venom apparatuses prepared as mentioned for Fig. 8.4 were stained with uranyl acetate and lead citrate and observed in a JEOL 1011 transmission electron microscopy. (A) General view of the ultrastructure of a venom gland of *A. japonica*. Scale bar represents 10 μm. (B) Detail of the ultrastructure of a secretory cell of *A. japonica*'s venom gland showing the abundant rough endoplasmic reticulum and an end apparatus lined with numerous microvillar digitations. Scale bar represents 5 μm. C and D Ultrastructure of venom secretory cells observed in *A. tabida*. Note the presence of large vacuoles, either clear or electron-dense, that occupy a large proportion of the cytoplasm, and the absence of end apparatuses as observed in *A. japonica*. Scale bars represent 2 (in C) or 5 μm (in D). CV, clear vacuole; DV, dense vacuole; EA, end apparatus; ER, endoplasmic reticulum; GL, gland's lumen; HS, hemolymph space; MV, microvilli; N, nucleus.

ductule is delimited by a chitin layer lined with an epithelial cell (duct cell) and communicates with the gland lumen, which is also lined with a thin cuticle (Fig. 8.5). Such a histological organization is similar to that of venom glands of social Hymenoptera of the genus *Vespa* and *Apis* (van Marle, 1977) and to some other parasitoid and solitary wasp species (Gnatzy and Volknandt, 2000; King and Ratcliffe, 1969; van Marle, 1977). It corresponds to "class III" insect epidermal glands as described by Noirot and Quennedey (1974) in which several cells are deployed along a cuticular duct which drains the secretion of the secretory cells outside the secretory epithelium.

The venom glands of *A. tabida* are also composed of a large central lumen surrounded by a single layer of secretory cells (Fig. 8.5). Secretions produced by these cells are unloaded into the gland lumen through similar ductules as in *A. japonica*. However, the ultrastructure of the secretory cells is very different to that of *A. japonica*. The most important difference concerns the absence of microvillar surfaces. A succession of large, transparent and electron-dense vacuoles can be observed in the cytoplasm of secreting cells, surrounded by abundant endoplasmic reticulum and Golgi apparatuses (Fig. 8.5). The ontogeny and functioning of these gland cells are not yet fully understood and the process by which venom secretions are synthesized, processed and unloaded still constitutes a persistent black box with promising developments for future investigations.

8.3. THE VENOM OF *A. TABIDA*

8.3.1. Physiological effects

Unlike other braconid endoparasitoids, *A. tabida* is naturally deprived of virus-like particles and polydnaviruses (Eslin et al., 1996; personal observations). *A. tabida* also lacks teratocytes, which are cells derived from the serosal membrane of the parasitoid egg and observed in several hosts endoparasitized by braconid species (Dahlman and Vinson, 1993). Both these factors are known to contribute greatly to the survival of the parasitoid as they can notably affect the immunity and development of the host (Gupta and Ferkovitch, 1998; Nogushi et al., 1995; Pennacchio et al., 1994; Qin et al., 2000; Strand and Dover, 1991). In the absence of such factors, the parasite's success mainly relies on surface features of *A. tabida* eggs and on the physiological effects of its venom.

The exochorion of *A. tabida*'s eggs is made of a fibrous layer responsible for their adhesiveness to the internal tissues of *Drosophila* larvae, thus protecting them from encapsulation by circulating hemocytes (Eslin and

Prévost, 2000; Eslin et al., 1996; Kraaijeveld and van Alphen, 1994). In addition, *A. tabida* induces a slight but significant reduction of hemolymph phenoloxidase activity (Moreau et al., 2000)—an essential enzyme of the host immune system (Nappi et al., 1991)—and a delayed development in parasitized hosts (Moreau et al., 2002).

Injections of *A. tabida* venom extracts into *Drosophila* larvae induced transient paralysis and mortality of the host in a dose-dependant manner (Moreau et al., 2002). Interestingly, the venom of the WOPV strain, whose virulence toward *D. melanogaster* larvae is low, induced higher mortality rates and a stronger paralyzing effect compared to the venom of the virulent A1 strain. These results were the first report of intraspecific variation in the ability of parasitoids to induce paralysis. The nature of the venom components that can cause such strain-specific effects is still unknown. Whether the transient paralyzing effect, initially described by van Alphen (1982) after parasitization, could be related to the mortality observed after venom injection is not yet established. Even though transient, this effect differs from what has been shown in other endoparasitoid Braconids, whose venoms are commonly considered nonparalyzing (Coudron, 1991). In contrast, several ectoparasitoid venoms have been reported to induce host developmental arrest (Coudron and Brandt, 1996; Rivers and Denlinger, 1995; Weaver et al., 1997) and most of them cause paralysis (for review, see Moreau and Guillot, 2005). The ability of *A. tabida* to delay development in parasitized hosts and to induce paralysis suggested that this species could share some properties with ectoparasitoid braconids even though its lifestyle was clearly endoparasitic. This view is in accordance with Dowton et al. (1998) who suggested on the basis of molecular phylogeny that the *Asobara* genus, which belongs to a predominantly ectoparasitic clade, would have reverted to endoparasitism. Thus, the study of the physiological effects of *A. tabida*'s venom led to unexpected results which provided new insights on the evolution of parasitic wasps in combination with data inferred from molecular phylogeny. It would be interesting to test if the venoms of other *Asobara* spp. have also retained the ancestral ability to paralyze their natural hosts.

8.3.2. Protein composition

Like numerous parasitoid Hymenoptera, the venom of *A. tabida* contains acidic proteins with isoelectric points ranking from 5.1 to 6.5. The most abundant venom proteins of *A. tabida* have approximate molecular masses of 200, 130, 110, 76, 69, 44, 30 and 18 kDa, and were designated as P200-P18. Two additional major proteins of 10 and 8.5 kDa were also detected in venom extracts (Fig. 8.6; Moreau et al., 2004). Three of the most

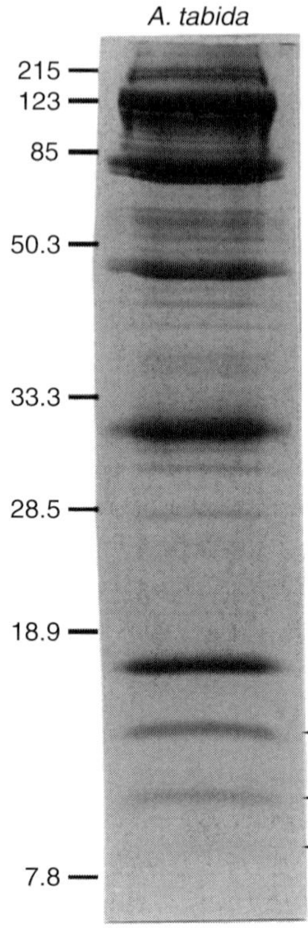

FIGURE 8.6 Protein analysis of the venom extracts of *A. tabida*. Venom apparatuses were dissected and their content allowed diffusing in PBS. Venom proteins were ran on a 15% sodium dodecyl sulfate polyacrylamide gel electrophoresis and stained with R250 Coomassie brilliant blue. (See Page 7 in Color Section at the back of the book.)

abundant venom proteins (P44, P30, P18) were identified as either subunits or precursors of an aspartylglucosaminidase (AtAGA) enzyme.

The API-ZYM system (Biomérieux) allowed us to detect several enzyme activities in *A. tabida* venom extracts and among them, mostly hydrolases such as phosphatase, arylamidase, lipase and protease (Fig. 8.7).

A complementary deoxyribonucleic acid (cDNA) library was recently constructed from messenger ribonucleic acid (mRNA) extracted from venom glands of *A. tabida*. The sequencing of the expressed sequence tags (ESTs) allowed the identification of proteins sharing homologies with several enzymes (Vinchon et al., unpublished results). An extensive work is currently underway to analyze the transcriptome of the venom glands of *A. tabida* fully and to identify the main venomous components.

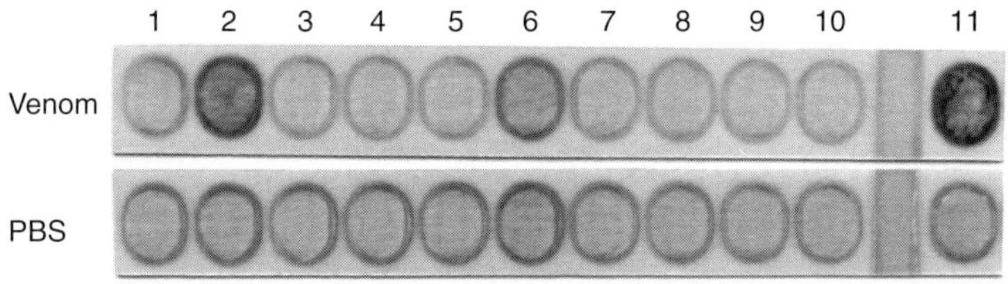

FIGURE 8.7 Detection of enzymatic activities in the venom of *A. tabida* using the API-ZYM system. The API-ZYM system (Biomérieux, Marcy l'Etoile, France) allowed the detection of strong alkaline phosphatase (well 2) and acid phosphatase (well 11) activities in venom extracts from *A. tabida*. A weak leucine arylamidase activity (well 6) was also detected at the limits of the detection threshold. The specific substrates adsorbed in the other wells were fitted to detect the following activities: well 1, negative control (no substrate); well 2, alkaline phosphatase; well 3, C4 esterase; well 4, C8 esterase/lipase; well 5, C14 lipase; well 6, L-Leucine arylamidase; well 7, L-valine arylamidase; well 8, L-cystine arylamidase; well 9, trypsin; well 10, α-chymotrypsin; well 11, acid phosphatase. The control line consisted of replacing venom by PBS for the tests.

8.3.3. The aspartylglucosaminidase (AtAGA)

The main component of the venom of *A. tabida* corresponds to an AtAGA enzyme (E.C. 3.5.1.26; Moreau et al., 2004). Based on analogies to other enzymes to which it is related, it has been suggested that this protein, once injected into the host, could be responsible for the production of aspartate, which is a known excitatory neurotransmitter of the *Drosophila* nervous system (Besson et al., 2000), a precursor of other amino acids and an intermediate of the citrate cycle. Therefore, AtAGA could potentially interfere with the host's neurophysiology or metabolism.

AtAGAs are typically lysosomal enzymes virtually present in almost all eucaryotic cells that possess lysosomes. The full length sequence of the main venomous enzyme of *A. tabida*, AtAGA, is now available (Vinchon et al., unpublished results). It clearly derives from a lysosomal gene and is specifically expressed by the venom gland cells and not by other tissues of the parasitoid wasp. The functional and molecular evolution of this enzyme might well represent a typical evolutionary case of a whole class of insect venom components and therefore, be worth investigating.

8.4. THE VENOM OF *A. JAPONICA*

8.4.1. Physiological effects

A. tabida eggs benefit from their chorion adhesive properties to escape from encapsulation in *D. melanogaster* larvae. Conversely, *A. japonica* eggs are never encapsulated although their eggs are nonadhesive. To prevent host defenses, the parasitoid has developed an aggressive strategy to regulate the immune system of its host. Several other effects are observed during oviposition by *A. japonica*. Since the *Asobara* genus seems to be devoid of any endosymbiotic virus, effects on the parasitized hosts must be closely and mainly dependant on the components of the fluids accompanying the egg during oviposition. Alterations of the host's hematopoietic organ occur concomitantly to the decrease of the hemocyte population and the phenoloxidase activity (Mabiala-Moundoungou et al., unpublished results).

A. japonica venom has been extracted from the glands and reservoir, and tested by injection into *D. melanogaster* larvae. The venom extracts from *A. japonica* induce host death within hours following injection, thus demonstrating a strong effect of the venom on the host physiology.

8.4.2. Protein composition

Venom extracts from *A. japonica and A. citri* which were analyzed for their protein content by sodium dodecyl sulfate polyacrylamide gel electrophoresis (SDS-PAGE) and silver staining revealed numerous polypeptides ranging from 16 to 150 kDa. Several proteins ranging from 30 to 80 kDa are not clearly identified (Fig. 8.8). The analysis of *A. japonica* venom proteins proved the absence of aspartylglucosaminidase (AGA), the main component of *A. tabida* venom: specific antibodies raised against the subunits of AtAGA showed no cross reaction with *A. japonica* venom proteins and no AGA activity has been detected in the venom extract from *A. japonica*.

As with *A. tabida* venom, several proteins were found belonging to protease and phospholipase families identified and characterized in the venom of several other insects and vertebrates (snakes).

8.5. EXPECTED PROSPECTS FROM STUDYING VENOMS IN THE *ASOBARA* GENUS

Secreted components of the wasps' venoms may have evolved differently depending on the habitat and ecological niche of the parasitoid species. However, the similarity of the venom gland proteins among Hymenoptera seems to indicate that the biosynthetic pathways have been particularly well conserved during evolutionary processes. Venom glands mostly synthesize the same class of molecules (acidic proteins) although being morphologically and functionally different. Interestingly, a number of enzymes of lysosomal origin have been described from the venoms of numerous Hymenoptera. The AtAGA is a characteristically lysosomal resident protein observed in *A. tabida* venom extract only, and not in the

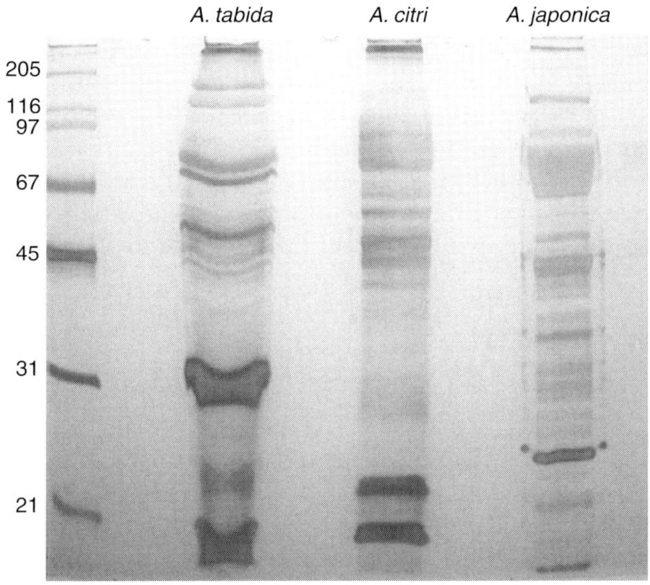

FIGURE 8.8 Analysis of *Asobara* venoms by sodium dodecyl sulfate polyacrylamide gel electrophoresis (SDS-PAGE). Samples of 5 μg of protein extracts collected from the venom glands of *A. tabida*, *A. japonica* and *A. citri* were subjected to SDS-PAGE under reducing conditions and 10% gels were submitted to silver staining. Molecular weight values of the markers are shown on the left. (See Page 8 in Color Section at the back of the book.)

venom of other studied *Asobara* spp. Classically, lysosomal enzymes are labeled with mannose-6-phosphate (M-6-P) which allows their recognition by M-6-P receptors and their correct addressing to lysosomes. Interestingly, AGA and other lysosomal-like proteins (identified by enzyme detection or cDNA construction) are commonly observed in hymenoptera venom glands.

In the *Asobara* genus, species geographically distant exhibit very different strategies of virulence. The European species *A. tabida* itself has two different strains that present two parasitism destinies. Also, neither the WOPV nor the A1 strain of *A. tabida* possesses venom exhibiting a physiological effect on *D. melanogaster* larvae as strong as that observed in terms of induced mortality upon injection of *A. japonica's* venom. Other *Asobara* spp. associated with the host *D. melanogaster* show intermediate effects (Mabiala-Moundoungou, personal communication). In order to investigate the molecular tools that parasitoids developed to overcome the defenses of their host, an extended analysis of the venom and ovary contents of female wasps from several *Asobara* spp. has been recently initiated. Protein separation and analysis, associated with the analysis of cDNA libraries from *A. tabida* and *A. japonica* venoms, will undeniably be of great interest to identify active molecules mediating the physiological effects of these parasitoids on their *Drosophila* hosts. The deadly molecules in *A. japonica* venom are under investigation. The comparison presently carried on of two other *Asobara* spp., namely *A. citri* and *A. persimilis*, should bring some insight into understanding the diversity of the molecular factors in venom involved in the host–parasitoid interactions.

ACKNOWLEDGMENTS

The authors wish to thank the Pr B. Arbeille and the members of the Department of Microscopy, University François-Rabelais (Tours, France) for acquisition of pictures in TEM.

REFERENCES

Asgari, S., 2006. Venom proteins from polydnavirus-producing endoparasitoids: their role in host–parasite interactions. Arch. Insect Biochem. Physiol. 61, 146–156.

Asgari, S., 2007. Endoparasitoid venom proteins as modulators of host immunity and development. In: Rivers, Yoder, (Eds.), Recent Advances in the Biochemistry, Toxicity, and Mode of Action of Parasitic Wasp Venoms. Research Signpost, Kerala, India, pp. 57–73.

Barrera, J.F., Gomez, J., Alauzet, C., 1994. Evidence for a marking pheromone in host discrimination by *Cephalonomia stephanoderis* (Hym.: Bethylidae). Entomophaga 36, 363–366.

Beard, R.L., 1978. Venoms of Braconidae. In: Bettini, (Ed.), Handbuch der Experimentellen Pharmakologie, vol. 48: Arthropod Venoms. Springer Verlag, Berlin, pp. 773–800.

Bender, J.C., 1943. Anatomy and histology of the female reproductive organs of *Habrobracon juglandis* (Ashmead) (Hymenoptera, Braconidae). Ann. Entomol. Soc. Am. 36, 537–545.

Besson, M.T., Soustelle, L., Birman, S., 2000. Selective high-affinity transport of aspartate by a *Drosophila* homologue of the excitatory amino acid transporters. Curr. Biol. 10, 207–210.

Blass, S., Ruthmann, A., 1989. Fine structure of the accessory gland of the female genital tract of the ichneumonid *Pimpla turionellae* (Insecta, Hymenoptera). Zoomorphology 108, 367–377.

Coudron, T.A., 1991. Host regulating factors associated with parasitic Hymenoptera. In: Hedin, (Ed.), Naturally Occuring Pest Bioregulators, vol. 449. American Chemical Society, Washington, DC, pp. 41–65 A.C.S. Symposium Series.

Coudron, T.A., Brandt, S.L., 1996. Characteristics of a developmental arrestant in the venom of the ectoparasitoid wasp *Euplectrus comstockii*. Toxicon 34, 1431–1441.

Dahlman, D.L., Vinson, S.B., 1993. Teratocytes: developmental and biochemical characteristics. In: Beckage, Thompson, Federici, (Eds.), Parasites and Pathogenes of Insects, Vol. 1. Parasites. Academic Press, San Diego, pp. 145–165.

Doury, G., Bigot, Y., Periquet, G., 1997. Physiological and biochemical analysis of factors in the female venom gland and larval salivary secretions of the ectoparasitoid wasp *Eupelmus orientalis*. J. Insect Physiol. 43, 69–81.

Dowton, M., Austin, A.D., Antolin, M.F., 1998. Evolutionary relationships among the Braconidae (Hymenoptera: Ichneumonoidae) inferred from partial 16S rDNA gene sequences. Insect Mol. Biol. 7, 129–150.

Edson, K.M., Barlin, M.R., Vinson, S.B., 1982. Venom apparatus of braconid wasps: comparative ultrastructure of reservoirs and gland filaments. Toxicon 20, 553–562.

Edson, K.M., Vinson, S.B., 1979. A comparative morphology of the venom apparatus of females Braconids (Hymenoptera: Braconidae). Can. Entomol. 111, 1013–1102.

Eslin, P., Giordanengo, P., Fourdrain, Y., Prévost, G., 1996. Avoidance of encapsulation in the absence of VLP by a braconid parasitoid of *Drosophila* larvae: an ultrastructural study. Can. J. Zool. 74, 2193–2198.

Eslin, P., Prévost, G., 2000. Racing against host's immunity defenses: a likely strategy for passive evasion of encapsulation in *Asobara tabida* parasitoids. J. Insect Physiol. 46, 1161–1167.

Gnatzy, W., Volknandt, W., 2000. Venom gland of the digger wasp *Liris niger*: morphology, ultrastructure, age-related changes and biochemical aspects. Cell Tissue Res. 302, 271–284.

Guillot, F.S., Joiner, R.L., Vinson, S.B., 1974. Host discrimination from Dufour's gland of a braconid parasitoid. Ann. Entomol. Soc. Am. 67, 720–721.

Gupta, P., Ferkovitch, S.M., 1998. Interaction of calyx fluid and venom from *Microplitis croceipes* (Braconidae) on developmental disruption of the natural host, *Helicoverpa zea*, and two atypical hosts, *Galleria mellonella* and *Spodoptera exigua*. J. Insect Physiol. 44, 713–719.

Howard, R.W., Baker, J.E., 2003. Morphology and chemistry of Dufour glands in four ectoparasitoids: *Cephalonomia tarsalis, C. waterstoni* (Hymenoptera:Bethylidae), *Anisopteromalus calandrae* and *Pteromalus cerealellae* (Hymenoptera:Pteromalidae). Comp. Biochem. Physiol. B 135, 153–167.

King, P.E., Ratcliffe, N.A., 1969. The structure and possible mode of functioning of the female reproductive system in *Nasonia vitripennis* (Hymenoptera: Pteromalidae). J. Zool. 157, 319–344.

Kraaijeveld, A.R., van Alphen, J.M., 1994. Geographical variation in resistance of the parasitoid *Asobara tabida* against encapsulation by *Drosophila melanogaster* larvae: the mechanism explored. Phys. Entomol. 19, 9–14.

Li, W.-D., Yu, R.-X., Chen, X.-X., He, J.-H., 2006. Venom gland of the ichneumonid *Diadromus collaris*: morphology, ultrastructure and age-related changes. Insect Sci. 13, 137–143.

Moreau, S.J.M., Cherqui, A., Dubois, F., Doury, G., Fourdrain, Y., Bulet, P., et al., 2004. Identification of an aspartylglucosaminidase-like protein in the venom of the parasitic wasp *Asobara tabida* (Hymenoptera: Braconidae). Insect Biochem. Mol. Biol. 34, 485–492.

Moreau, S.J.M., Dingremont, A., Doury, G., Giordanengo, P., 2002. Effects of parasitism by *Asobara tabida* (Hymenoptera: Braconidae) on the development, survival and activity of *Drosophila melanogaster* larvae. J. Insect Physiol. 48, 337–347.

Moreau, S.J.M., Doury, G., Giordanengo, P., 2000. Intraspecific variation in the effects of parasitism by *Asobara tabida* on phenoloxidase activity of *Drosophila melanogaster* larvae. J. Invert. Pathol. 76, 151–153.

Moreau, S.J.M., Guillot, S., 2005. Advances and prospects on biosynthesis, structures and functions of venom proteins from parasitic wasps. Insect Biochem. Mol. Biol. 35, 1209–1223.

Nakamatsu, Y., Tanaka, T., 2003. Venom of ectoparasitoid, *Euplectrus* sp. near *plathypenae* (Hymenoptera: Eulophidae) regulates the physiological sate of *Pseudaletia separate* (Lepidoptera: Noctuidae) host as a food resource. J. Insect Physiol. 49, 149–159.

Nappi, A.J., Carton, Y., Frey, F., 1991. Parasite-induced enhancement of hemolymph tyrosinase activity in a selected immune reactive strain of *Drosophila melanogaster*. Arch. Insect Biochem. Physiol. 18, 159–168.

Nogushi, H., Hayakawa, Y., Downer, R.G.H., 1995. Elevation of dopamine levels in parasitized insect larvae. Insect Biochem. Mol. Biol. 25, 197–201.

Noirot, C., Quennedey, A., 1974. Fine structure of insect epidermal glands. Ann. Rev. Entomol. 19, 61–80.

Pennacchio, F., Vinson, S.B., Tremblay, E., Ostuni, A., 1994. Alteration of ecdysone metabolism in *Heliothis virescens* (F.) (Lepidoptera: Noctuidae) larvae induced by *Cardiochiles nigriceps* Viereck (Hymenoptera: Braconidae) teratocytes. Insect Biochem. Mol. Biol. 24, 383–394.

Prévost, G., Eslin, P., Doury, G., Moreau, S.J.M., Guillot, S., 2005. *Asobara*, braconid parasitoids of *Drosophila* larvae: unusual strategies to avoid encapsulation without VLPs. J. Insect Physiol. 51, 171–179.

Qin, Q., Gong, H., Ding, T., 2000. Two collagenases are secreted by teratocytes from *Microplitis mediator* (Hymeoptera: Braconidae) cultured *in vitro*. J. Invert. Pathol. 76, 79–80.

Rivers, D.B., Denlinger, D.L., 1995. Venom-induced alterations in fly lipid metabolism and its impact on larval development of the ectoparasitoid *Nasonia vitripennis* (Walker) (Hymenoptera: Pteromalidae). J. Invert. Pathol. 66, 104–110.

Robertson, P.L., 1968. A morphological and functional study of the venom apparatus in representatives of some major groups of Hymenoptera. Aust. J. Zool. 16, 133–166.

Soliman, H.S., 1941. Studies in the structure of *Microbracon hebetor* Say. Bull. Soc. Entomol. Egyp. 25, 1–97.

Strand, M.R., Dover, B.A., 1991. Developmental disruption of *Pseudaletia includens* and *Heliothis virescens* larvae by the calyx fluid and venom of *Microplitis demolitor*. Arch. Insect Biochem. Physiol. 18, 131–145.

Ueno, T., Tanaka, T., 1996. Self-host discrimination by a parasitic wasp: the role of short-term memory. Animal Behav. 52, 875–883.

Van Alphen, J.J.M., 1982. Foraging Behaviour of *Asobara tabida*, a Larval Parasitoid of *Drosophila*. University of Leiden, The Netherlands [thesis].

Van Marle, J., 1977. Structure and histochemistry of the venom glands of the wasps *Microbracon hebetor* Say and *Philanthus triangulum* F. Toxicon 15, 529–539.

Van Marle, J., Piek, T., 1986. Morphology of the venom apparatus. In: Piek, (Ed.), Venoms of the Hymenoptera. Academic Press, London, pp. 17–44.

Vinson, S.B., Guillot, F.S., 1972. Host marking: source of a substance that results in host discrimination in insect parasitoids. Entomophaga 17, 241–245.

Weaver, R.J., Marris, G.C., Olieff, S., Mosson, J.H., Edwards, J.P., 1997. Role of ectoparasitoid venom in the regulation of haemolymph ecdysteroid titres in a host noctuid moth. Arch. Insect Biochem. Physiol. 35, 169–178.

SECTION 3

STRATEGIES AND EVOLUTION OF PARASITOID VIRULENCE AND HOST RESISTANCE

CHAPTER 9

Strategies of Avoidance of Host Immune Defenses in *Asobara* Species

Geneviève Prévost, Géraldine Doury, Alix D.N. Mabiala-Moundoungou, Anas Cherqui, and **Patrice Eslin**

Contents

Abstract

Eggs and larvae of endophagous parasitoids face the host's immunity reaction once they penetrate the insect host's hemocele. In order to overcome the host's immune barrier, endoparasitoids have developed various strategies. Conformer parasitoids hide and/or get protected from the attack by the host's immunity cells without interfering with the host's immune system. Differently, regulator parasitoids directly attack the host's hemocytes, therefore totally inhibiting the immunity reaction of encapsulation in the parasitized

Laboratoire de Biologie des Entomophages, EA 3900 BioPI, Université de Picardie-Jules Verne, 33 rue Saint Leu, 80039 Amiens cedex, France

Advances in Parasitology, Volume 70
ISSN 0065-308X, DOI: 10.1016/S0065-308X(09)70009-0

host. Female wasps may also discriminate immunoreactive hosts from nonreactive, permissive ones before laying an egg. These different strategies coexist within the same genus of the braconids *Asobara*, endoparasitoids of *Drosophila* larvae. The physiological mechanisms underlying the conformer and regulator strategies in *Asobara* are exposed. The factors which may contribute to the diversity of the means developed by *Asobara* parasitoids to overcome the hosts' immunity defenses are discussed.

9.1. INTRODUCTION

Drosophila species are distributed world wide and among them, *Drosophila melanogaster*, the very cosmopolitan species, is potentially parasitized by many larval parasitoids of different families, genus and species. This provides us with a unique opportunity to study host–parasitoid relationships between different wasp species and *D. melanogaster* as the reference host, and to compare the means different parasitoids have developed to succeed in their parasitic life.

Parasitoids of *Drosophila* have long been used as models to study the immunity defense reactions of insect host larvae against large parasites, with both humoral and cellular aspects of the so-called reaction of encapsulation being described in *D. melanogaster* (Nappi and Streams, 1969; Nappi and Vass, 1993, 2001; Nappi et al., 1991; Rizki and Rizki, 1980). Until recently when it was shown that an inert foreign body can be used to observe the hemocytic reaction of encapsulation of *Drosophila* larvae (Eslin and Doury, 2006), the formation of the melanized cellular capsules had been described using larvae parasitized by endoparasitoids, mostly of *Leptopilina* genus (Carton and Nappi, 1997, 2001; Carton et al., 2008; Nappi et al., 1995; Russo et al., 1996). From these studies a broad knowledge has also been built up on the physiology, biochemistry and genetics of the virulence of *Leptopilina* parasitoids (Carton and Boulétreau, 1985; Carton et al., 1992, 2008; Dubuffet et al., 2007; Dupas et al., 2003; Labrosse et al., 2003, 2005; Poirié et al., 2009; Chapter 4 by Nappi et al.; Chapter 6 by Dubuffet et al.). It is more recent that the physiological and immunity aspects of host–parasitoid relationships have been studied using *Asobara* as larval parasitoids of *D. melanogaster* (Eslin and Prévost 1996, 2000; Eslin et al., 1996; Moreau et al., 2002, 2003; Nappi, 1981; Prévost et al., 2005).

The *Asobara* genus is represented by a complex of species, several of which have been well studied. The species *Asobara tabida* occurs over most Europe (Carton et al., 1986; Kraaijeveld and van der Wel, 1994) and was recently observed in the Iberian Peninsula (van Alphen, personal communication). It is also found in North America. Its sibling species, *A. rufescens*, which has a wide European distribution (Kraaijeveld et al., 1994), was also found as far South as Spain and Portugal. Other *Asobara* species which have been studied include *A. citri*, a species originating from Africa, and *A. japonica*, whose range is limited to Japan (Ideo et al., 2008; Mitsui et al., 2007). *A. persimilis* from Australia and *A.* near *orientalis* from Indonesia were more recently collected and both are presently under investigation. One striking aspect of the biology of *Asobara* species is the diversity of the means different species use to deal with their hosts' immunity defenses.

In the early 1980s, Vinson and Iwantsch (1980a,b) described the host adequation and host regulation as two key steps of the success of endoparasitoid development. Since then, an abundant literature showed that parasitoids have evolved an amazing diversity of mechanisms to manipulate host physiology, therefore creating an environment favorable for their own development (for review, see Beckage and Gelman, 2004). Lawrence (1986, 1990) suggested that parasitoids could be divided into two categories on the basis of their interactions with their hosts: "regulator" parasitoids, which would trigger disruption of the host physiology, and "conformer" parasitoids, which would not redirect the host development. A similar classification can apply to immunity relationships, with parasitoids either actively disrupting/depressing/suppressing their host's immunity system (regulators), or using a "passive" strategy to avoid the immune reaction (conformers) (Moreau, 2003; Prévost et al., 2005; Schmidt et al., 2001; Strand and Pech, 1995; Vass and Nappi, 2000; Vinson, 1990).

Conformer versus regulator strategies are here discussed on the basis of the physiological ways parasitoids of the *Asobara* genus developed to either avoid or overcome their host immunity defenses.

9.2. CONFORMER VERSUS REGULATOR STRATEGY

Nonpermissive hosts often eliminate endoparasitoids by encapsulation, which involves adhesion of cellular layers of hemocytes to the parasitoid egg or larva, usually associated with blackening of the hemocytic capsule due to melanization (Nappi, 1981; Nappi and Vass, 1993, 2001; Russo et al., 1996). Larval endoparasitoids evade encapsulation and other host defenses passively and/or by inhibiting the host's immune system (Schmidt et al., 2001; Strand and Pech, 1995).

9.2.1. The conformer strategy in *Asobara* parasitoids

Conformer strategies include oviposition or development in host tissues that are inaccessible to host hemocytes. Many plastygastrids, for example, oviposit in the host's gut or ganglia where they keep out of reach of the host immunity cells, melanization and cytotoxic by-products present in the host hemolymph (Strand and Pech, 1995). Other parasitoids may display physical properties at the surface of their eggs that protect them from encapsulation (Davies and Vinson, 1986) or can evade nonself recognition using antigenic mimicry (Asgari and Schmidt, 1994; Asgari et al., 1998; Hayakawa and Yazaki, 1997).

The species *A. tabida* has developed an original way to evade encapsulation. As in most endoparasitoid species, eggs are laid in the host hemocele and therefore exposed to attack by the host hemocytes. Study of the hemocyte population in the hemolymph of parasitized *Drosophila* larvae proved that *A. tabida* eggs are well recognized by the host's immune system. Oviposition by *A. tabida* is followed by a burst of lamellocytes in the host hemolymph, a hemocytic reaction which typically indicates that the host immune system is responding to the presence of the parasite (Eslin and Prévost, 1996, 1998; Chapter 7 by Eslin et al.). In the A1 strain of *A. tabida*, the exochorion of the parasitic egg possesses adhesive properties such that it can attach to almost any host tissue floating in the hemocele (fat body, digestive tube, tracheal cells; Fig. 9.1A; Eslin et al., 1996; Kraaijeveld and van Alphen, 1994; Monconduit and Prévost, 1994). Movements of the host larva probably contribute to create many contact areas between the parasitic egg (at first floating free in the host hemocele) and the host tissues, therefore resulting to the embedment of the

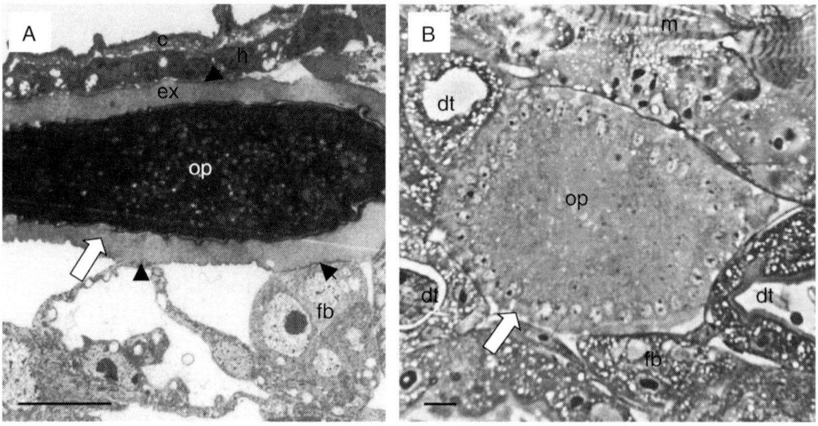

FIGURE 9.1 (A) Electron micrograph of a section of *A. tabida* (A1 strain) egg (white arrow) in the hemocele of a *D. melanogaster* first-instar host larva 3 h after parasitization. The surface of the exochorion (ex) sticks to the host tissues (black arrow head). (B) Light micrograph showing the parasitic egg (white arrow) completely embedded in the tissues of a first instar host larva 12 h after parasitization. c, larval cuticle; h, hypodermal tissue; fb, fat body; dt, digestive tract; m, muscle; ex, exochorion; op, ooplasm; scale bars: 10 μm.

parasitoid within the host tissues (Fig. 9.1B). Embedment can be so well completed that the presence of a parasitoid egg may not be detectable in the dissected host. It is therefore easily understandable that attachment of the parasitoid egg's chorion to the host tissues protects the parasite from any contact with the host's cellular and humoral defenses, that is, spreading hemocytes associated with the concomitant cytotoxic products like melanin. *A. tabida* eggs benefit an efficient protection from adhering to the host tissues. Proof of this is given by the partially embedded eggs, the unattached area becoming covered with host hemocytes and/or melanin (Fig. 9.2). More proof is given by the existence of a nonvirulent strain of *A. tabida*, the WOPV strain from The Netherlands. WOPV wasps lay nonsticky eggs floating free in the host hemocele that are always readily encapsulated in *D. melanogaster* (Kraaijeveld, 1994). The fact that encapsulation always occurs for nonadhesive (WOPV) eggs also suggests that in *A. tabida* there is no alternative mechanism but attachment to the host tissues to protect the parasitoid.

Differences between the WOPV strain from The Netherlands and our A1 reference strain (originating from the Rhone Valley, France) demonstrate that geographical variations exist in the virulence of *A. tabida* toward *D. melanogaster* (Kraaijeveld, 1994), although the genetics of this trait has not been established yet. Preliminary experiments suggested that the genetic determinism of egg adhesiveness could be complex (Prévost, unpublished results).

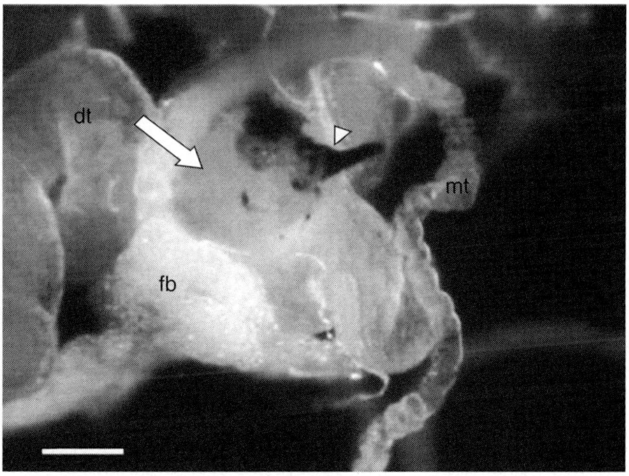

FIGURE 9.2 *A. tabida* egg (arrow) almost totally embedded in *Drosophila* host tissues observed under a binocular stereomicroscope. The unattached area (arrow head) of the egg is surrounded by a partially melanized cellular capsule. fb, fat body; dt, digestive tract; mt, malpighian tubules; scale bars: 100 μm. *From Prévost et al. (2005).*

It is possible that this physiological trait can be influenced by environmental factors. Because the host represents the primary environment of the parasitoid egg after oviposition in the hemocele, the host species itself could play a role in the egg's adhesive property. However, a series of experiments conducted on six species of the *melanogaster* subgroup (Eslin and Prévost, 1998) did not show any evidence of different degrees of attachment between *A. tabida* eggs and the host tissues in the different host species. Nevertheless, avoidance of encapsulation by the A1 strain of *A. tabida*, which is correlated to its ability to get embedded and "hide" before the host mounts a hemocytic reaction, tended to vary considerably between host species. Avoidance of encapsulation by *A. tabida* was shown to reach 100% in *D. sechellia*, with no more than 20% in *D. simulans*. Intermediate values were observed for the other tested *Drosophila* spp., namely *D. teissieri* (where 35% of the parasitoid eggs escaped from encapsulation), *D. yakuba* (75%), *D. mauritania* (80%) and *D. melanogaster* (95%) (Fig. 9.3; Eslin and Prévost, 2000). However, avoidance of encapsulation by *A. tabida* here obviously depends on whether or not the host has the capacity to mount layers of hemocytes around the parasitoid egg before it can be protected, rather than on any particular parasitoid property. In other words, it is the host's potential to mount hemocyte capsules quickly that determines the issue of the encapsulation reaction. This potentiality was clearly shown to be correlated with the hemocyte load in *Drosophila* larvae, suggesting that host and parasitoid were facing a race that

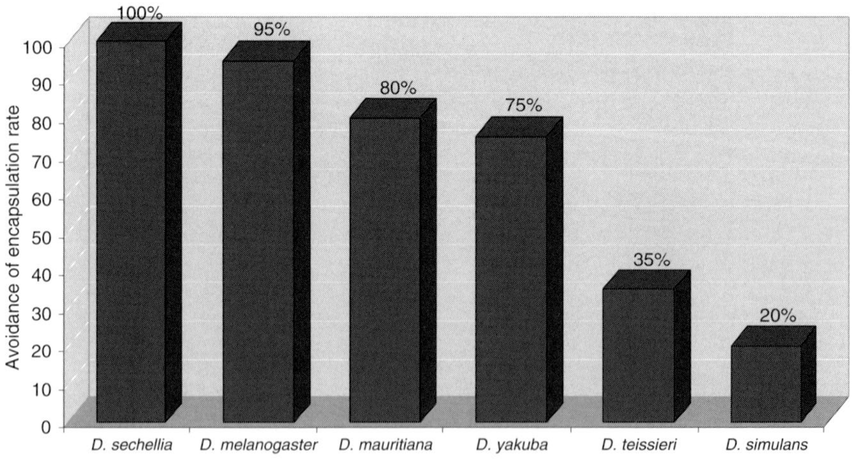

FIGURE 9.3 Rate of avoidance of encapsulation by *A. tabida* eggs in third-instar (6 days old) larvae of *Drosophila* from the *melanogaster* subgroup, determined by dissection 96 h after parasitization by the protected strain (A1). *Adapted from Eslin and Prévost (1998).*

A. tabida would win if the pool of circulating hemocytes remained too low in the host's hemocele (Eslin and Prévost, 1998, 2000).

Another environmental factor that may influence the capacity of *A. tabida* to avoid encapsulation is temperature. In Europe, *A. tabida* has been commonly found in The Netherlands (Kraaijeveld, 1994; Kraaijeveld and van Alphen, 1994), and in the northern part of the Rhone Valley in France (Allemand et al., 1999; Fleury et al., 2004; Chapter 1 by Fleury et al.). Although *A. tabida* has occasionally been observed in more southern and/or warmer regions like Spain or Greece (van Alphen, personal communication), its occurrence clearly tends to diminish with increasing temperatures. In the laboratory, 20 °C is usually considered the maximum temperature for the best success of *A. tabida* development. Preliminary experiments recently suggested that in *D. melanogaster*, failure of development at a higher temperature is associated with a higher rate of encapsulation (Zanchi et al., unpublished results). Whether the encapsulation reaction is causing the death of the parasitoid is not established yet. However, this result is in agreement with the hypothesis that the population of the *Drosophila* host's circulating hemocytes would build up and mature faster at a higher temperature, therefore creating conditions in which *A. tabida* would loose the race between the protection from embedment and attack by the host's hemocytes. Nevertheless, there is no actual evidence that *A. tabida* would attach and get embedded faster when the temperature rises.

Avoidance of encapsulation does not necessarily mean success of parasitoid development. Actually, results obtained with the several *Drosophila* species of the *melanogaster* subgroup clearly demonstrated that once the parasitoid overcome the host immunity defenses, success of development was not guaranteed. The concept of host adequation, developed by Vinson and Iwantsch (1980a), designated the physiological and biochemical conditions provided by hosts to endophagous parasitoids to complete their development and emerge as adult wasps successfully. These conditions were proven to have not been met in a significant proportion of the hosts from different *Drosophila* species infested by *A. tabida*, even when the parasitoid overcome immunity defenses. It is interesting to note that *D. simulans*, the most encapsulating species (80% encapsulation), is not the worst host since nearly 18% of the parasitoids could emerge as adult wasps. This is the same proportion of adult wasps as the one obtained with *D. mauritiana* which encapsulates only 20% of the parasitoids (Eslin and Prévost, 1998). Conversely, *D. yakuba*, which encapsulates *A. tabida* at a rate of only 25%, is a nonadequate host because it permits less than 2% of wasp emergence. Also, our *D. melanogaster* strain, encapsulating 5% of *A. tabida* eggs, was adequate to the development of less than 60% of the parasitoids (Fig. 9.4). These results show that in

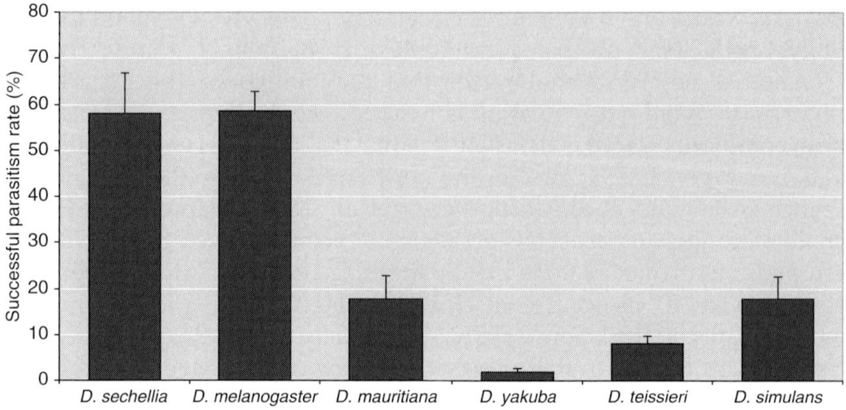

FIGURE 9.4 Successful parasitism rate (SPR) in six *Drosophila* species parasitized by the protected strain (A1) of *A. tabida*. Results were recorded after the parasitoids had completed their development. Percentages were estimated from 30 series of 20 larvae in each *Drosophila* species. *Adapted from Eslin and Prévost (1998).*

many cases (for instance in *D. mauritiana*, *D. yakuba* and *D. melanogaster*), the nonencapsulated parasitoid dies in the course of development.

A. tabida was shown to induce a significant reduction of larval weight gain and an increase in larval development time in *D. melanogaster* parasitized larvae. However, similar effects were also recorded in larvae carrying an encapsulated, nondeveloping *A. tabida* egg. These were interpreted as the consequences of the energy costs of parasitism imposed on the *Drosophila* host larva, rather than a regulative effect of the parasite, leading to its successful development (Moreau et al., 2002). Therefore, despite transient physiological changes in larvae parasitized by *A. tabida*, results are in agreement with the idea that *A. tabida* acts as a ''conformer'' parasitoid, with little investment—if any—in redirecting host physiology to its own benefit once the egg is laid in the host hemocele.

How can a conformer parasitoid ensure the success of development of its progeny? One strategy is to make the best decision at the time of parasitization. *A. tabida* female wasps obviously are good ''decision makers'' able to discriminate more or less good hosts. For instance *D. subobscura*, known as the nonencapsulating host (Eslin and Doury, 2006; Chapter 7 by Eslin et al.), is also the preferred one when *A. tabida* females are offered the choice between *D. subobscura* and *D. melanogaster* larvae (van Alphen and Janssen, 1982). As observed by van Alphen and Drijver (1982), *A. tabida* can also take advantage of parasitizing younger hosts which are believed to possess a less efficient cellular defense than older ones (Eslin and Prévost, 2000). Also, *A. tabida* females may choose to lay their eggs in hosts already parasitized by *L. boulardi* (a Figitidae clearly on the side of regulator parasitoids), a strategy of kleptoparasitim (Kraaijeveld, 1999) that allows their progeny to benefit from the physiological effects of *L. boulardi* on the host, if *A. tabida* is to win the battle that will take place between the larvae of these two solitary parasitoids. These behavioral traits of *A. tabida* wasps are well described in an abundant literature and in this volume (for synthesis, see Chapter 2 by Thiel and Hoffmeister; Chapter 10 by Kraaijeveld and Godfray).

9.2.2. The regulator strategy in *Asobara* parasitoids

Host regulation designates the many effects parasitoids can have on their host and which benefit their own development. It evokes developmental disruption of the host usually via hormonal or neurohormonal pathways, like endocrine signaling, which coordinates development of the parasitoid with that of the host so that the two partners molt in synchrony (Beckage and Gelman, 2004). It also covers all the effects on the host immunity system (Carton et al., 2008; Pennacchio and Strand, 2006; Schmidt et al., 2001; Strand and Pech, 1995), the first physiological barrier that endoparasitoids encounter after they break the cuticle of their host. A large part of our knowledge concerning the disruptive effects of parasitoids on their host's physiology comes from studies on Ichneumonoidae, in particular these ichneumonid and braconid species parasitizing Lepidoptera hosts, most of which can actively suppress the immune responses of their larval hosts (Schmidt et al., 2001; Vinson, 1990). Effects are usually directed toward the circulating hemocytes and in particular, the capsule forming hemocytes in which spreading ability is inhibited.

Similar phenomena can be observed in parasitoids of *Drosophila*. Several well-studied Figitidae of the *Leptopilina* genus (i.e., *L. boulardi*, *L. heterotoma* and *L. victoriae*) were shown to disrupt the hemocytic reaction of encapsulation by specifically affecting the lamellocytes (Carton and Kitano, 1979; Chiu et al., 2006; Morales et al., 2005; Rizki and Rizki, 1994; Rizki et al., 1990), the hemocytes that are responsible for the formation of cellular layers around foreign bodies during the encapsulation process. *L. boulardi* alters the morphology of the host's circulating lamellocytes (Labrosse et al., 2005; Rizki et al., 1990) while the two species *L. heterotoma* and *L. victoriae* promote their apoptosis and cellular lysis (Chiu et al., 2006; Morales et al., 2005; Rizki et al., 1990). The mechanism by which the other figitid *Ganaspis xanthopoda* affects the *Drosophila* immune system is believed to be similar to those of *L. heterotoma* and *L. victoriae* (Chiu et al., 2000).

Like these figitid parasitoids, the braconid *A. citri* provokes an overall suppression of the *D. melanogaster* hosts' encapsulation ability. The inhibition of capsule formation is extended to the entire host larva, such that the parasitoid egg and any other foreign body, like an injected oil drop or supernumerary parasitoids, are protected (Moreau et al., 2003). Differently to those of *A. tabida*, *A. citri* eggs float freely in the host hemolymph and thus, seem to lack any adhesive property. Nevertheless, *A. citri* is very rarely encapsulated in *D. melanogaster*, showing that it has developed efficient means to disrupt the host's cellular defenses. It was shown from the hemocyte counts that *Drosophila* larvae parasitized by *A. citri* possessed fewer circulating hemocytes than either unparasitized ones or larvae parasitized by *A. tabida* (Moreau et al., 2003). The amounts of plasmatocytes and lamellocytes involved in capsule formation were particularly reduced, but no cell pathology or cell lysis was observed among the circulating hemocytes. The lymph gland is the main center of hemocyte production, and the source of the lamellocytes needed for encapsulation (Lanot et al., 2001; Sorrentino et al., 2002). A study of the hematopoietic organs of *D. melanogaster* larvae parasitized by *A. citri* was thus conducted. The study revealed that the size of the anterior lobes of the lymph glands (the hematopoietic organ) was strongly reduced (Fig 9.5A and C), while the posterior lobes were more developed than in unparasitized larvae (Moreau et al., 2003, 2005). In *D. melanogaster* larvae parasitized by *A. tabida*, both anterior and posterior lobes were more developed than in control, unparasitized larvae (Fig. 9.5A and B). Since *A. tabida* proved to cause no inhibition of the host's cellular immunity reaction and ability to mount hemocytic capsules, increased sizes of both

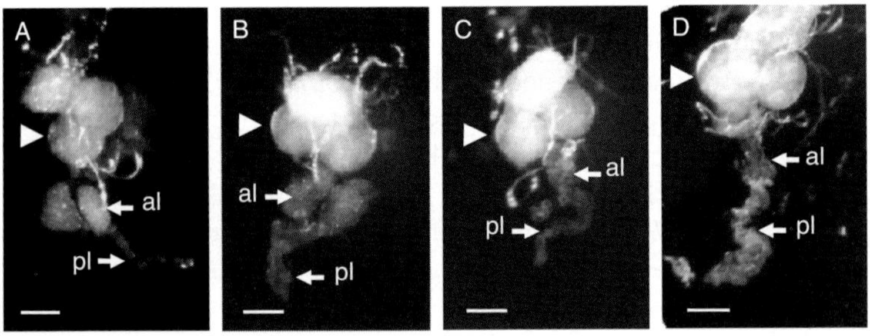

FIGURE 9.5 Morphology of hematopoietic organs from third-instar (6 days old) larvae of *D. melanogaster*: unparasitized (A), parasitized (96 h after parasitization) by *A. tabida* (B), *A. citri* (C) or *A. japonica* (D). Hematopoietic organs are observed under a binocular stereomicroscope and found near the brain (white arrow head). This organ is composed of two prominent anterior lobes (al) and several posterior lobes (pl). When larvae are parasitized by *A. citri* or *A. japonica*, they clearly show altered anterior lobes. In contrast, parasitization by *A. tabida* is followed by a slight increase of the size of anterior lobes compared to control unparasitized larvae. Scale bars: 100 μm. *From Moreau et al. (2003) and Mabiala* et al. *(unpublished results).*

the anterior and posterior lobes were considered to reflect a primary step of the host's immune response to the presence of the parasitoid. Therefore, in *D. melanogaster* parasitized by *A. citri*, the larger size of the posterior lobes of the lymph gland could not account for one of the pathological effects induced by the parasite. Conversely, it was supposed that the severe disruption of the anterior lobes of the lymph glands was the major cause of the low hemocyte load in the hemolymph of larvae parasitized by *A. citri* (Moreau et al., 2003). Electronic microscopy of hemocytes in the lymph gland of *D. melanogaster* larvae parasitized by *A. citri* clearly showed altered cells in the anterior lobes, supporting the hypothesis that the host's hematopoietic organ is one main target of the immunosuppressive effect of the parasitoid (Prévost et al., 2005).

In *D. melanogaster* parasitized by parasitoids of the *Leptopilina* genus, hemocytes in the hematopoietic organ are also the targets of the parasites (Chiu and Govind, 2002). However, the reported effects of *Leptopilina* species differ in several aspects from what is observed with *A. citri*. First, cell lysis provoked by *Leptopilina* species has been described as apoptosis, while electronic microscopy suggested that the effect of *A. citri* on the host's lymph gland was necrosis (Prévost et al., 2005). Second, cell lysis has been observed in circulating hemocytes of *D. melanogaster* larvae parasitized by *Leptopilina* and *Ganaspis* species (Chiu and Govind, 2002; Russo et al., 2001), while the effects of *A. citri* are targeted on the host's hematopoietic organ, only (Moreau et al., 2003).

The Japanese parasitoid *A. japonica* was recently investigated. Results showed that this species unambiguously ranges among "regulator" parasites. Effects on the immunity system and the encapsulation ability of *D. melanogaster* larvae are similar to those previously described with *A. citri* (Mabiala-Moudoungou et al., unpublished results). *A. japonica* is responsible for the overall suppression of the host's encapsulation ability with a marked effect on the host's hematopoietic organ (Fig. 9.5D). Like with *A. citri*, it is considered that the host's hematopoietic organ is one main target of the parasitoid, and that this effect accounts for the suppression of the host's ability to mount hemocytic capsules (Mabiala-Moudoungou et al., unpublished results).

Preliminary observations conducted on *A. persimilis*, the Australian *Asobara* species, also suggested a regulator strategy targeting the host's hematopoietic organ.

Parasitism by the three species *A. tabida*, *A. citri* and *A. japonica* provokes a decrease of phenoloxidase (PO) activity in the hemolymph of *D. melanogaster* larvae (Moreau et al., 2002, 2003). The same effect has been observed with the figitid parasitoids of *Leptopilina* genus (Colinet et al., 2007). The timing and the importance of the variation may vary between parasitoid species, but the lowered PO activity in the host's hemolymph seems to be commonly associated with the infestation by a larval parasitoid in *D. melanogaster* larvae. Although it is possible that this effect contributes to the avoidance of the host's immune response, it does not seem to be a good indicator of the status—either conformer or regulator—of the parasitoid species.

Another expression of the regulative effects of *Asobara* species on the physiology of the host *D. melanogaster* is the transient paralysis induced upon parasitization. This effect has been measured, or at least observed, with the three well-studied species *A. tabida*, *A. citri* and *A. japonica* (Mabiala-Moundoungou et al., unpublished results; Moreau et al., 2002). As previously reported by Moreau et al. (2002), a total paralysis, which may be followed by a transient immobility, is unusual in larvae parasitized by endoparasitoids, compared to what is known with ectoparasitoid species (Doury et al., 1997). This question is discussed in Section 9.3 and in Chapter 8 by Moreau et al.

9.3. ARMS DEVELOPED BY *ASOBARA* PARASITOIDS TO REGULATE OR EVADE HOST IMMUNITY DEFENSES

9.3.1. The arms of regulation

Tools developed by endophagous parasitoids to regulate the host's physiology and immunity come from either the female wasp's reproductive apparatus and the associated glands, or the parasitic egg or larva itself. In many species of the ichneumonid and braconid families, symbiotic polydnaviruses (PDVs) or virus-like particles (VLPs; Beckage, 1998; Beckage and Gelman, 2004; Pennacchio and Strand, 2006; Schmidt and Schumann-Feddersen, 1989; Strand and Pech, 1995) can act as infecting agents. PDVs multiply in the calyx cells of the female wasp's ovaries while VLPs can be produced either in the ovaries or the venom apparatus (Barratt et al., 1999; Suzuki and Tanaka, 2006). Once injected (along with the parasitoid egg) into the host hemocele, PDVs can specifically infect host tissues, while both PDVs and VLPs may enter the circulating hemocytes.

PDVs were never found in parasitoids of *Drosophila*, while VLPs were observed in the venom of all studied *Leptopilina* species (Dupas et al., 1996; Labrosse et al., 2003; Morales et al., 2005; Rizki and Rizki, 1990). Although the nature of VLPs which do not contain DNA is not established, VLPs of *L. heterotoma* have been reported in the cytoplasm of the host's lamellocytes, where they induce morphological changes (Rizki and Rizki, 1994). In *L. boulardi*, there is variability in the morphology and the number of VLPs carried in the wasp's venom, but the role these particles play in the virulence of the parasitoid is not fully understood (Dupas et al., 1996; Labrosse et al., 2003; Chapter 6 by Dubuffet et al.).

Nevertheless, the factor responsible for the immunosuppressive effect of *L. boulardi* on *D. melanogaster* has been isolated from the wasp's venom and is now clearly identified. The so-called LbGAP protein enters the host's plasmatocytes and lamellocytes and affects the cytoskeleton of lamellocytes (Colinet et al., 2007), therefore inhibiting *D. melanogaster* ability to form cellular capsules (Chapter 6 by Dubuffet et al.).

No PDV or VLP was ever reported in any of the *Asobara* species, while they were found in most of the studied braconid endoparasitoids. In many braconids, venom components are necessary to enhance the effects of PDVs (Stoltz, 1986; Stoltz et al., 1988). A limited number of studies also suggest that the venom of parasitoid species devoid of symbiotic viruses and VLP may perturb the host's immune defenses (Cai et al., 2004; Richards and Parkinson, 2000). In *Asobara* parasitoids, venom is the only factor identified so far, exerting an active, regulative effect on *Drosophila* hosts (Mabiala-Moundoungou et al., unpublished results; Chapter 8 by Moreau et al.).

In *D. melanogaster* larvae parasitized by *Asobara* parasitoids, two major "pathologies" associated with parasitism were considered responsible for the host's incapacity to mount cellular capsules. One of them is the destruction of the anterior lobes of the larval host's hematopoietic organ, an effect which is strong enough to be clearly visible under a stereomicroscope (Fig. 9.5C and D). This effect was reported with both *A. citri* (Moreau et al., 2003) and *A. japonica* (Mabiala-Moundoungou et al., unpublished results), although disruption of the host's lymph gland was more pronounced with *A. japonica*. The second important change is the drop of PO activity in the hemolymph of parasitized larvae, another effect of *Asobara* parasitoids which can account, at least partially, for the inhibition of the encapsulation reaction. The effect of an endoparasite on the host's PO system is rather common to many host–parasitoid interactions, but attacking the anterior lobes of the host's hematopoietic organ (with no concomitant destruction of the circulating hemocytes) seems to be specific to the regulation effects of *Asobara* species. However, none of these effects on host defense systems could be proved to be directly produced by the venom of *Asobara* female wasps.

Conversely to the inhibition of host immunity defenses, paralysis of host larvae could be undoubtedly attributed to the effect of the female wasps' venoms in the three tested *Asobara* species. Manual injection of wasp venom into unparasitized *D. melanogaster* larvae gives rise to different degrees of paralysis depending on the *Asobara* species. However, the segregation between *Asobara* species according to their either conformer or regulator strategy to avoid host defenses does not apply to the paralytic effect. Both *A. tabida* and *A. citri* possess the less harmful venoms, while *A. japonica* venom can be considered a "lethal weapon," since all larvae die within a few hours following injection (Mabiala-Moundoungou et al., unpublished results; Chapter 8 by Moreau et al.). Whether these paralytic effects can be associated with any other regulation of host physiology and immunity is unknown.

Most active venoms from endoparasitoids affect the host physiology. However, the here described paralytic effects by *Asobara* wasps' venoms on the host *D. melanogaster* differ from what is commonly reported with endoparasitoid braconids, whose venoms are usually nonparalyzing. This intriguing feature of *Asobara* parasitoids suggests that these species may share some properties with ectoparasitic braconids (e.g., affecting developmental programming and inducing paralysis; Coudron, 1991; Coudron and Brandt, 1996; Doury et al., 1997), even though their lifestyle is endoparasitic (Moreau and Guillot, 2005; Prévost et al., 2005; Chapter 8 by Moreau et al.).

9.3.2. The arms of evasion

It resorts from the investigation of several species of the *Asobara* genus that active inhibition of host defenses could be the main rule among *Asobara* species (Mabiala-Moundoungou et al., unpublished results), while "passive" evasion is clearly established in the only *A. tabida* species. *A. tabida* avoids the attack by the host hemocytes thanks to the adhesiveness of the eggs, a feature which allows them to quickly attach and get embedded within the host tissues before layers of host hemocytes start building up and form a hermetic melanized capsule around the parasite. An indication of the critical role played by the adhesiveness of *A. tabida* eggs is given by the comparison between the A1 French strain, with "sticky" eggs, and the WOPV Dutch strain, with "nonsticky" ones. The ultrastructural study shows that the attachment of A1 eggs to *D. melanogaster* organs results from the large zone of coalescence between the egg exochorion and the basement membranes of the surrounding host tissues (Fig. 9.6A and C). In order to provide an efficient protection to the parasitoid, the adhesion of the eggs to the host's tissues needs to last up until the end of the parasitoid embryologic development. This can be achieved only if the integrity of the structure of the egg exochorion is maintained. It was shown that in the A1 eggs, the exochorionic structure remains, therefore permitting a durable adhesion to the host tissues. Conversely, the outer fibrous layer of WOPV eggs' exochorion becomes flaky after a few hours in the host hemocele (Fig. 9.6B and D). After initiating some attachment to the surrounding host tissues, WOPV eggs eventually detach, therefore becoming exposed to the encapsulating hemocytes.

The biochemical analysis of the egg's chorion components revealed that one protein of 80 kDa—the so-called P80 protein—is present at the surface of A1 eggs, while missing in the exochorion of WOPV ones

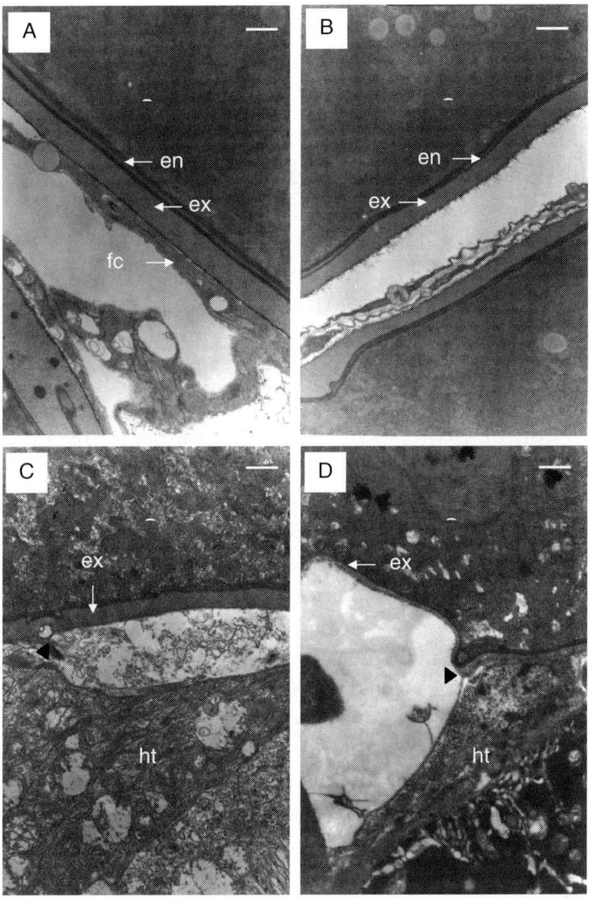

FIGURE 9.6 Electron micrographs of *A. tabida* eggs and comparison of the ultrastructural features between chorions of A1 (A and C) and WOPV (B and D) eggs. The eggs were observed inside females' ovaries (A and B) and 6 h after oviposition (C and D) in *D. melanogaster* larvae. Black arrow heads show the adhesion zone (chorions of the eggs/basal membranes of host's tissues); fc, follicular cells; ht, host tissues; en, endochorion; ex, exochorion; O, oocyte; scale bars: 1 μm.

(Prévost et al., 2005). This chorionic protein, which could play a major role in the stickiness of *A. tabida* eggs, may be considered one main agent of the parasitoid strategy to avoid encapsulation.

In braconid species, teratocytes are giant cells derived from a serosal membrane which envelops the developing embryo (Pennacchio and Strand, 2006). Teratocytes which circulate in the host hemolymph are believed to participate in the nutrition of the developing parasite. So far, teratocytes were never observed associated with *Asobara* species, another trait which contributes to distinguish the *Asobara* genus from other endophagous braconids.

9.4. CONCLUDING REMARKS AND PROSPECTS

The conformer and regulator strategies in the *Asobara* genus raise the question of their relative cost for the parasitoids. *A. citri* and *A. japonica* attacking the host's lymph gland must produce in their venom, ovaries or embryos, an active factor responsible for the impairment of the host's cellular defense system. Conversely, *A. tabida* exerts no effect on the host's immunity system but it does produce chorionic proteins allowing the eggs to attach and be protected from encapsulation. The WOPV strain of *A. tabida* which does not produce "sticky" eggs is unable to avoid encapsulation in *D. melanogaster*. However, this lack of egg protection is counterbalanced by a better acuity of the female wasps to select *D. subobscura* (nonencapsulating) larvae as hosts (Kraaijeveld, 1994; personal observations). This shows that avoiding rather than disrupting host defenses requires physiological and behavioral traits which can account for an "active" strategy to overcome immune barriers of the hosts. Our study on the *Asobara* species, as well as studies covering a large number and diversity of parasitoid species (Moreau, 2003; Pennacchio and Strand, 2006; Schmidt et al., 2001; Siva-Jothy et al., 2005; Strand and Pech, 1995), tend to demonstrate that the concepts of "passive" and "active" strategies to overcome host defenses do not reflect the cost of the behavioral, physiological and molecular tools developed by parasitoids to either regulate or avoid host immunity defenses.

It is worth note that there is an amazing array of molecular factors developed by endoparasitoids to overcome the host immunity defenses (Beckage and Gelman, 2004). Whether the nature and diversity of these molecules are more likely to reflect the lifestyle of the parasitoids, or their habitat and ecological niche, or the genus or family of their host, or rather their ancestral origin, is not well understood yet. This is probably due to the fact that our knowledge in this area is still scarce and that the identification of the molecules mediating the virulence of parasitoids has just started. The recently identified LbGAP factor from the figitid *L. boulardi* (Colinet et al., 2007) is one example showing that identifying the molecular tools of the virulence in parasitoids opens a promising area of research.

Parasitoids may inherit their virulence factors from regulatory molecules present in ancestral species. How this basal toolkit of gene products has changed with the developmental strategies of the parasitoid species needs to be investigated. For this approach, the model *Asobara* is of particular interest because it shows some variability between *Asobara* species—and even within species—in the means the parasitoids have developed to circumvent host defenses. In addition, the *Asobara* genus is atypical because it shares some properties with ectoparasitoid braconids, like the paralyzing effects of the wasps' venoms (Mabiala-Moundougou et al., unpublished results; Chapter 8 by Moreau et al.). The hypothesis that the *Asobara* genus belongs to a predominantly ectoparasitic clade and may have recently reverted to endoparasitism (Dowton et al., 1998) would explain why *Asobara* species present several peculiar traits compared to what is known in other endophagous braconids. Thus, these wasps may be of great interest for studying the evolutionary relationship between ecto- and endoparasitic species.

It is also possible that among the large toolkit of gene products inherited from Apocrita Hymenoptera, the virulence factors which are retained by one given family, genus, or even species of parasitoids, is partially fortuitous and is affected by other factors than the host ranges. This could explain the large diversity of the virulence strategies met in parasitoids which are either phylogenetically related or exploiting the same hosts.

In conclusion, larval parasitoids of *Drosophila*, and among them *Asobara* species, proved to be a particularly interesting model which will be worth exploiting to pursue different approaches of the virulence strategies developed by endophagous parasites.

REFERENCES

Allemand, R., Fleury, F., Lemaitre, C., Boulétreau, M., 1999. Dynamique des populations et interactions compétitives chez deux espèces de *Leptopilina*, parasitoïdes de drosophiles, dans la vallée du Rhône (Hymenoptera: Figitidae). Ann. Soc. Entomol. Fr. 35, 97–103.

Asgari, S., Schmidt, O., 1994. Passive protection of eggs from the parasitoid, *Cotesia rubecula*, in the host, *Pieris rapae*. J. Insect Physiol. 40, 789–795.

Asgari, S., Theopold, U., Welby, C., Schmidt, O., 1998. A protein with protective properties against the cellular defense reactions in insects. Dev. Biol. 95, 3690–3695.

Barratt, B.I.P., Evans, A.A., Stoltz, D.B., Vinson, S.B., Easingwood, R., 1999. Virus-like particles in the ovaries of *Microctonus aethiopoides* Loan (Hymenoptera: Braconidae), a parasitoid of adult weevils (Coleoptera: Curculionidae). J. Invert. Pathol. 73, 182–188.

Beckage, N.E., 1998. Modulation of immune responses to parasitoids by polydnaviruses. Parasitology 116, 57–64.

Beckage, N.E., Gelman, D.B., 2004. Wasp parasitoid disruption of host development: implications for new biologically based strategies for insect control. Ann. Rev. Entomol. 49, 299–330.

Cai, J., Ye, G.-Y., Hu, C., 2004. Parasitism of *Pieris rapae* (Lepidoptera: Pieridae) by a pupal endoparasitoid, *Pteromalus puparum* (Hymenoptera: Pteromalidae): effects of parasitization and venom on host hemocytes. J. Insect Physiol. 50, 315–322.

Carton, Y., Boulétreau, M., 1985. Encapsulation ability of *Drosophila melanogaster*: a genetic analysis. Dev. Comp. Immunol. 9, 211–219.

Carton, Y., Kitano, H., 1979. Changes in the haemocyte population of *Drosophila* larvae after single and multiple parasitization by *Cothonaspis* (Parasitic Cynipidae). J. Invert. Pathol. 34, 88–89.
Carton, Y., Nappi, A., 1997. *Drosophila* cellular immunity against parasitoids. Parasitol. Today 13, 218–226.
Carton, Y., Nappi, A., 2001. Immunogenetics aspects of the cellular immune response of *Drosophila* against parasitoids. Immunogenetics 5, 157–164.
Carton, Y., Boulétreau, M., van Alphen, J.J.M., van Lenteren, J.C., 1986. The *Drosophila* parasitic wasps. In: Ashburner, Thomson, (Eds.), ''The Genetics and Biology of *Drosophila*.'' Vol. 3C. Academic Press, New York/London, pp. 347–394.
Carton, Y., Frey, F., Nappi, A.J., 1992. Genetic determinism of the cellular immune reaction in *Drosophila melanogaster*. Heredity 69, 393–399.
Carton, Y., Poirié, M., Nappi, A.J., 2008. Insect immune resistance to parasitoids. Insect Sci. 15, 67–87.
Chiu, H., Govind, S., 2002. Natural infection by virulent parasitic wasps induces apoptotic depletion of hematopoietic precursors. Cell Death Diff. 9, 1379–1381.
Chiu, H., Morales, J., Govind, S., 2006. Identification and immuno-electron microscopy localization of p40, a protein component of immunosuppressive virus-like particles from *Leptopilina heterotoma*, a virulent parasitoid wasp of *Drosophila*. J. Gen. Virol. 87, 461–470.
Chiu, H., Sorrentino, R.P., Govind, S., 2000. Suppression of the *Drosophila* cellular immune response by *Ganaspis xanthopoda*. Adv. Exp. Med. Biol. 484, 161–167.
Colinet, D., Schmitz, A., Depoix, D., Crochard, D., Poirié, M., 2007. Convergent use of RhoGAP toxins by eukaryotic parasites and bacterial pathogens. PLoS Pathog. 3, e203.
Coudron, T.A., 1991. Host regulating factors associated with parasitic Hymenoptera. In: Hedin, (Ed.), ''Naturally Occuring Pest Bioregulators.'' 449, A. C. S. Symposium Series. American Chemical Society, Washington, DC, pp. 41–65.
Coudron, T.A., Brandt, S.L., 1996. Characteristics of a developmental arrestant in the venom of the ectoparasitoid wasp *Euplectrus comstockii*. Toxicon 34, 1431–1441.
Davies, D.H., Vinson, S.B., 1986. Passive evasion by eggs of braconid parasitoid, *Cardiochiles nigriceps*, of encapsulation *in vitro* by haemocytes of host *Heliothis virescens*. Possible role of fibrous layer in immunity. J. Insect Physiol. 32, 1003–1010.
Doury, G., Bigot, Y., Periquet, G., 1997. Physiological and biochemical analysis of factors in the female venom gland and larval salivary secretions of the ectoparasitoid wasp *Eupelmus orientalis*. J. Insect Physiol. 43, 69–81.
Dowton, M., Austin, A.D., Antolin, M.F., 1998. Evolutionary relationships among the Braconidae (Hymenoptera: Ichneumonoidae) inferred from partial 16S rDNA gene sequences. Insect Mol. Biol. 7, 129–150.
Dubuffet, A., Dupas, S., Frey, F., Drezen, J.-M., Poirié, M., Carton, Y., 2007. Genetic interactions between the parasitoid wasp *Leptopilina boulardi* and its *Drosophila* hosts. Heredity 98, 21–27.
Dupas, S., Brehélin, M., Frey, F., Carton, Y., 1996. Immune suppressive virus-like particles in a *Drosophila* parasitoid: significance of their intraspecific morphological variations. Parasitology 113, 207–212.
Dupas, S., Carton, Y., Poirié, M., 2003. Genetic dimension of the coevolution of virulence-resistance in *Drosophila*–parasitoid wasp relationships. Heredity 90, 83–89.
Eslin, P., Doury, G., 2006. The fly *Drosophila subobscura*: a natural case of innate immunity deficiency. Dev. Comp. Immunol. 30, 977–983.
Eslin, P., Prévost, G., 1996. Variation in *Drosophila* concentration of haemocytes associated with different ability to encapsulate *Asobara tabida* larval parasitoid. J. Insect Physiol. 42, 549–555.

Eslin, P., Prévost, G., 1998. Hemocyte load and immune resistance to *Asobara tabida* are correlated in species of *Drosophila melanogaster* subgroup. J. Insect Physiol. 44, 807–816.

Eslin, P., Prévost, G., 2000. Racing against host's immunity defenses: a likely strategy for passive evasion of encapsulation in *Asobara tabida* parasitoids, J. Insect Physiol. 46, 1161–1167.

Eslin, P., Giordanengo, P., Fourdrain, Y., Prévost, G., 1996. Avoidance of encapsulation in the absence of VLP by a braconid parasitoid of *Drosophila* larvae: an ultrastructural study. Can. J. Zool. 74, 2193–2198.

Fleury, F., Ris, N., Allemand, R., Fouillet, P., Carton, Y., Boulétreau, M., 2004. Ecological and genetic interactions in *Drosophila*-parasitoids communities: a case study with *D. melanogaster, D. simulans* and their common Leptopilina parasitoids in south-eastern France. Genetica 120, 181–194.

Ideo, S., Watada, M., Mitsui, H., Kimura, M.T., 2008. Host range of *Asobara japonica* (Hymenoptera: Braconidae), a larval parasitoid of drosophilid flies. Entomol. Sci. 11, 1–6.

Hayakawa, Y., Yazaki, K., 1997. Envelope protein of parasitic wasps's symbiont virus, polydnavirus, protects the wasp eggs from cellular immune reactions by the host insect. Eur. J. Biochem. 246, 820–826.

Kraaijeveld, A.R., 1994. Local Adaptations in a Parasitoid–Host System – a Coevolutionary Arms Race. University of Leiden, The Netherlands [PhD thesis].

Kraaijeveld, A.R, 1999. Kleptoparasitism as an explanation for paradoxical oviposition decisions of the parasitoid *Asobara tabida*. J. Evol. Biol. 12, 129–133.

Kraaijeveld, A.R., van der Wel, N.N., 1994. Geographical variation in reproductive success of the parasitoid *Asobara tabida* in larvae of several *Drosophila* species. Ecol. Entomol. 19, 221–229.

Kraaijeveld, A.R., Voet, S., van Alphen, J.J.M., 1994. Geographical variation in habitat choice and host suitability in the parasitoid *Asobara rufescens*. Entomol. Exp. Appl. 72, 109–114.

Labrosse, C., Carton, Y., Dubuffet, A., Drezen, J.-M., Poirié, M., 2003. Active suppression of *D. melanogaster* immune response by long gland products of the parasitic wasp *Leptopilina boulardi*. J. Insect Physiol. 49, 513–522.

Labrosse, C., Stasiak, K., Lesobre, J., Grangeia, A., Huguet, E., Drezen, J.-M., et al., 2005. A RhoGAP protein as a main immune suppressive factor in the *Leptopilina boulardi* (Hymenoptera, Figitidae)–*Drosophila melanogaster* interaction. Insect Biochem. Mol. Biol. 35, 93–103.

Lanot, R., Zachary, D., Holder, F., Meister, M., 2001. Postembryonic hematopoiesis in *Drosophila*. Dev. Biol. 230, 243–257.

Lawrence, P.O., 1986. Host–parasitoid hormonal interactions: an overview. J. Insect Physiol. 32, 295–298.

Lawrence, P.O., 1990. The biochemical and physiological effects of insect hosts on the development and ecology of their parasites: an overview. Arch. Insect Biochem. Physiol. 13, 217–228.

Mitsui, H., Van Achterberg, K., Nordlander, G., Kimura, M.T., 2007. Geographical distributions and host associations of larval parasitoids of frugivorous Drosophilidae in Japan. J. Nat. Hist. 41, 1731–1738.

Monconduit, H., Prévost, G., 1994. Avoidance of encapsulation by *Asobara tabida*, a larval parasitoid of *Drosophila* species. Norw. J. Agr. Sci. 16, 301–309.

Morales, J., Chiu, H., Oo, T., Plaza, R., Hoskins, S., Govind, S., 2005. Biogenesis, structure, and immune-suppressive effects of virus-like particles of a *Drosophila* parasitoid, *Leptopilina victoriae*. J. Insect Physiol. 51, 181–195.

Moreau, S.J.M., 2003. Relationships between parasitoid Hymenoptera and the immune system of their hosts: the "active" and "passive" mechanisms redefined. Ann. Soc. Entomol. Fr. 39, 305–314.

Moreau, S.J.M., Guillot, S., 2005. Advances and prospects on biosynthesis, structures and functions of venom proteins from parasitic wasps. Insect Biochem. Mol. Biol. 35, 1209–1223.

Moreau, S.J.M., Dingremont, A., Doury, G., Giordanengo, P., 2002. Effects of parasitism by *Asobara tabida* (Hymenoptera: Braconidae) on the development, survival and activity of *Drosophila melanogaster* larvae. J. Insect Physiol. 48, 337–347.

Moreau, S.J.M., Eslin, P., Giordanengo, P., Doury, G., 2003. Comparative study of the strategies evolved by two parasitoids of the *Asobara* genus to avoid immune response of the host, *Drosophila melanogaster*. Dev. Comp. Immunol. 27, 273–282.

Moreau, S.J.M., Guillot, S., Populaire, C., Doury, G., Prévost, G., Eslin, P., 2005. Conversely to its sibling *Drosophila melanogaster*, *D. simulans* overcomes the immunosuppressive effects of the parasitoid *Asobara citri*. Dev. Comp. Immunol. 29, 205–209.

Nappi, A.J., 1981. Cellular immune response of *Drosophila melanogaster* against *Asobara tabida*. Parasitology 83, 319–324.

Nappi, A.J., Streams, F.A., 1969. Haemocytic reactions of *Drosophila melanogaster* to the parasites *Pseudocoila mellipes* and *P. hochei*. J. Insect Physiol. 15, 1551–1556.

Nappi, A.J., Vass, E., 1993. Melanogenesis and the generation of cytotoxic molecules during insect cellular immune reactions. Pigment Cell Res. 6, 117–126.

Nappi, A.J., Vass, E., 2001. Cytotoxic reactions associated with insect immunity. In: Berck, Sugumaran, Cooper, (Eds.), "Phylogenetic Perspectives on the Vertebrate Immune System." Kluwer/Plenum, New York, pp. 329–348.

Nappi, A.J., Carton, Y., Frey, F., 1991. Parasite induced enhancement of haemolymph tyrosinase activity in a selected immune reactive strain of *Drosophila melanogaster*. Arch. Insect Biochem. Physiol. 18, 159–168.

Nappi, A.J., Vass, E., Frey, F., Carton, Y., 1995. Superoxide anion generation in *Drosophila* during melanotic encapsulation of parasites. Eur. J. Cell Biol. 68, 450–456.

Pennacchio, F., Strand, M.R., 2006. Evolution of developmental strategies in parasitic hymenoptera. Ann. Rev. Entomol. 51, 233–258.

Poirié, M., Carton, Y., Dubuffet, A., 2009. Virulence strategies in parasitoid Hymenoptera as an example of adaptive diversity. Compt. Rend. Biol. 332, 311–320.

Prévost, G., Eslin, P., Doury, G., Moreau, S.J.M., Guillot, S., 2005. *Asobara*, braconid parasitoids of *Drosophila* larvae: unusual strategies to avoid encapsulation without VLPs. J. Insect Physiol. 51, 171–179.

Richards, E.H., Parkinson, N.M., 2000. Venom from the endoparasitic wasp *Pimpla hypochondriaca* adversely affects the morphology, viability, and immune function of hemocytes from larvae of the tomato moth, *Lacanobia oleracea*. J. Invert. Pathol. 76, 33–42.

Rizki, T.M., Rizki, R.M., 1980. Properties of the larval hemocytes of *Drosophila melanogaster*. Experientia 36, 1223–1226.

Rizki, R.M., Rizki, T.M., 1990. Parasitoid virus-like particles destroy *Drosophila* cellular immunity. Proc. Natl. Acad. Sci. USA 87, 8388–8392.

Rizki, T.M., Rizki, R.M., 1994. Parasitoid-induced cellular immune deficiency in *Drosophila*. Ann. N. Y. Acad. Sci. 712, 178–194.

Rizki, T.M., Rizki, R.M., Carton, Y., 1990. *Leptopilina heterotoma* and *L. boulardi*: strategies to avoid cellular defense responses of *Drosophila melanogaster*. Exp. Parasitol. 70, 466–475.

Russo, J., Brehélin, M., Carton, Y., 2001. Haemocyte changes in resistant and susceptible strains of *D. melanogaster* caused by virulent and avirulent strains of the parasitic wasp *Leptopilina boulardi*. J. Insect Physiol. 47, 167–172.

Russo, J., Dupas, S., Frey, F., Carton, Y., Brehélin, M., 1996. Insect immunity: early events in encapsulation process of parasitoid (*Leptopilina boulardi*) eggs in reactive and non reactive strains of *Drosophila*. Parasitology 112, 135–142.

Schmidt, O., Schumann-Feddersen, I., 1989. Role of virus-like particles in parasitoid–host interactions in insects. In: Harris, J.R (Ed.), "Subcellular Biochemistry." 15, Plenum Publishing Corporation, New York, pp. 91–119.

Schmidt, O., Theopold, U., Strand, M.R., 2001. Innate immunity and its evasion and suppression by hymenopteran endoparasitoids. BioEssays 23, 344–351.
Siva-Jothy, M.T., Moret, Y., Rolff, J., 2005. Insect immunity: an evolutionary ecology perspective. Adv. Insect Physiol. 32, 1–48.
Sorrentino, R.P., Carton, Y., Govind, S., 2002. Cellular immune response to parasite infection in the drosophila lymph gland is developmentally regulated. Dev. Biol. 243, 65–80.
Stoltz, D.B., 1986. Interaction between parasitoid derived products and host insects: an overview. J. Insect Physiol. 32, 347–350.
Stoltz, D.B., Guzo, D., Belland, E.R., Lucarotti, C.J., MacKinnon, E.A., 1988. Venom promotes uncoating *in vitro* and persistence *in vivo* of a braconid polydnavirus. J. Gen. Virol. 69, 903–907.
Strand, M.R., Pech, L.L., 1995. Immunological basis for compatibility in parasitoid–host relationships. Ann. Rev. Entomol. 40, 31–56.
Suzuki, M., Tanaka, T., 2006. Virus-like particles in venom of *Meteorus pulchricornis* induce host hemocyte apoptosis. J. Insect Physiol. 52, 602–613.
Van Alphen, J.J.M., Drijver, R.A.B., 1982. Host selection by *Asobara tabida* Nees (Hymenoptera: Alysiinae), a larval parasitoid of fruit inhabiting *Drosophila* species. I. Host stage selection with *Drosophila melanogaster* as host species. Neth. J. Zool. 32, 194–214.
Van Alphen, J.J.M., Janssen, A.R.M., 1982. Host selection by *Asobara tabida* Nees (Hymenoptera: Alysiinae), a larval parasitoid of fruit inhabiting *Drosophila* species. II. Host species selection. Neth. J. Zool. 32, 215–231.
Vass, E., Nappi, A.J., 2000. Developmental and immunological aspects of *Drosophila*–parasitoid relationships. J. Parasitol. 86, 1259–1270.
Vinson, S.B., 1990. How parasitoids deal with the immune system of their hosts: an overview. Arch. Insect Biochem. Physiol. 13, 3–27.
Vinson, S.B., Iwantsch, G.F., 1980a. Host suitability for insect parasitoids. Ann. Rev. Entomol. 25, 397–419.
Vinson, S.B., Iwantsch, G.F., 1980b. Host regulation by insect parasitoids. Q. Rev. Biol. 55, 143–165.

CHAPTER 10

Evolution of Host Resistance and Parasitoid Counter-Resistance

Alex R. Kraaijeveld* and **H. Charles J. Godfray**†

Contents

Abstract

By their nature, parasitoids will exert a selection pressure on their hosts to evolve a mechanism through which to resist parasitoid attack. In turn, such a resistance mechanism will lead to parasitoids evolving counter-resistance. In this chapter, we present an overview of the research on the (co)evolutionary interaction between *Drosophila* and their parasitoids, with the main focus on the cellular immune response of *D. melanogaster*, and the counter-resistance mechanism of one of its main parasitoids, *Asobara tabida*. A key aspect of this interaction is the existence of genetic variation: in the field, host resistance and parasitoid counter-resistance vary, both between and within populations. Host resistance and parasitoid counter-resistance are costly, and both these costs

* University of Southampton, School of Biological Sciences, Southampton SO16 7PX, United Kingdom
† Department of Zoology, University of Oxford, Oxford OX1 3PS, United Kingdom

Advances in Parasitology, Volume 70
ISSN 0065-308X, DOI: 10.1016/S0065-308X(09)70010-7

turn out to be density dependent. These tradeoffs can explain the existence of genetic variation. We briefly touch upon behavioral aspects of the interaction and the parasites and pathogens that the parasitoids themselves suffer from. We end this chapter by considering the data coming from gene chip experiments: early indications suggest that the genes involved in the actual immune response against parasitoids are mostly different from the genes involved in the evolution of resistance.

10.1. INTRODUCTION

Parasitoids cannot exist without hosts. Given that parasitoids need to parasitize hosts and that, by definition, they kill their host as part of their normal lifecycle, the effect of parasitism by a parasitoid is very severe for a host. At the very least, the host's fecundity is likely to be reduced and, if the parasitoid kills the host before it becomes reproductively active, parasitoid attack equals genetic death for the parasitized host. This means that, unless attack rates by parasitoids are rare, there will be a strong selection pressure on the host to develop a resistance mechanism that either prevents parasitism taking place, or kills the parasitoid egg or larva before it can do further damage to the host. In turn, the evolution of a resistance mechanism in the host will exert a selection pressure on the parasitoid for a counter-resistance mechanism that somehow avoids or overcomes the host's resistance mechanism.

Because parasitoid–host interactions result in the death of either of the two, the coevolution of host resistance and parasitoid counter-resistance is more antagonistic than the typical parasite–host interaction. In this chapter, we aim to summarize the existing data on the reciprocal antagonistic evolution of resistance in *Drosophila* and counter-resistance in the parasitoids attacking *Drosophila*. Our focus will be on *D. melanogaster* and *Asobara tabida* in Europe, as this species pair has been a focus of studying coevolution between parasitoids and host, and most of the work on these two species has been carried out in that part of the world. However, we will discuss data on other (host and parasitoid) species as and when appropriate.

In some of the parasitoid literature, including that focusing on *Drosophila* parasitoids, the ability of the parasitoid to overcome the host's resistance mechanism has been referred to as parasitoid "virulence." Virulence is usually defined as the fitness effect a parasite has on its host. Because a parasitoid either kills its host or is killed by it, parasitoid "virulence" is essentially either 1 or 0. Because virulence when applied to parasites has a different meaning then when applied to parasitoids, we avoid using the term parasitoid ''virulence'' and instead talk about parasitoid ''counter-resistance.''

10.2. *DROSOPHILA MELANOGASTER* AND ITS PARASITOIDS

D. melanogaster primarily feeds on fermenting substrates, such as rotting fruits. In Europe, besides *D. melanogaster*, the most common *Drosophila* species in fermenting substrates are its sibling *D. simulans*, species of the *obscura* group, of which *D. subobscura* and *D. obscura* are the most abundant, and *D. immigrans*. The parasitoids most commonly found parasitizing *Drosophila* larvae on fermenting substrates in Europe are the braconid *Asobara tabida* (Carton et al., 1986; Kraaijeveld and van der Wel, 1994) and the figitids (= eucoilids) *Leptopilina boulardi* and *L. heterotoma* (= *Pseudeucoila bochei* in older literature; Carton et al., 1986). Of these three species, *A. tabida* and *L. heterotoma* occur over most of Europe, whereas *L. boulardi* is restricted to the Mediterranean. In Africa, *Drosophila* larvae in fermenting substrates mainly belong to the *melanogaster* group (*D. melanogaster*, *D. simulans* and *D. yakuba* being the most abundant) where they are attacked by species in the *L. boulardi* complex. However, less is known about the exact species composition of *Drosophila* and its parasitoids on fermenting substrates in Africa than in Europe. Data from other continents are largely anecdotal.

Once larvae have pupated, they are susceptible to attack by pupal parasitoids. In Europe, the most common parasitoids attacking *Drosophila* pupae belong to the chalcidoid genera *Pachycrepoideus* and *Spalangia* (Carton et al., 1986). Unlike the parasitoids attacking *Drosophila* larvae, these pupal parasitoids are not *Drosophila* specialists, but attack pupae from a wide range of dipteran species (Nøstvik, 1954). Also, whereas the larval parasitoids are endoparasitoids (their eggs are oviposited inside the larva), the pupal parasitoids are ectoparasitoids, as they lay their eggs inside the puparium, but outside the actual pupa. *Pachycrepoideus* is often regarded as a facultative hyperparasitoid, as it will parasitize and develop in puparia that contain a fly pupa as well as puparia that contain the pupa of a larval parasitoid (van Alphen and Thunnissen, 1983). No parasitoid has ever been reported attacking *Drosophila* eggs or attacking (or emerging from) adult flies (Carton et al., 1986).

The principal mechanism through which *D. melanogaster* larvae defend themselves against parasitism by parasitoids is immunological and consists of a cellular phase followed by a humoral phase (Meister and Lagueux, 2003; Nappi, 1981; Rizki and Rizki, 1984). When the parasitoid egg is oviposited in the host's hemocoel, it is recognized as nonself (exactly how this nonself recognition works is still not clear). Lamellocytes (large flattened hemocytes) envelop the parasitoid egg until it is covered by several layers. Typically, this cellular phase of the immune reaction is completed in a day or so. Then, enzymes are released in the hemocoel from another type of hemocytes, the crystal cells. These enzymes initiate the humoral phase of the immune reaction by initiating and regulating a series of biochemical reactions referred to as the prophenoloxidase (PPO) cascade. The end product of this cascade is melanin, a black substance that is deposited on the lamellocyte-encapsulated parasitoid egg during the next few days. Once the egg is completely covered by melanin, it dies, due to a lack of oxygen and nutrients and/or as a result of necrotizing compounds emanating from the melanin. Although cellular immunity plays a role against microbial pathogens, the immune reaction against such parasites is mainly based on antimicrobial peptides (Hoffmann, 2003; Lemaitre et al., 1997; Wang and Ligoxygakis, 2006).

The immune reaction briefly described above for *D. melanogaster* also occurs in other *melanogaster* group species such as *D. simulans* and *D. yakuba*. Eslin and Prévost (1998) showed that across species in the *melanogaster* group, there is a positive correlation between the ability to encapsulate eggs of *A. tabida* and the number of free-floating hemocytes, suggesting that hemocyte load is an important parameter in a host's ability to defend itself against parasitoid attack (though the analysis did not control for phylogeny). Outside the *melanogaster* group, *D. obscura* is able to encapsulate and melanize parasitoid eggs to a very low degree (Havard et al., 2009; Kraaijeveld and van der Wel, 1994; Chapter 7 by Eslin et al.). However, there are other species which are known to be resistant to parasitism by parasitoids, such as *D. phalerata* and *D. immigrans*. Melanized capsules have never been reported in these species, and the nature of their resistance mechanism is unknown. In addition, *D. immigrans* larvae, which are larger than larvae of any of the other species, appear to have a skin thick enough to ward off most parasitism attempts by *A. tabida* and *Leptopilina* species (Ideo et al., 2008; personal observation).

What is intriguing is that *D. subobscura*, one of the most abundant species on fermenting substrates, does not appear to have any resistance mechanism against parasitoid attack. Larvae of this species are readily accepted by parasitoids for oviposition, and are commonly attacked by parasitoids in the field, yet only very rarely survive parasitoid attack. They appear not to have the ability to produce lamellocytes and this lack of the crucial type of hemocyte appears to make them unable to launch an effective immune reaction against parasitoid eggs (Eslin and Doury, 2006). At the end of this chapter, we will briefly revisit the question as to why a host would not invest in an immune system despite being commonly parasitized.

Pupae of *Drosophila* have not been reported to mount an immune reaction upon parasitism by parasitoids. As the pupal parasitoids attacking *Drosophila* are ectoparasitoids, an immune response against parasitoid eggs, as seen in *Drosophila* larvae, would not do much good. Whether *Drosophila* pupae have no immune reaction against other parasites, such as microbial pathogens, is unknown.

Parasitoid larvae are never encapsulated by their *Drosophila* host, presumably because they are too mobile to become encapsulated by lamellocytes. This means that parasitoids need to prevent their eggs from becoming encapsulated and melanized. Different parasitoids species have evolved different counter-resistance mechanisms. *A. tabida* has eggs which have a chorion with proteinaceous filaments. These filaments make the egg "stick" to host tissue such as fat body, guts, etc. Due to this "stickiness," the egg ends up embedded in host tissue, where it cannot be reached easily by lamellocytes and thus avoids complete melanization. Because the egg is not fully covered in melanin, the parasitoid larva can continue development and escape from the partial capsule when it hatches from the egg. *L. heterotoma* and *L. boulardi* have a much more active counter-resistance mechanism which consists of essentially blocking the immune reaction right from the beginning. Together with the egg, females inject venom and virus-like particles (VLPs) in the host. *L. heterotoma*'s VLPs enter the host's hemocytes and cause them to apoptose (Rizki and Rizki, 1994). The VLPs from *L. boulardi* differ morphologically from those of *L. heterotoma* and do not cause apoptosis. Rather, together with the venom they appear to block lamellocyte recruitment and alter lamellocyte morphology (Labrosse et al., 2003, 2005; Russo et al., 2001). In addition, *L. boulardi* eggs "stick" to host tissue to some degree (Rizki et al., 1990), but much less so than those of *A. tabida*. Eslin et al. (1996) specifically looked for VLPs in *A. tabida*, but could not detect any.

10.3. GEOGRAPHIC VARIATION

Given that there is geographic variation in the host species that *Drosophila* parasitoids typically encounter, and the rate of parasitism that these hosts experience, it is expected that there is geographic variation in both host resistance and parasitoid counter-resistance.

To determine whether there is indeed geographic variation in the ability of *D. melanogaster* larvae to encapsulate parasitoid eggs, and if so, whether this variation shows a specific geographic pattern, the resistance against *A. tabida* of larvae from 41 populations collected from the field across Europe was measured. In all measurements, *A. tabida* females from a single strain (Sospel, originating from southern France) were used. The results showed that, across Europe, there is a large amount of variation among wild populations of *D. melanogaster* in the ability to encapsulate the eggs of *A. tabida*. Larvae with the highest encapsulation ability (20–64%) are found in central European and central Mediterranean populations whereas larvae from populations collected in north-western Europe and the south-western (Spain, Portugal) and south-eastern Mediterranean (Greek islands, Turkey, Cyprus, Israel) have on average a much lower ability to encapsulate *A. tabida* eggs (0–31%; Kraaijeveld and van Alphen, 1995). Part of this geographic pattern (the north–south component) can be explained by the relative abundance of *D. subobscura*, which is much more common at higher latitudes. As mentioned above, *D. subobscura* does not appear able to launch an immune response against parasitoid eggs. Parasitoids from northern parts of Europe will have the opportunity to oviposit in a host that does not defend itself, while parasitoids from southern Europe have no choice but to oviposit in *D. melanogaster*. This means that the selection pressure on *D. melanogaster* to evolve a highly developed immune system is much stronger in southern than in northern Europe. This is what we indeed see, at least partly; why *D. melanogaster* larvae from the south-western and south-eastern Mediterranean have a lower encapsulation ability than larvae from the central Mediterranean remains unclear.

Larvae from a subset of 28 populations were also exposed to *L. boulardi* females and their ability to encapsulate eggs of this parasitoid species was determined as described for *A. tabida* (the *L. boulardi* strain used for this originated from Tasagil, Turkey). Again, a substantial amount of variation in encapsulation ability was found among these populations (0–24%), but this time there was no discernible geographic pattern (Kraaijeveld and van Alphen, 1995). As mentioned above, *L. boulardi* only occurs in the Mediterranean, but *D. melanogaster* larvae from parts of Europe where *L. boulardi* does not occur did not have a lower ability to encapsulate its eggs than larvae from parts of Europe where the species does occur. However, *L. heterotoma* occurs over much of Europe and as this species counter-resistance mechanism resembles that of *L. boulardi* (even though the VLPs of both species are different), the encapsulation ability measured could very well be against *Leptopilina* VLPs in general rather than those of *L. boulardi* specifically.

In addition to the lack of a clear geographic pattern in encapsulation ability against *L. boulardi* (or possibly *Leptopilina* in general) in European *D. melanogaster* populations, there was no correlation between populations in their ability to encapsulate eggs from *A. tabida* and *L. boulardi* (Kraaijeveld and van Alphen, 1995). We will return to this issue, but it does suggest that the immune system of *D. melanogaster* has components specific to particular counter-resistance mechanisms and that an ability to deal with "sticky" eggs is no guarantee for being able to deal with venom and/or VLPs.

In addition to the geographic variation *between* populations mentioned above, there is also variation *within* populations. Carton and Boulétreau (1985) collected 22 females from a single population of *D. melanogaster* in the field. They established isofemale lines from these females and exposed larvae from these isofemale lines to *L. boulardi*. The encapsulation abilities of these lines ranged from just above 0% to almost 100%, showing that substantial additive genetic within-population variation for encapsulation ability can exist in the field.

As parasitoids searching for *Drosophila* larvae on fermenting substrates in northern parts of Europe have the option to oviposit in larvae of a nonencapsulating species (*D. subobscura*), whereas parasitoids in southern Europe have no option but to use *D. melanogaster*, we would expect parasitoids from southern Europe to be under much stronger selection pressure for a well-developed counter-resistance mechanism. Kraaijeveld and van Alphen (1994) collected 27 *A. tabida* populations from the field across Europe and measured their level of counter-resistance by letting them parasitize *D. melanogaster* larvae. Dissection of parasitized larvae and scoring of encapsulated and nonencapsulated eggs was done as described above and, mirroring the previous experiments, a single host strain was used (originating from Hamburg, Germany). The geographic pattern showed a clear north–south divide with, as expected, southern parasitoids being much better at preventing encapsulation of their eggs than northern parasitoids. Also, the degree of egg "stickiness" (measured by estimating the proportion of the egg chorion that is attached to and embedded in host tissue) was highly and positively correlated with the level of counter-resistance (Kraaijeveld and van Alphen, 1994), providing further support for egg "stickiness" being the counter-resistance mechanism used by *A. tabida*.

In addition to the 27 populations collected from the wild, mentioned above, counter-resistance levels were also measured in a small number (six) of *A. tabida* populations collected from urban sites in the Netherlands and the UK. For two sites (a tropical bird shop in Den Haag and a fruit market in Leeds) it was known that *D. melanogaster* was the dominant *Drosophila* species and that *D. subobscura* was present in low numbers or rare (Atkinson and Shorrocks, 1977; Kraaijeveld, 1994). The level of counter-resistance was significantly higher in the urban strains (24–61%) than in the strains collected from natural habitats in these two countries (13–22%), and the level of egg "stickiness" approached that found in Mediterranean strains (Kraaijeveld, 1994). Although only based on a handful of strains, this higher level of counter-resistance in north-western European urban strains compared to those from nearby natural habitats lends further support to counter-resistance evolving in *A. tabida* when *D. melanogaster* larvae are their main host.

The one part of Europe where, despite several attempts, *A. tabida* was never found during the collections of strains discussed above, is the Iberian peninsula. Instead, its sibling species *A. rufescens* was repeatedly found attacking *Drosophila* larvae in fermenting substrates in Spain and Portugal (Kraaijeveld et al., 1994). In other parts of Europe, *A. rufescens* is a parasitoid of *Scaptomyza pallida* and other drosophilids in decaying plant material (Vet and Janse, 1984) and it has not been reported from fermenting fruits outside the Iberian Peninsula. Comparison of a Dutch and a Portuguese strain of *A. rufescens* showed that the Portuguese parasitoids had a much higher survival probability in *D. melanogaster* than the Dutch parasitoids (Kraaijeveld et al., 1994). This is yet again consistent with the idea that the immune system of *D. melanogaster* exerts a selection pressure on the counter-resistance mechanism of parasitoids attacking it.

The use of a single parasitoid strain to measure host resistance, and a single host strain to measure parasitoid counter-resistance, rests on the assumption that both resistance and counter-resistance are graded one-dimensional traits, with the outcome (i.e., whether the fly or parasitoid survives) a function of the difference between the two. In other words, there is no local adaptation or genotype x genotype interaction. Kraaijeveld and Godfray (2001) found that, across 20 *A. tabida* strains, more than 56% of the variation in survival of the parasitoid in its sympatric *D. melanogaster* strain could be predicted from the resistance and counter-resistance measurements obtained from exposing host strains to a single parasitoid strain and parasitoid strains to a single host strain as discussed above. This strongly suggests that local adaptation plays at most a minor role in the interaction between *A. tabida* and *D. melanogaster*. The egg "stickiness" mechanism used by *A. tabida* to prevent encapsulation by its host indeed appears to be a graded trait rather than one which allows for genetic matching.

L. boulardi uses VLPs that enter the host's hemocytes. Carton (1984), when comparing the survival of *L. boulardi* (originating from Guadeloupe, France, Italy, Tunisia and Brazil) in sympatric and allopatric *D. melanogaster* strains, show that several of the parasitoid strains cause higher mortality in allopatric than in sympatric larvae. They interpret this as evidence for local adaptation, but a closer inspection of the data shows that parasitoid emergence from surviving host larvae is not different between sympatric and allopatric hosts, suggesting that the difference in host mortality is not an adaptive parasitoid trait. Clearly, more data are needed to show whether or not local adaptation and genetic specificity play a role in the interaction between *L. boulardi* and *D. melanogaster*.

10.4. EXPERIMENTAL EVOLUTION OF RESISTANCE AND COUNTER-RESISTANCE

The results presented in the previous section are consistent with selection pressure by parasitoids leading to an increase in host resistance and selection pressure by hosts leading to an increase in parasitoid counter-resistance. However, these data are mostly of a correlative nature, and correlation is not necessarily causation. So does a host population indeed evolve higher levels of resistance when exposed to parasitoids and does a parasitoid population indeed evolve higher levels of counter-resistance when reared on highly resistant hosts? Experimental evolution in the laboratory can give an answer to these questions.

Kraaijeveld and Godfray (1997) collected a large *D. melanogaster* population from the field in the Netherlands. The encapsulation ability of this field population against *A. tabida* was around 5%, typical for a north-western European field population. Selection for increased resistance was carried out by splitting this base population into four subpopulations, and subsequently splitting each of these subpopulations into a selection line which was exposed to parasitism by *A. tabida*, and a paired control line, which was treated in the same way as its paired selection line, apart from exposure to parasitoids. When the larvae of the selection lines had pupated, the pupae were individually checked under the microscope. Only pupae which contained an encapsulated parasitoid egg were chosen for the next generation; pupae showing no signs of parasitoid encapsulation were discarded. The same number of pupae was then picked at random from the paired control line and care was taken in the experimental setup that the effective population size was kept above 50. Encapsulation ability in the selection lines rose from the initial 5% to 50–60% in five generations, after which it leveled off; all four selection lines showed the same rate of change in encapsulation ability over time. Encapsulation ability of the control line did not change over the course of the experiment (eight generations). Fellowes et al. (1998a) used an identical experimental setup, including the same base population of *D. melanogaster*, to select for increased resistance to *L. boulardi*. In this case, encapsulation ability in the selection lines increased from just 0.5% to around 45% in five generation. As in the previous experiment, the response leveled off after five generations, all selection lines showed the same rate of increase, and encapsulation ability of the control lines showed no change during the experiment. These two sets of experimental evolution show that, given a strong selection pressure (and remember that only larvae which had successfully encapsulated the parasitoid egg were allowed into the next generation), resistance of *D. melanogaster* larvae against parasitoids can evolve rapidly. The level of heritability was hard to estimate accurately, but was found to be approximately 20%, which was surprisingly high.

As encapsulation ability across species of the *melanogaster* group is correlated with numbers of hemocytes circulating in the hemocoel (Eslin and Prévost, 1998; see above), the question is whether this across-species relationship is also found at the within-species level. Focusing on the lines selected for increased resistance to *A. tabida* and their controls, Kraaijeveld et al. (2001b) bled second-instar larvae on a hemocytometer and counted relative numbers of hemocytes (without distinguishing between different classes of hemocytes). The results showed that larvae from the selection lines had about double the number of hemocytes floating in the hemocoel than larvae from the control lines. Thus, evolution of increased resistance to *A. tabida* in these lines was correlated with a substantial increase in the numbers of circulating hemocytes.

The *A. tabida* strain used to select for increased resistance was the same as the one used for assessing geographic variation (Sospel, originating from southern France). Larvae from the control and selection lines were also exposed to parasitism by females from three other parasitoid strains (originating from north-western Europe, the Mediterranean and Canada). In all cases, the selection larvae showed a significantly higher encapsulation rate as the control larvae (Kraaijeveld and Godfray, 1999). This supports the theory that the counter-resistance mechanism of *A. tabida* is a graded trait, as resistance that evolved against one parasitoid genotype is effective against three other genotypes from around the world.

But what about cross-resistance across parasitoid species, especially if parasitoid species use very different counter-resistance mechanisms? Does resistance evolved against one parasitoid species increase resistance against another parasitoid species? Fellowes et al. (1999a) tackled this question by exposing larvae from the lines selected for resistance against *A. tabida* and *L. boulardi* against "the other" parasitoid species. Larvae selected for resistance against *A. tabida* showed no increase in resistance to *L. boulardi*. In contrast, larvae selected for resistance to *L. boulardi* had also become much more resistant against *A. tabida*, even though they had not been exposed to this species before. Both sets of selection lines showed an increase in resistance to *L. heterotoma*. The results lend further support to the idea that the resistance mechanism of *D. melanogaster* has parasitoid-species-specific components. Exposure to *A. tabida* appears to select for aspects of the immune system that play a role in resistance to parasitoids in general (such as an increase in circulating hemocytes), whereas a further component is needed for resistance specifically to the VLPs of *L. boulardi*.

Kraaijeveld et al. (2001a) subjected *A. tabida* to experimental evolution for increased counter-resistance in a way similar to that described above for host resistance. The base population originated from southern England and was first split into five subpopulations. Each subpopulation was subsequently split in a selection line which was reared on *D. melanogaster*, and a control line which was reared on the nonencapsulating *D. subobscura*, giving a total of five selection lines and five paired control lines. The selection procedure was simpler than the one employed for host resistance, as all parasitoids emerging from *D. melanogaster* must have escaped encapsulation as an egg. For the first seven generations, a weakly encapsulating *D. melanogaster* strain was used for the selection lines, and little difference was found in the level of counter-resistance of selection and control lines against this strain after seven generations (67% vs. 56%). Subsequently, a strongly encapsulating strain was used for an additional 10 generations. After these 17 generations of experimental evolution, the level of counter-resistance of the selection lines was significantly higher than that of the control lines (37% vs. 11%). Coupled with a change in counter-resistance was a change in egg "stickiness": eggs from selection females were found much more embedded in host tissue than eggs from control females. This result offers more support for egg "stickiness" being the prime counter-resistance mechanism of *A. tabida*. A final small piece of support comes from a selection experiment in which *A. tabida* was specifically selected for increased egg "stickiness." For this, 10 females were allowed to oviposit in *D. subobscura* and kept. All parasitized larvae were dissected and the degree of embedding of each egg was scored. The two females with the highest degree of egg embedding were then allowed to form the next generation. After 15 generations, eggs were indeed much more embedded than at the start of the experiment and the level of counter-resistance in *D. melanogaster* had increased from 20% to 60% (Kraaijeveld, 1994). However, as the experiment was not replicated and did not have a proper control, this result must be regarded with care.

In the experiments mentioned above there was no opportunity for coevolution to take place. Either the host or the parasitoid was able to evolve in response to, respectively, a parasitoid or host strain which itself was reared separately. In order to allow host resistance and parasitoid counter-resistance to evolve simultaneously, Green et al. (2000) set up six replicated sets of three cages. *D. melanogaster* was allowed twice weekly to lay eggs on slices of banana in one cage after which the banana slices were exposed to *A. tabida* for 1 week in the second cage. Subsequently, the banana slices were kept in the third cage; emerging flies were released in the fly cage and emerging parasitoids in the parasitoid cage. A further three sets of cages, in which the larvae were not exposed to parasitoids, was set up as controls. The experiment ran for 5 months, which is equivalent to approximately 10 fly generations or five parasitoid generations. Host resistance indeed increased in the sets of cages exposed to parasitoids, but not by nearly as much as in the selection experiments described above; parasitoid counter-resistance did not change. The fact that host resistance evolved at a lower rate in these coevolution experiments than in the earlier selection experiments can be explained by the fact that in the earlier selection experiments only hosts which had been parasitized and had successfully encapsulated the parasitoid egg formed the next generation. In the coevolution experiment, hosts which were not parasitized in the first place also contributed to the next generation, thereby diluting the strength of the selection pressure. In the parasitoid selection experiment, little effect was seen after seven generations of relatively weak selection, and a clear effect was only seen after 10 generations of strong selection. Given that, it is not surprising that five generations of relatively weak selection did not lead to an increase in parasitoid counter-resistance.

10.5. COSTS OF RESISTANCE AND COUNTER-RESISTANCE

Within-population genetic variation in host resistance has been observed in the field (Carton and Boulétreau, 1985; see above) and the experiments described in the section above confirm that *D. melanogaster* and *A. tabida* populations are genetically variable for resistance and counter-resistance, respectively. Focusing first on host resistance, why would one find substantial genetic variation in a trait which is so closely linked to fitness? One explanation for this is that resistance is costly and that this cost exerts a selection pressure *against* high resistance. Variation in time and space in parasitism rates, coupled with a tradeoff between resistance and other fitness parameters, could result in the relative strengths of the selection pressures for and against high resistance varying both temporally and spatially. As a result, high resistance never reaches fixation and genetic variation is maintained. The same line of reasoning can be used to explain variation in counter-resistance.

When talking of costs of resistance, it is important to keep in mind that there are two types of cost. First, there is the cost of *actual resistance*, resulting from energy and resources being spent when the immune response is launched against the parasitoid egg after parasitism. This type of cost is assessed by comparing individuals that have successfully defended themselves against parasitism with individuals which have not been parasitized, although care must be taken in interpreting the results, as costs may be confused with direct pathogenic effects of parasitism. In *D. melanogaster*, surviving parasitoid attack has indeed been shown to incur costs. Parasitized larvae have a lower competitive ability (Tiën et al., 2001) than unparasitized larvae. After pupation, larvae which have successfully encapsulated the parasitoid egg have thinner puparial walls and an increased risk of being attacked by pupal parasitoids (Fellowes et al., 1998b). Adult flies which succeeded in encapsulating the parasitoid egg as larvae are smaller than flies which were not parasitized, and are more susceptible to desiccation and starvation (Hoang, 2001). Additionally, females have a lower fecundity and males a lower mating success (Carton and David, 1983; Fellowes et al., 1999b). However, not paying these costs of resistance will of course result in genetic death, so incurring them will always be in the evolutionary interests of the organism.

The more interesting cost from an evolutionary perspective is the cost of *having the ability to resist*. This type of cost involves energy and resources invested in the immune system prior to being parasitized, that is, in anticipation of future parasitism. Whether this type of cost should be paid depends on the risk of parasitism. When parasitism is common, individuals who invest in an immune system have a higher overall fitness. However, when parasitism is rare, these individuals are at a disadvantage compared to individuals who have not invested in an immune system, as they have paid the cost of something they do not need.

A powerful method to detect tradeoffs is artificial selection, in which a specific trait is selected for, and the existence of correlated responses in other traits then measured.

Kraaijeveld and Godfray (1997) compared the lines selected for resistance to *A. tabida* with their control lines for a range of standard life history parameters (e.g., survival rate, developmental time, adult size), but found no differences between control and selection flies. As tradeoffs are more likely to be detected when resources are limited, they set up a second set of experiments, in which larvae from the control and selection lines were reared in food patches together with larvae of a strain that could be distinguished phenotypically from them (an eye color mutant in this case). The amount of food in the patches was varied from plentiful to severely limited. When food was plentiful, no difference between control and selection larvae was found, but when food was severely limited, larvae from the selection lines had a lower relative competitive ability than larvae from the control lines. To explore further the cause of the reduced competitive ability of larvae selected for parasitoid resistance, Fellowes et al. (1999c) compared the feeding rate of larvae selected for resistance to both *A. tabida* and *L. boulardi* with their respective controls. Feeding rate is known to be an important parameter of competitive ability in *Drosophila* (Joshi and Mueller, 1988; Mueller, 1988a,b) and can easily be measured by putting a larva in a drop of liquid yeast suspension and counting the number of retractions of the cephalopharyngeal feeding apparatus (the mouth hooks). Larvae selected for increased parasitoid resistance had a reduced feeding rate, independent of whether they had increased resistance to *A. tabida* or *L. boulardi*. Thus, resistance to parasitoids tradeoffs with larval competitive ability. Evolution of increased resistance leads to an increase in the numbers of circulating hemocytes, but this comes at a cost of a reduction in larval feeding rate. As discussed above, Green et al. (2000) found that encapsulation ability increased more slowly in a cage setup than in the earlier selection experiments. Apart from the explanation offered above (i.e., dilution of the selection pressure due to unparasitized flies contributing to the next generation), the cost of resistance may very well have played a role in slowing down the rate of evolution of resistance, as competition for food was not prevented in the cages.

Why hemocyte numbers would trade off with feeding rate is unclear. Several, not mutually exclusive, explanations could underlie this tradeoff. First, it may simply be a redirection of resources from a trophic to an immune function. Secondly, as both the hematopoietic organ and the head musculature originate from the same part of the embryo (Tepass et al., 1994), the tradeoff may be the result of a shift in the balance of embryonic tissue allocated to these two parts of the future larva. Finally, it could be that the increase in circulating hemocytes increases the viscosity of the hemolymph, thereby reducing the rate at which oxygen and nutrients can reach the head muscles.

Is the tradeoff between resistance and competitive ability symmetrical? Sanders et al. (2005) explored this question with an experiment in which larvae were reared at either high or low levels of competition. The lines selected for increased resistance to *A. tabida* were pooled and served as the base population for this experiment; first the base population was split into five subpopulations, each of which was subsequently split in a line reared under a high level of competition (200 larvae on a banana slice) and a line reared under a low level of competition (50 larvae on a banana slice). After eight generations of selection, larvae from the high competition lines were indeed superior competitors than larvae from the low competition lines, although they did not differ in feeding rate. Contrary to expectation, larvae from the high competition lines appeared to have a slightly but significantly *higher* level of resistance to *A. tabida* than larvae from the low competition lines (55% vs. 41%) and a correspondingly higher number of circulating hemocytes. The reason why larvae reared under high levels of competition evolved higher parasitoid resistance, despite never having been exposed to parasitoids, is unclear. One explanation may be that the risk of larvae wounding each other with their mouth hooks is higher under crowded conditions. As wound healing and encapsulation of parasitoid eggs partially share physiological pathways (Galko and Krasnow, 2004; Lackie, 1988), selection for increased wound healing capability under crowded conditions may have a slightly increased parasitoid resistance as a by-product.

In *A. tabida*, there is unlikely to be a cost of *actual counter-resistance*. The egg ''stickiness'' mechanism appears to be passive and not require additional energy and/or resources to be expended once the egg is oviposited in the host. To detect costs of the *ability to counter-resist*, Kraaijeveld et al. (2001a) compared parasitoids from the *A. tabida* lines described above that had been selected for increased counter-resistance to their controls. A comparison of a range of life history parameters (e.g., survival in *D. subobscura*, development time, fat content and egg load) showed no difference between selection and control parasitoids. In the host, costs of resistance were only found when food was limited. A parasitoid equivalent of this situation occurs when more than one parasitoid larva is present in a host. As only one parasitoid can develop from a host, parasitoid larvae present in the same host will have to fight for possession of that host. The time it takes for the parasitoid egg to hatch is important in determining the winner of a fight for the host (van Strien-van Liempt, 1983; Visser et al., 1992): the first parasitoid out of the egg has a higher probability to kill the other egg(s) or smaller larva(e). Kraaijeveld et al. (2001a) allowed *D. subobscura* larvae to be attacked by parasitoids from the selection and control lines and dissected the larvae at various subsequent time intervals in order to determine the time window during which the parasitoid eggs hatched inside the host. Eggs from selection-line parasitoids hatched on average 2.5 h later than eggs from control parasitoids. This delay in egg development is likely to be caused by the eggs being embedded in host tissue and therefore experiencing a lower rate of oxygen and nutrients reaching the egg. Although a delay of 2.5 h may not seem much given that parasitoid egg-to-adult development time is about 1 month, a small difference in hatch rate can make a major difference when parasitoid larvae are fighting for possession of the host (van Strien-van Liempt, 1983; Visser et al., 1992).

Within-population variation in resistance also appears to have a sexual dimension. Kraaijeveld et al. (2008) exposed larvae from a single *D. melanogaster* strain to *A. tabida*, sexed the larvae just prior to pupation, and dissected them 5 days after parasitism to score encapsulated and nonencapsulated eggs. Male larvae had a significantly lower encapsulation ability than female larvae (51% vs. 65%). This sex difference in resistance to parasitoids suggests that the optimal level of immunity might not be the same in males and females, something which is predicted under certain circumstances by life history theory.

10.6. BEHAVIOR RELATED TO RESISTANCE AND COUNTER-RESISTANCE

Whenever variation in resistance to parasitoids exists within *D. melanogaster* populations, it would be beneficial for a female's offspring if she mates with males that have genes coding for high resistance to parasitoids. Of course, for mate preference to occur, a female must be able to detect the genetic resistance status of a male. Rolf and Kraaijeveld (2003) offered around 300 females a choice of a pair of males, one of which was from a line selected for resistance to *A. tabida*, the other was from its paired control line. Males were color-marked to allow identification and the triad was observed until the first mating took place. Males from the selection lines were more successful in obtaining matings. In a separate set of experiments, females were offered either a male from a selection line or a male from a control line and the time until mating was measured. In these experiments, the time until mating took place was less with males from the selection lines than with males from the control lines. The proximate reason for why females prefer males from the selection lines is as yet unclear. Males from selection and control lines may differ behaviorally (e.g., in their courtship song) and/or they may differ in cuticular hydrocarbons and thus smell differently to females.

When a larva is successful in encapsulating the parasitoid egg, the blackened capsule is often visible through the abdominal wall of the adult fly. Kraaijeveld et al. (1997) tested whether females show a preference for capsule-bearing males over males which had not been parasitized. A Perspex cage, containing 25 capsule-bearing males, 25 unparasitized males and 25 virgin females was observed continuously (replicated 15 times). Mating pairs were removed and the type of the male determined. No female preference was found for males of either type. As males usually face a female when courting, she may not be able to observe clearly the abdomen. Alternatively, although capsule-bearing males reveal they have the genes to resist parasitoids, they show at the same time that they behaved in a way that made them susceptible to parasitoid attack and this may outweigh the benefit of the resistance genes.

There are several potential ways in which *Drosophila* larvae could alter their behavior to increase their probability of successfully encapsulating a parasitoid egg. *A. tabida* females leave a kairomone on the substrate when searching (van Alphen and Galis, 1983). This could potentially be used by host larvae to indicate the presence of searching females in the environment and lead to an upregulation of the immune system in anticipation of imminent parasitism. Robertson and Kraaijeveld (unpublished observation) let larvae crawl for 3 h on either a patch previously visited by *A. tabida* or an unvisited patch. Subsequently, all larvae were exposed to parasitoid attack and dissected to score encapsulation rate; no difference was found between the two groups of larvae. Like most physiological processes in invertebrates, the efficacy of the immune reaction of *Drosophila* is temperature dependent. Locusts are known to increase their body temperature after pathogen infection by actively moving to warmer places (Blanford et al., 2002). To test whether this phenomenon, referred to as "behavioral fever," also occurs in *Drosophila* larvae parasitized by a parasitoid, Croxson and Kraaijeveld (unpublished observation) established a temperature gradient of 10 °C (range 15–25 °C) over bananas. Subsequently, they released either parasitized or unparasitized larvae in the center of the bananas and allowed the larvae to crawl freely across the temperature gradient. Five days later, the bananas were cut into five sections, which were kept separately until fly/parasitoid emergence. Although it was predicted that parasitized larvae would move more toward the warm end of the banana than unparasitized larvae, there was in fact no difference in the distribution of parasitized and unparasitized larvae over the banana sections. Both sets of experiments described above were small-scale pilot experiments, but neither gave any indication that, following parasitism, *D. melanogaster* larvae change their behavior to maximize their immune system's opportunity of dealing with the parasitoid egg.

There is a link between larval behavior and susceptibility to pupal parasitoids. As mentioned above, *Drosophila* pupae have no immune reaction against parasitoid eggs and variation in the thickness of the puparial wall, a barrier that parasitoids must overcome prior to oviposition, does not appear to influence risk of parasitism (Kraaijeveld and Godfray, 2003). Rearing larvae on banana slices in large Petri dishes filled with vermiculite, allows them to freely choose their pupation site. Larvae that pupate away from the banana slice have a higher probability to be parasitized by searching *Pachycrepoideus* females than those that pupate on the slice (Kraaijeveld and Godfray, 2003).

When searching *A. tabida* females encounter a mixture of *D. subobscura* and *D. melanogaster* larvae, the *D. subobscura* larvae are virtually always accepted for oviposition. Whether the *D. melanogaster* larvae are also accepted depends on the geographic origin of the parasitoid, and thus on the probability that its eggs will avoid encapsulation. Kraaijeveld et al. (1995) showed that *A. tabida* females from southern European populations, which have a high probability of survival in *D. melanogaster* almost always accept this host for oviposition, whereas females from more northern populations, which have a lower probability of survival in *D. melanogaster*, tend to reject it.

Rolff and Kraaijeveld (2001) showed that variation in host choice behavior in *A. tabida* has a genetic basis. They offered a mixture of *D. subobscura* and *D. melanogaster* larvae to females from the lines selected for increased counter-resistance in *D. melanogaster* and their paired control lines. Females selected for increased counter-resistance accepted *D. melanogaster* more often than females from the control lines. Rearing *A. tabida* on *D. melanogaster* could exert a direct selection pressure on host choice behavior (females rejecting *D. melanogaster* will not contribute to the next generation), or the evolution of host choice behavior could be a pleiotropic effect of the evolution of counter-resistance.

A. tabida's host choice behavior appears to have evolved to take adaptive host choice decisions when offered a range of host species differing in the survival probability they offer to the parasitoid egg (Kraaijeveld et al., 1995; van Alphen and Janssen, 1982). In this light, the decisions taken when offered *D. melanogaster* and *D. simulans* appears at first maladaptive. *D. simulans* is a stronger encapsulator than *D. melanogaster* and the survival probability of *A. tabida* eggs in this host species is virtually nil (Kraaijeveld and van der Wel, 1994). When offered a mixture of *D. melanogaster* and *D. simulans*, parasitoid females do distinguish between the two host species, but still readily accept *D. simulans* (Kraaijeveld, 1999). *L. boulardi* is well able to prevent encapsulation by *D. simulans* and using a set of strains from Ormos Panagias (Greece),

Kraaijeveld (1999) showed that the survival rate of *A. tabida* in *D. simulans* larvae previously parasitized by *L. boulardi* is 15% compared to the virtually nil survival rate in *D. simulans* when on its own. *L. boulardi*'s survival rate of 88% when alone in *D. simulans* larvae is reduced to 35% in multiparasitized hosts. In other words, *A. tabida* acts as a kleptoparasitoid of *L. boulardi* in *D. simulans*, profiting from *L. boulardi*'s VLPs knocking out the host's immune system. An optimal foraging model showed that the option to act as a kleptoparasitoid increases the value of *D. simulans* to a level where it should at least be partially accepted (Kraaijeveld, 1999).

10.7. PARASITOIDS AS HOSTS

It has already been mentioned that the pupal parasitoid *Pachycrepoideus* can be regarded as a facultative hyperparasitoid, as it will also parasitize *Drosophila* puparia which contain the pupa of a larval parasitoid. In addition, *A. tabida*, *L. heterotoma* and *Pachycrepoideus* can be infected by the entomopathogenic fungus *Beauveria bassiana* (personal observation) and *A. tabida* and *Pachycrepoideus* are also susceptible to infection by the microsporidian *Tubulinosema kingi* (Franzen et al., 2006; Futerman et al., 2006). No data exist on the incidence of infection by these pathogens in the field and nothing is known about any immune responses that parasitoids may mount against their own parasites and pathogens.

In the field, *A. tabida* is known to be infected with the bacterium *Wolbachia* and *L. boulardi* with a virus. The often fascinating interaction between these pathogens and their parasitoid hosts are the focus of other chapters in this book (Chapter 12 by Vavre et al.; Chapter 13 by Varaldi et al.), but one aspects of *Wolbachia* deserves mention in this chapter as it relates directly to host resistance and parasitoid counter-resistance. Fytrou et al. (2006) established *Wolbachia*-free strains of *D. simulans* and *L. heterotoma* through curing with tetracycline, and then compared the encapsulation rates of infected and uninfected larvae parasitized by infected and uninfected parasitoids. The results showed that the bacterium suppressed both the host's immune reaction as well as the parasitoid's counter-resistance. Although the underlying mechanisms are as yet unknown, these results do show that species other than the host and parasitoid themselves have the ability to influence levels of resistance and counter-resistance.

10.8. GENETICS AND GENOMICS

A full understanding of the (co)evolution of host resistance and parasitoid counter-resistance will ultimately need to include consideration of the genes involved. Although a number of genes are known to be involved in immunity against microbial pathogens in *D. melanogaster*, very little was known until recently about the genes involved in resistance against parasitoids. Poirié et al. (2000) showed that different genes were involved in resistance to *A. tabida* and *L. boulardi* (Rat and Rlb, respectively), although the precise role of these genes in the immune reaction is still unclear. More recently, the immune reaction against *A. tabida* and the two *Leptopilina* species was investigated at a genomic level, using Affymetrix expression microarrays (Schlenke et al., 2007; Wertheim et al., 2005). These studies compared genome-wide gene expression profiles of parasitized and unparasitized larvae and found several hundred genes which were expressed differentially in parasitized and unparasitized larvae. Wertheim et al. (2005) found that most of the genes they identified had not previously been associated with immune function. Several groups of genes were expressed together and shared functional annotations. For example, a group of genes involved in proteolysis and peptidolysis was upregulated during the encapsulation/melanization phase of the immune response. The *Drosophila* genome codes for various types of pattern recognition receptors (PRRs) for recognizing invading organisms, such as lectins and receptors for microbial peptides. The immune response against *A. tabida* showed differential expression. During the first few hours after *A. tabida* attack two of the 20 peptidoglycan recognition proteins (*PGRP-LB* and *PGRP-SB1*) were upregulated, which may be a response to microbial infections following the puncturing of the cuticle by the parasitoid. Of the 30 C-type lectins in the *Drosophila* genome, one (*lectin-24A*) showed a striking increase in expression during the encapsulation phase. More overlap was found in genes involved in the resistance to *A. tabida* and *L. boulardi* than in the genes involved in the resistance to *A. tabida* and *L. heterotoma* (Schlenke et al., 2007). *L. boulardi* appears to provoke a relatively normal immune response that is only sabotaged at the final stage, while *L. heterotoma* seems to cause a near-complete lack of a transcriptional immune response. *L. boulardi* induced a similar massive upregulation of the same *lectin-24A* gene during the first 2–5 h after infection, while the lectin was not upregulated at all after attack by *L. heterotoma*. Lectins are thought to be also important in changing cell adhesion properties, and it is therefore plausible that *lectin-24A* is a key player in recruiting the hemocytes to the parasitoid egg and the subsequent formation of the multilayered cellular encasing of the egg.

However, these studies focused on genes involved in the *actual immune response*. These are not necessarily the genes which are involved in the *variation in the ability to launch an immune response*, which is observed both between and within populations. It is this latter group of genes which is important for the evolution of resistance. A first step in identifying the genes involved in the *variation* in resistance was taken by Wertheim et al. (unpublished observations). They compared the genome-level gene expression profiles of larvae which were selected for increased resistance to *A. tabida* with that of their paired control larvae at different time points (from young egg to second instar). As the focus was on genes involved in building up (rather than using) the immune system, all larvae were unparasitized. Analysis of the data is in progress at the time of writing this chapter, but early indications suggest that selection and control larvae differ in the expression profiles of several hundred genes, and that there is little overlap between these genes and the genes identified as being involved in the actual immune response.

Nothing is known as of yet of genes involved in variation in counter-resistance in *A. tabida* or the *Leptopilina* spp. However, the recent sequencing of three species of *Nasonia* (http://www.hgsc.bcm.tmc.edu/projects/nasonia; last accessed July 01, 2009), coupled with the increasing ability to analyze the genomics of nonmodel organisms means that obtaining an understanding of the genetics and genomics of parasitoid counter-resistance means progress is likely in this area.

10.9. CONCLUDING REMARKS

At a geographic scale, coevolution between *D. melanogaster* and its parasitoids is consistent with Thompson's (1994, 2005) geographic mosaic model. Coevolutionary hot spots for the interaction between *D. melanogaster* and *A. tabida* occur in the central Mediterranean whereas coevolutionary cold spots occur in north-western Europe. We believe the dynamics at a geographic level are driven by variation in the abundance of *D. subobscura*, leading to parasitoids in the cold spots having the option to use this nonencapsulating host for oviposition, whereas in the hot spots they have no choice but to oviposit in *D. melanogaster*. In the hot spots, this results in reciprocal selection for increased resistance and counter-resistance.

At the population level, variation within populations appears to be mainly driven by the costs of (counter-)resistance. Fluctuating and opposing selection pressures maintain genetic variation and the balance of the opposing selection pressures drive the (counter-)resistance of a population to the level that is optimal in terms of fitness at any point in space or time.

In both *D. melanogaster* and *A. tabida*, increased (counter-)resistance evolves at the expense of larval competitive ability. This means that the costs of both traits are density dependent. A reduction in competitive ability will be close to selective neutrality when population sizes are small and food is plentiful, whereas it will be detrimental to fitness at high population sizes when competition for food is severe. Sasaki and Godfray (1999) modeled a general (though inspired by the *Drosophila*–parasitoid system) host–parasitoid interaction where both host resistance and parasitoid counter-resistance were costly and found that the course of coevolution between host resistance and parasitoid counter-resistance was largely determined by the relative costs of the two traits. In some parts of parameter space, when resistance is more costly for the host than counter-resistance is for the parasitoid, hosts do not invest in an immune system at all, but basically gamble on not being found. Such a scenario might apply to *D. subobscura* and explain why this host species lacks the ability to launch an immune response against parasitoids

The *Drosophila*–parasitoid system has proven to be very valuable in integrating a range of aspects of host–parasitoid interactions, from evolutionary ecology (e.g., costs) to physiology, behavior and ultimately genetic/genomics. We believe it is a valuable model system for understanding the evolution of resistance and counter-resistance in host–parasite interactions.

REFERENCES

Atkinson, W., Shorrocks, B., 1977. Breeding site specificity in the domestic species of *Drosophila*. Oecologia 29, 223–232.

Blanford, S., Thomas, M.B., Langewald, J., 2002. Behavioural fever in the Senegalese grasshopper, *Oedaleus senegalensis*, and its implications for biological control using pathogens. Ecol. Ent. 23, 9–14.

Carton, Y., 1984. Analyse expérimentale de trois niveaux d'interaction entre *Drosophila melanogaster* et le parasite *Leptopilina boulardi* (sympatrie, allopatrie, xénopatrie). Génét. Sél. Evol. 16, 417–430.

Carton, Y., Boulétreau, M., 1985. Encapsulation ability of *Drosophila melanogaster*: a genetic analysis. Dev. Comp. Immunol. 9, 211–219.

Carton, Y., David, J.R., 1983. Reduction of fitness in *Drosophila* adults surviving parasitism by a cynipid wasp. Experientia 39, 231–233.

Carton, Y., Boulétreau, M., van Alphen, J.J.M., van Lenteren, J.C., 1986. The *Drosophila* parasitic wasps. In: Ashburner, Carson, Thompson, (Eds.), The Genetics and Biology of Drosophila, Vol. 3e. Academic Press, London, pp. 347–394.

Eslin, P., Doury, G., 2006. The fly *Drosophila subobscura*: a natural case of innate immune deficiency. Dev. Comp. Immunol. 30, 977–983.

Eslin, P., Prévost, G., 1998. Hemocyte load and immune resistance to *Asobara tabida* are correlated in species of the *Drosophila melanogaster* subgroup. J. Ins. Phys. 44, 807–816.

Eslin, P., Giordanengo, P., Fourdrain, Y., Prévost, G., 1996. Avoidance of encapsulation in the absence of VLP by a braconid parasitoid of *Drosophila* larvae: an ultrastructural study. Can. J. Zool. 74, 2193–2198.

Fellowes, M.D.E., Kraaijeveld, A.R., Godfray, H.C.J., 1998a. Trade-off associated with selection for increased ability to resist parasitoid attack in *Drosophila melanogaster*. Proc. R. Soc. Lond. B 265, 1553–1558.

Fellowes, M.D.E., Kraaijeveld, A.R., Godfray, H.C.J., 1999a. Cross-resistance following artificial selection for increased defence against parasitoids in *Drosophila melanogaster*. Evolution 53, 966–972.

Fellowes, M.D.E., Kraaijeveld, A.R., Godfray, H.C.J., 1999b. The relative fitness of *Drosophila melanogaster* (Diptera, Drosophilidae) that have successfully defended themselves against the parasitoid *Asobara tabida* (Hymenoptera, Braconidae). J. Evol. Biol. 12, 123–128.

Fellowes, M.D.E., Kraaijeveld, A.R., Godfray, H.C.J., 1999c. Association between feeding rate and defence against parasitoids in *Drosophila melanogaster*. Evolution 53, 1302–1305.

Fellowes, M.D.E., Masnatta, P., Kraaijeveld, A.R., Godfray, H.C.J., 1998b. Pupal parasitoid attack influences the relative fitness of *Drosophila* that have encapsulated larval parasitoids. Ecol. Ent. 23, 281–284.

Franzen, C., Futerman, P.H., Schroeder, J., Salzberger, B., Kraaijeveld, A.R., 2006. An ultrastructural and molecular study of *Tubulinosema kingi* Kramer (Microsporidia: Tubulinosematidae) from *Drosophila melanogaster* (Diptera: Drosophilidae) and its parasitoid *Asobara tabida* (Hymenoptera: Braconidae). J. Invert. Path. 91, 158–167.

Futerman, P.H., Layen, S.J., Kotzen, M.L., Franzen, C., Kraaijeveld, A.R., Godfray, H.C.J., 2006. Fitness effects and transmission routes of a microsporidian parasite infecting *Drosophila* and its parasitoids. Parasitology 132, 479–492.

Fytrou, A., Schofield, P.G., Kraaijeveld, A.R., Hubbard, S.F., 2006. *Wolbachia* suppresses both host defence and parasitoid counter-defence. Proc. R. Soc. Lond. B 273, 791–796.

Galko, M.J., Krasnow, M.A., 2004. Cellular and genetic analysis of wound healing in *Drosophila* larvae. PLoS Biol. 8, e329.

Green, D.M., Kraaijeveld, A.R., Godfray, H.C.J., 2000. Evolutionary interactions between *Drosophila melanogaster* and its parasitoid *Asobara tabida*. Heredity 85, 450–458.

Havard, S., Eslin, P., Prévost, G., Doury, G., 2009. Encapsulation ability: are all *Drosophila* species equally armed? An investigation in the *obscura* group. Can. J. Zool. in press.

Hoang, A., 2001. Immune response to parasitism reduces resistance of *Drosophila melanogaster* to desiccation and starvation. Evolution 55, 2353–2358.

Hoffmann, J.A., 2003. The immune response of *Drosophila*. Nature 426, 33–38.

Ideo, S., Watada, M., Mitsui, H., Kimura, M.T., 2008. Host range of *Asobara japonica* (Hymenoptera: Braconidae), a larval parasitoid of drosophilid flies. Entomol. Science 11, 1–6.

Joshi, A., Mueller, L.D., 1988. Evolution of higher feeding rate in *Drosophila* due to density dependent natural selection. Evolution 42, 1090–1093.

Kraaijeveld, A.R., 1994. Local Adaptations in a Parasitoid–Host System – a Coevolutionary Arms Race? University of Leiden, The Netherlands [PhD thesis].

Kraaijeveld, A.R., 1999. Kleptoparasitism as an explanation for paradoxical oviposition decisions of the parasitoid *Asobara tabida*. J. Evol. Biol. 12, 129–133.

Kraaijeveld, A.R., Godfray, H.C.J., 1997. Trade-off between parasitoid resistance and larval competitive ability in *Drosophila melanogaster*. Nature 389, 278–280.

Kraaijeveld, A.R., Godfray, H.C.J., 1999. Geographic patterns in the evolution of resistance and virulence in *Drosophila* and its parasitoids. Am. Nat. 153, S61–S74.

Kraaijeveld, A.R., Godfray, H.C.J., 2001. Is there local adaptation in *Drosophila*–parasitoid interactions? Evol. Ecol. Res. 3, 107–116.

Kraaijeveld, A.R., Godfray, H.C.J., 2003. Potential life history costs of parasitoid avoidance in *Drosophila melanogaster*. Evol. Ecol. Res. 5, 1251–1261.

Kraaijeveld, A.R., van Alphen, J.J.M., 1994. Geographical variation in resistance of the parasitoid *Asobara tabida* against encapsulation by *Drosophila melanogaster* larvae: the mechanism explored. Physiol. Ent. 19, 9–14.

Kraaijeveld, A.R., van Alphen, J.J.M., 1995. Geographic variation in encapsulation ability of *Drosophila melanogaster* larvae and evidence for parasitoid-specific components. Evol. Ecol. 9, 10–17.

Kraaijeveld, A.R., van der Wel, N.N., 1994. Geographical variation in reproductive success of the parasitoid *Asobara tabida* in larvae of several *Drosophila* species. Ecol. Ent. 19, 221–229.

Kraaijeveld, A.R., Barker, C.L., Godfray, H.C.J., 2008. Stage-specific sex differences in *Drosophila* immunity to parasites and pathogens. Evol. Ecol. 22, 217–228.

Kraaijeveld, A.R., Emmett, D.A., Godfray, H.C.J., 1997. Absence of direct sexual selection for parasitoid encapsulation in *Drosophila melanogaster*. J. Ecol. Biol. 10, 337–342.
Kraaijeveld, A.R., Hutcheson, K.A., Limentani, E.C., Godfray, H.C.J., 2001a. Costs of counter-defenses to host resistance in a parasitoid of *Drosophila*. Evolution 55, 1815–1821.
Kraaijeveld, A.R., Limentani, E.C., Godfray, H.C.J., 2001b. Basis of the trade-off between parasitoid resistance and larval competitive ability in *Drosophila melanogaster*. Proc. R. Soc. Lond. B 268, 259–261.
Kraaijeveld, A.R., Nowee, B., Najem, R.W., 1995. Adaptive variation in host selection behaviour of *Asobara tabida*, a parasitoid of *Drosophila* larvae. Funct. Ecol. 9, 113–118.
Kraaijeveld, A.R., Voet, S., van Alphen, J.J.M., 1994. Geographical variation in habitat choice and host suitability in the parasitoid *Asobara rufescens*. Ent. Exp. Appl. 72, 109–114.
Labrosse, C., Carton, Y., Dubuffet, A., Drezen, J.M., Poirié, M., 2003. Active suppression of *D. melanogaster* immune response by long gland products of the parasitic wasp *Leptopilina boulardi*. J. Ins. Physiol. 49, 513–522.
Labrosse, C., Eslin, P., Doury, G., Drezen, J.M., Poirié, M., 2005. Haemocyte changes in *D. melanogaster* in response to long gland components of the parasitoid wasp *Leptopilina boulardi*: a Rho-GAP protein as an important factor. J. Ins. Physiol. 51, 161–170.
Lackie, A.M., 1988. Haemocyte behaviour. Adv. Ins. Physiol. 21, 85–177.
Lemaitre, B., Reichhart, J.-M., Hoffmann, J.A., 1997. *Drosophila* host defense: differential induction of antimicrobial peptide genes after infection by various classes of microorganisms. Proc. Natl. Acad. Sci. USA 94, 14614–14619.
Meister, M., Lagueux, M., 2003. *Drosophila* blood cells. Cell. Microbiol. 5, 573–580.
Mueller, L.D., 1988a. Density-dependent population growth and natural selection in food-limited environments: the *Drosophila* model. Am. Nat. 132, 786–809.
Mueller, L.D., 1988b. Evolution of competitive ability in *Drosophila* by density-dependent natural selection. Proc. Natl. Acad. Sci. USA 85, 4383–4386.
Nappi, A.J., 1981. Cellular immune response of *Drosophila melanogaster* against *Asobara tabida*. Parasitology 83, 319–324.
Nøstvik, E., 1954. Biological studies of *Pachycrepoideus dubius* Ashmead (Chalcidoidea: Pteromalidae), a pupal parasite of various Diptera. Oikos 5, 195–204.
Poirié, M., Frey, F., Hita, M., Huguet, E., Lemeunier, F., Periquet, G., Carton, Y., 2000. *Drosophila* resistance genes to parasitoids: chromosomal location and linkage analysis. Proc. R. Soc. Lond. B 267, 1417–1421.
Rizki, T.M., Rizki, R.M., 1984. The cellular defense system of *Drosophila melanogaster*. In: King, Akai, (Eds.), Insect Ultrastructure. Vol. 2. Plenum Publishing Co, New York, pp. 579–603.
Rizki, T.M., Rizki, R.M., 1994. Parasitoid-induced cellular immune deficiency in *Drosophila*. Ann. NY Acad. Sci. 712, 178–194.
Rizki, T.M., Rizki, R.M., Carton, Y., 1990. *Leptopilina heterotoma* and *L. boulardi*: strategies to avoid cellular defense responses of *Drosophila melanogaster*. Exp. Parasitol. 70, 466–475.
Rolff, J., Kraaijeveld, A.R., 2001. Host preference and survival in selected lines of a *Drosophila* parasitoid. Asobara tabida. J. Evol. Biol. 14, 742–745.
Rolff, J., Kraaijeveld, A.R., 2003. Selection for parasitoid resistance alters mating success in *Drosophila*. Proc. R. Soc. Lond. B (Suppl.) 270, S154–S155.
Russo, J., Brehelin, M., Carton, Y., 2001. Haemocyte changes in resistant and susceptible strains of *D. melanogaster* caused by virulent and avirulent strains of the parasitic wasp *Leptopilina boulardi*. J. Ins. Physiol. 47, 167–172.
Sanders, A.E., Scarborough, C., Layen, S.J., Kraaijeveld, A.R., Godfray, H.C.J., 2005. Evolutionary change in parasitoid resistance under crowded conditions in *Drosophila melanogaster*. Evolution 59, 1292–1299.
Sasaki, A., Godfray, H.C.J., 1999. A model for the coevolution of resistance and virulence in coupled host–parasitoid interactions. Proc. R. Soc. Lond. B 266, 455–463.

Schlenke, T.A., Morales, J., Govind, S., Clark, A.G., 2007. Contrasting infection strategies in generalist and specialist wasp parasitoids of *Drosophila melanogaster*. PLoS Pathog. 3, e158.
Tepass, U., Fessler, L., Aziz, A., Hartenstein, V., 1994. Embryonic origin of hemocytes and their relationship to cell death in *Drosophila*. Development 120, 1829–1837.
Thompson, J.N., 1994. The Coevolutionary Process. University of Chicago Press, Chicago.
Thompson, J.N., 2005. The Geographic Mosaic of Coevolution. University of Chicago Press, Chicago.
Tiën, N.S.H., Boyle, D., Kraaijeveld, A.R., Godfray, H.C.J., 2001. Competitive ability of parasitized *Drosophila* larvae. Evol. Ecol. Res. 3, 747–757.
van Alphen, J.J.M., Galis, F., 1983. Patch time allocation and parasitization efficiency of *Asobara tabida*, a larval parasitoid of *Drosophila*. J. Anim. Ecol. 52, 937–952.
van Alphen, J.J.M., Janssen, A.R.M., 1982. Host selection by *Asobara tabida* Nees (Braconidae; Alysiinae) a larval parasitoid of fruit inhabiting *Drosophila* species. II. Host species selection. Neth. J. Zool. 32, 194–214.
van Alphen, J.J.M., Thunnissen, I., 1983. Host selection and sex allocation by *Pachycrepoideus vindemiae* Rondani (Pteromalidae) as a facultative hyperparasitoid of *Asobara tabida* Nees (Braconidae; Alysiinae) and *Leptopilina heterotoma* (Thomson) (Cynipoidea; Eucoilidae). Neth. J. Zool. 33, 497–514.
van Strien-van Liempt, W.T.F.H., 1983. The competition between *Asobara tabida* Nees von Esenbeck, 1834 and *Leptopilina heterotoma* (Thomson, 1862) in multiparasitized hosts. I. The course of competition. Neth. J. Zool. 33, 125–163.
Vet, L.E.M., Janse, C., 1984. Fitness of two sibling species of *Asobara* (Braconidae; Alysiinae), larval parasitoids of Drosophilidae in different microhabitats. Ecol. Ent. 9, 345–354.
Visser, M.E., Luyckx, B., Nell, H.W., Boskamp, G.J.F., 1992. Adaptive superparasitism in solitary parasitoids: marking of parasitized hosts in relation to the pay-off from superparasitism. Ecol. Entomol. 17, 76–82.
Wang, L., Ligoxygakis, P., 2006. Pathogen recognition and signalling in the *Drosophila* innate immune response. Immunobiology 211, 251–261.
Wertheim, B., Kraaijeveld, A.R., Schuster, E., Blanc, E., Hopkins, M., Pletcher, S.D., et al., 2005. Genome-wide gene expression in response to parasitoid attack in *Drosophila*. Genome Biol. 11, R94.

CHAPTER 11

Local, Geographic and Phylogenetic Scales of Coevolution in *Drosophila*–Parasitoid Interactions

S. Dupas,[*,†] **A. Dubuffet,**[‡] **Y. Carton,**[*] and **M. Poirié**[§]

Contents

[*] IRD, UR072 Laboratoire Evolution, Génomes et Spéciation/UPR9034, CNRS 91198 Gif-sur-Yvette cedex, France/Université Paris-Sud 11, 91405 Orsay cedex, France
[†] Pontificia Universidad Católica del Ecuador, Facultad de Ciencias Exactas y Naturales, Quito, Ecuador
[‡] Institute of Integrative and Comparative Biology, Faculty of Biological Sciences, University of Leeds, United Kingdom
[§] UMR Interactions Biotiques et Santé Végétale, Institut Agrobiotech, 06 903 Sophia Antipolis, France. INRA, UMR 1301/CNRS UMR 6243/Université Nice Sophia Antipolis, 28, avenue de Valrose, 06103 Nice Cedex 2, France

Advances in Parasitology, Volume 70
ISSN 0065-308X, DOI: 10.1016/S0065-308X(09)70011-9

Abstract

In this chapter, we describe the geographically widespread genetic fixation of traits involved in *Drosophila*–parasitoid immune interactions and the situations where such fixation is not observed. We then discuss how the three classes of coevolutionary dynamics that can occur at the local scale (coevolutionary escalation, coevolutionary alternation and coevolutionary polymorphism), the geographic mosaic of selection, and the phylogenetic constraints may explain such evolutionary patterns and drive diversification in the interactions. Most *Drosophila* parasitoid traits involved in virulence are host-species specific. Directional selection (coevolutionary escalation) on such traits can lead to their fixation or on the contrary maintain their polymorphism if these traits are associated with fitness costs. When hosts targeted by different host-specific virulence systems coexist, fluctuations in selective pressures on these systems, together with the ability of *Drosophila* parasitoids to select the most susceptible host for parasitization, can lead to coevolutionary alternation. Finally, we discuss the potential for parasitoid diversification in relation with the fact that most observed geographic situations, for different parasitoid clades, correspond to coevolutionary cold spots, due to fixation of virulence in parasitoid taxa.

11.1. INTRODUCTION

Most *Drosophila* parasitoids are larval solitary endoparasitoids that lay their eggs inside the host, consume the host tissues in the course of their development—leading to the death of the host—and emerge as free adults from the pupa. Immune interactions are thus highly important in determining which of the host or the parasitoid survives infestation. Besides, the prevalence of *Drosophila* parasitoids can reach 90% in field populations (Fleury et al., 2004), and so the traits governing the issue of the interaction are expected to evolve under high coevolutionary selective pressures. Of course, selection and thus coevolution can only occur in the presence of genetic variation in the traits involved in the outcome of the interaction (Sorci, 1997). Such genetic variation has been reported in *Drosophila* spp. and their associated parasitoids for host behavioral traits associated to the rate of infestation, as the rover-sitter phenotype (Carton and Sokolowski, 1992; Sokolowski, 1980) or for traits involved in host immune resistance or parasitoid virulence (Chapter 6 by Dubuffet et al.). When variation of such traits was reported, the observed heritability was above 0.2 (Carton and Boulétreau, 1985; Carton et al., 1989; Fellowes et al., 1998; Kraaijeved et al., 1998), which is a high value for fitness-associated traits (Mousseau and Roff, 1987). Under strong selective pressures, a rapid evolution of traits involved in the outcome of interactions is thus expected and such evolution has been documented using population cages experiments (Dupas and Boscaro, 1999; Fellowes et al., 1998; Green et al., 2000; Kraaijeveld and Godfray, 1997; Kraaijeveld et al., 2001).

The variable issues of *Drosophila*–parasitoid coevolution can first be interpreted at a local scale. Thompson (2005) described three distinct classes of local coevolutionary dynamics that differ in the primary form of selection acting on the system: coevolving polymorphism, coevolutionary escalation and coevolutionary alternation. Coevolving polymorphism is expected when host resistance and parasitoid virulence are efficient only on a restricted number of genotypes of parasitoids and hosts respectively, that is, in case of specific interactions between host and parasitoid genotypes (as in matching alleles or gene-for-gene models). Polymorphism is then maintained by negative frequency dependent selection, which favors the rare genotypes. Alternatively, if traits involved in the interaction are not genotype specific and selection is directional, coevolutionary escalation is expected, which can lead either to a typical ''arms-race'' (each partner evolving a new ''arm'' in response to the one selected in the other partner) or to the maintenance of polymorphism if the traits are costly. Finally, if the parasite can chose to lay eggs in a least resistant host, coevolutionary alternation is expected which leads to fluctuating patterns of selection in the different partners.

The local coevolutionary dynamics can be further modified by the geographic mosaic of coevolution. This kind of coevolution has three attributes: (1) occurrence of a geographic variation in the selective pressures acting on each partner (geographic selection mosaic); (2) presence of coevolutionary ''hot spots'' (where coevolution occurs at a local scale) and ''cold spots'' (local absence of evolution in at least one partner) and (3) occurrence of gene flow between localities. The geographic mosaics and the local dynamics can ultimately favor speciation in one or both partners (Thompson, 2005). Of course, occurrence of coevolution may also depend on phylogenetic constraints acting on ancestral traits, which can lead to fixation of virulence or resistance during long evolutionary time scales.

In this chapter, we present available data on patterns of immune interactions between *Drosophila* spp. and parasitoids of the *Leptopilina* and *Asobara* genus at the local, geographic and phylogenetic scale, and we discuss which kind of local coevolutionary dynamics better explains them. Then, we show how the geographic mosaic of selection and the phylogenetic constraints may influence evolutionary patterns and drive diversification in these interactions.

11.2. THE LOCAL COEVOLUTIONARY DYNAMICS

11.2.1. Coevolutionary escalation

Coevolutionary escalation results from directional selection toward higher levels of investment in defense (resistance) in the host, and in counterdefense (virulence) in the parasite. It can be considered as the typical "arms race" coevolution. In this model, polymorphism is only transient (one coevolving trait is selected until fixation) unless there are tradeoffs between traits involved in virulence or resistance and other fitness traits. Fixation of virulence traits is suspected in *D. melanogaster*/ *L. heterotoma* interactions (Dupas, unpublished data) and in most *D. melanogaster*/*L. boulardi* interactions, based on the success of the parasitoid whatever the population or host reference strain used (Dupas et al., 2003, Fig. 11.1). In *D. subobscura*/*A. tabida* interactions, the host is always successfully parasitized (Kraaijeveld and Godfray, 1999) which might be due, as well, to a high virulence in the parasitoid, or to a low resistance in the host. Two different simulation models by Fellowes and Travis (2000) and Sasaki and Godfray (1999) provided a hypothesis regarding the absence of successful defense mechanisms in *D. subobscura*. They predict that when the cost of resistance is too high, the host would give up investment in higher resistance: depending on the percentage of

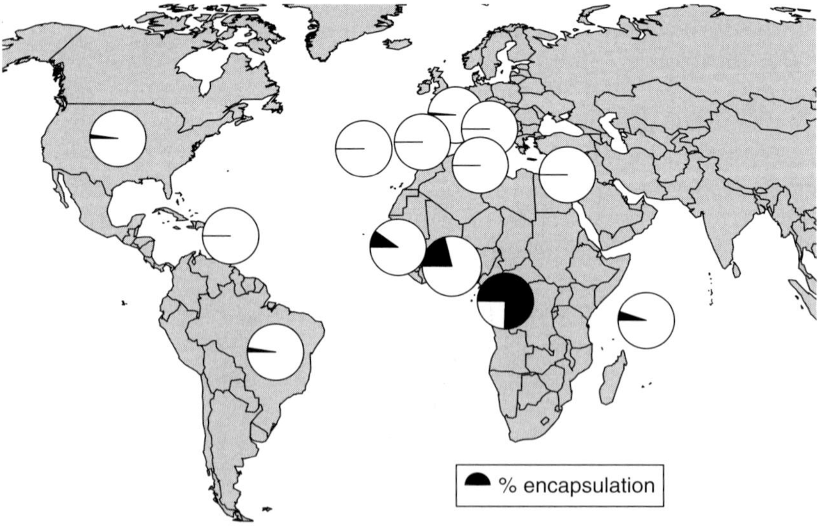

FIGURE 11.1 Outcome of immune interactions in *D. melanogaster*/*L. boulardi* sympatric interactions. The black portion of the circles represents the percentage of encapsulated eggs (failure of parasitism). Populations with the higher encapsulation rates are respectively Brazzaville (Congo) and Lamto (Ivory Coast). *Note*: From Dupas et al. (2003).

parasitized individuals, the loss of part of the population might indeed be preferable. This model could also account for the absence of *D. melanogaster* resistance to *L. heterotoma* and to the Mediterranean "ISm" type of *L. boulardi* [1] (see a detailed description of the ISm and ISy *L. boulardi* types in Chapter 6 by Dubuffet et al.). Another explanation for fixation of virulence and absence of resistance in most populations would be the imbalance of costs between virulence and resistance traits. For instance, *L. boulardi* (ISm) virulence against *D. melanogaster* seems to have no cost (Dupas and Boscaro, 1999) while experiments by Fellowes et al. suggest that *D. melanogaster* resistance selected against *L. boulardi* lines can be costly (Fellowes et al., 1998).

The situation is quite different for *D. melanogaster/L. boulardi* interactions in tropical Africa where virulence traits are not fixed and resistance is observed. As shown in Figure 11.1, parasitoids originating from populations from Lamto (Ivory Coast) and particularly from Congo (Brazzaville) are encapsulated at a high level by *D. melanogaster* sympatric populations (Dupas et al., 2003). *L. boulardi* is able to develop in several species of the *melanogaster* subgroup of Drosophilidae but it is considered as a specialist on *D. simulans* and *D. melanogaster* in Mediterranean areas. In tropical Africa, it faces a gradient from one additional host species (*D. yakuba*) in African coasts, to four additional host species (*D. yakuba, D. teissieri, D. erecta* and *D. orena*) in inland Cameroon (Lachaise et al., 1988). In these regions, selective pressures for virulence against *D. melanogaster*, which is host specific (see Chapter 6 by Dubuffet et al.), might then be reduced, and virulence toward this species counter-selected due to trade offs. This situation would be in agreement with predictions of coevolutionary escalation. The evolution of virulence toward *D. melanogaster* was analyzed using population cages experiments and no cost was detected (Dupas and Boscaro, 1999). However, the parasitoids used were not directly issued from African populations but had been produced by crosses between the ISm (Mediterranean) and ISy (Tropical) *L. boulardi* strains whose virulence systems clearly differ (see Chapter 6 by Dubuffet et al.). Besides, weak levels of costs that nevertheless would allow polymorphism to be maintained, are difficult to detect experimentally (Frank, 1993).

D. yakuba is one of the alternative species encountered by *L. boulardi* in Africa and it is targeted using a system of virulence genetically distinct from the one used against *D. melanogaster* (Dupas and Carton, 1999). *D. yakuba/L. boulardi* interactions are characterized by a polymorphism

[1] The *L. boulardi* ISm strain originates from Mediterranean areas. It is highly virulent against *D. melanogaster* and always encapsulated by *D. yakuba*, a tropical species. There is no described resistance of *D. melanogaster* flies against this type of parasitoid. The ISy strain originates from Congo. It is successful on "susceptible" strains of *D. melanogaster* and *D. yakuba* (both species are found in its area of origin) but is encapsulated by "resistant" populations or strains of both host species.

both in resistance and virulence (Dubuffet et al., 2007; Figs. 11.2 and 11.3) but in this case, virulence seems to be costly (Dupas and Boscaro, 1999). In population cage experiments as the ones reported above, virulence against *D. yakuba* dropped down when the parasitoid was reared on *D. melanogaster* while virulence against *D. melanogaster* was maintained when the parasitoid was reared on *D. yakuba*. The higher cost of virulence against *D. yakuba* compared to virulence against *D. melanogaster* is in agreement with the adaptive budget theory which predict that the cost of overcoming resistance should be higher when resistance itself involves more cost (Dupas and Boscaro, 1999). Indeed, *D. yakuba* is a species that is suspected to invest more in immune defenses than *D. melanogaster*, as suggested by its higher hemocyte load and its better ability to encapsulate a large number of host strains and species (Dupas, unpublished data; Eslin, 1996). Another level in terms of investment in resistance-virulence may thus have been reached in the arms race between *L. boulardi* and *D. yakuba*.

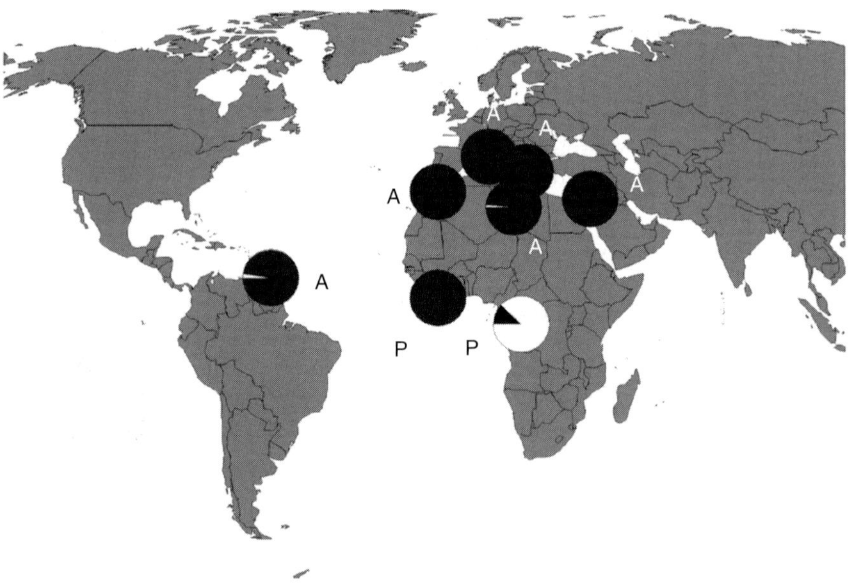

FIGURE 11.2 Rates of immune suppression of *D. yakuba* by *L. boulardi*. The black portion of the circles represents the percentage of encapsulated eggs. The letters in rectangles notify the presence (P) or absence (A) of *D. yakuba* in the locality. The host strain used for the experiments was Dy1880-D. *Note*: From Dupas and Boscaro, 1999.

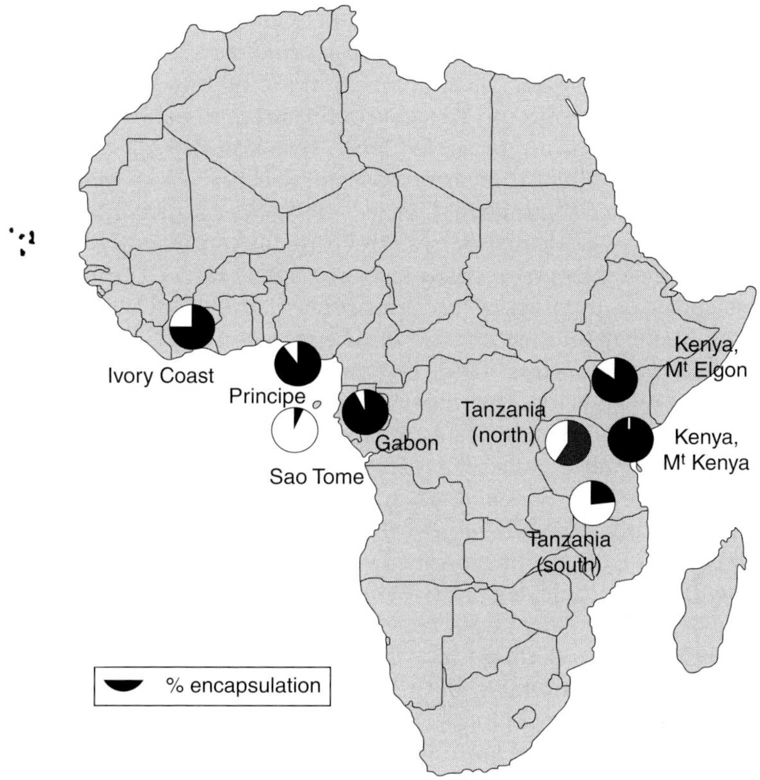

FIGURE 11.3 Distribution of *D. yakuba* resistance in tropical Africa. The resistance level (represented by the black portion of the pie chart) was estimated from the encapsulation rate following experimental infestation of *D. yakuba* populations by the reference ISy strain of *L. boulardi*. *Note*: From Dubuffet et al. (2007).

11.2.2. Coevolutionary alternation

Coevolutionary alternation involves a parasitoid species and at least two host species. The parasitoid is selected for increased preference toward the most susceptible host, which leads to higher infestation rates and thus higher selection pressures for resistance in this host. The preferred host then evolves more resistance while the nonpreferred host(s) evolves lower resistance until a point where the nonpreferred host(s) become less resistant than the other and it becomes advantageous for the parasitoid to switch host preference. Such alternation in host choice leads to fluctuations in resistance and virulence traits.

Both *L. boulardi* and *A. tabida* were shown to select the most suitable host species in choice situations (Carton et al., 1987; Dubuffet et al., 2006; Kraaijeveld et al., 1995). Dutch populations of *A. tabida* are more likely to accept *D. subobscura*, which is totally susceptible, than *D. melanogaster* which is partially resistant to this parasitoid species. In this system, some, but not all conditions for alternation are met. Indeed, *D. melanogaster* resistance is lower in the presence of the alternative host *D. subobscura*, suggesting that as a nonpreferred host, it is counter-selected for resistance. However, alternation cannot occur because *D. subobscura* never evolved resistance (Kraaijeveld et al., 1995). Resistance may have not evolved because parasitism pressures are not sufficiently strong and resistance mechanisms are costly, as suggested above. The fact that this species lacks of a category of hemocytes (lamellocytes) known to be involved in encapsulation of parasitoid eggs, might also be an important factor (Doury and Eslin, 2006). In *Leptopilina* parasitoids, the ability to choose to lay eggs on the most suitable/least resistant host species has also been demonstrated. Across most of its distribution range, in Mediterranean areas, where it faces at most two major host species, *D. melanogaster* and *D. simulans*, *L. boulardi* prefers to lay eggs in the most susceptible host, *D. melanogaster* (Carton et al., 1987). However, alternation is again not observed since resistance toward Mediterranean *L. boulardi* parasitoids has not evolved in *D. melanogaster*: *L. boulardi* populations facing only these two host species are never encapsulated by *D. melanogaster* (Dupas and Boscaro, 1999). However, in tropical Africa, *L. boulardi* interacts both with *D. yakuba* and *D. melanogaster*, which are both polymorphic in encapsulation ability toward sympatric parasitoid populations (Dubuffet et al., 2007). The ISy strain that originates from this area is similarly attracted by both host species but preferentially lays eggs in *D. yakuba* (Dubuffet et al., 2006). This behavior could have been selected at a time where the proportion of susceptible genotypes was higher in the local *D. yakuba* population than in the local *D. melanogaster* population. Polymorphism for host selection behavior as well as genetic variation for parasitoid virulence and host resistance, which are prerequisites for the coevolutionary alternation model to occur, might thus have existed in at least this *L. boulardi* population. Sampling efforts are now needed to collect and study populations from the same area as the ISy strain in order to determine if this model applies to at least some *Drosophila*–parasitoid interactions.

11.2.3. Coevolutionary polymorphism

Coevolutionary polymorphism results from a selection favoring rare genotypes in at least one partner (Thompson, 2005; Woolhouse et al., 2002). This polymorphism has been associated with mechanisms of interaction of the "key-lock" type. The compatibility of the key with the lock would determine either the success of host immune defenses or inversely, the success of parasitoid virulence mechanisms. In the first case, rare genotypes are favored in the parasite, while in the second, they are favored in the host. This leads to the maintenance of polymorphism in both partners with the cycling of allele frequencies, without fixation of a single host or parasite genotype.

Such kind of coevolution implies that interactions are genotype specific, with any of one partner genotypes outperformed by another genotype in at least one of the interactions with the other partner's genotypes. The few studies that partially addressed the question whether such situation occurs in host–parasitoid interactions did not reveal matching genotype effects for resistance or virulence factors. Green et al. (2000) and Kraaijeveld (2001) showed that the issue of sympatric interactions between *A. tabida* and *D. melanogaster* was explained by the additive interaction between the resistance of the host (estimated using a reference parasitoid strain) and the virulence of the parasitoid (estimated using a reference host strain) suggesting genotype specificity was not important. In *L. boulardi*, host-genotype specificity is rarely encountered. Most populations are "host-species specific," being highly successful on *D. melanogaster* and totally inefficient on *D. yakuba*, whatever the genotype of these hosts. Only the Congolese population (and the ISy line issued from this population) has been shown so far to have a success on *D. melanogaster* and *D. yakuba* depending on their respective genotypes (Dubuffet et al., 2007). However, we still do not know whether different virulence genotypes co-occur within the Congolese population, neither if "resistant genotypes" in the host populations are susceptible to these other parasitoid genotypes. Further investigations will be necessary to determine whether coevolutionary polymorphism could occur in this locality.

11.3. THE COMPONENTS OF THE GEOGRAPHIC MOSAIC OF COEVOLUTION

As stressed by Thompson (2005, 2009), local coevolutionary dynamics can be modified by the overall geographic mosaic of coevolution. Geographic mosaic of coevolution is driven by three components of geographic structure: selection mosaics, coevolutionary hot spots and traits remixing (Thompson 2005, 2009). Some of the geographic components of *Drosophila*–parasitoid interactions have been described by Kraaijeveld and Godfray (1999). Here, we intend to explain why *Drosophila* parasitoids should be considered as choice models to test the predictions of Thompson.

11.3.1. The geographic selection mosaic

Selection mosaic occurs when the natural selection pressures on interactions vary among different communities. This can be due to the environmental context in which a given host–parasitoid interaction evolves.

For instance, a host genotype can be highly resistant to one parasite genotype at a given temperature but totally susceptible at another temperature (host-genotype-by-parasitoid-genotype-by-environment (G_H x G_P x E) interactions). So far, the existence of G_H x G_P x E interactions has never been taken into account in studies on host parasitoid interactions. However, they have now been reported in many host–parasite interactions and the fact that several environmental factors modify the outcome of host–parasitoid interactions is generally admitted (Delpuech, 1993; Fellowes et al., 1999; Fytrou et al., 2006).

The geographic mosaic of selection can also originates from a geographic variation on the occurrence of the interaction. The presence of alternative species, for instance, does not act directly on the issue of a particular interaction, but it modifies the selective pressure acting on this interaction by modifying the frequency of its occurrence in the field. As described previously, *L. boulardi* virulence toward *D. melanogaster* drops down in tropical Africa where other host species of the *melanogaster* subgroup are present in addition to *D. melanogaster* and *D. simulans* (Dupas and Boscaro, 1999). Parasitoids from this area might be less specialized, having different virulence mechanisms, specific but not totally efficient, against each of these host species (Dupas and Boscaro, 1999; see Chapter 6 by Dubuffet et al.). This would create a geographic mosaic of coevolution with hot spots of reciprocal selection in the areas where the host varies genetically for resistance, and cold spots of reciprocal selection in the areas where the host does not resist.

11.3.2. Coevolutionary hot spots and cold spots

Polymorphism for virulence in *L. boulardi* occurs only in tropical Africa that can be considered as a hot spot of coevolution embedded in cold spots where the parasitoid always wins (Dupas et al., 2003). As stressed by van Alphen (2002), hot spots of coevolution are far less common than cold spots in host–parasitoid immune interactions. The fixation of resistance or virulence traits can be explained by the high heritability for variation of these traits, the strong selective pressures associated with the high infestation rates in the field, and the necessity parasitoids have to kill their host to survive and reproduce. Lapchin and Guillemaud (2005) explained the high levels of parasitoid virulence (or low levels of host resistance) by the asymmetry of host–parasitoid interactions. All parasitoids need a host but not vice versa and a more or less important part of the host population remains unparasitized. Accordingly, lower levels of genetic variation for virulence than for resistance seem to be a general pattern in *Drosophila*–parasitoid systems. In *L. boulardi*, fixation of virulence traits might be prevented in tropical Africa because the host community is more diversified than in Mediterranean areas, and the presence of a given host species likely more variable in time and space.

In the model of Lapchin and Guillemaud (2005), specialization is also favored in the parasitoid since the parasitoid can choose its host, but the host cannot choose the parasite. This leads to more important selection pressures on parasitoid virulence traits specific for the host compared to what observed for host immune resistance traits specific for the parasite. The parasitoid traits should then have higher evolutionary rates than the host ones.

11.3.3. Trait remixing

The mixture of coevolutionary cold spots and hot spots makes local *Drosophila*–parasitoid coevolutionary interactions particularly susceptible to gene flow from localities with monomorphic populations to localities with polymorphic ones. Significant population structuring was observed among African *D. melanogaster* populations using microsatellite loci (Dieringer et al., 2005), but the most important parameter for maintenance or not of hot spots of coevolution, that is, the gene flow between parasitoid populations, has not yet been estimated. The importance of trait remixing in parasitoid–host interactions thus remains an open research for the future.

11.4. HYPOTHESIS OF COEVOLUTIONARY DIVERSIFICATION

The major reason why coevolution should promote diversification is that fluctuating selection in species interactions should favor specialized lineages that respond more rapidly and survive more easily to coevolutionary arms races (Kawecki, 1998). In addition, sympatric coevolutionary models lead more readily to speciation when assuming a link between reproduction genes and adaptation genes. Empirical examples of coevolutionary driven sympatric speciation are also those where interactions are associated to reproductive isolation mechanisms such as sexual conflicts (Arnquvist, 2000; Rice et al., 2005) or host habitat selection (Smith and Benkman, 2007). In allopatry, coevolutionary driven hypotheses of speciation are related to the geographic mosaic theory of coevolution. Geographic populations differ in the range of species they interact with and the diverging coevolutionary interactions promote speciation (Thompson, 2005). Both sympatric and allopatric models of speciation may be supported by data on *Drosophila*–parasitoid interactions. Host specificity of the virulence systems has been demonstrated in *L. boulardi* and parasitoids ability to choose their host in relation to their ability to develop inside that host is strongly suspected (Carton et al., 1987; Dubuffet et al., 2006). The frequent mating on the host patch in this species (Fauvergue et al., 1999) may also promote an association between virulence, host choice and mate choice thereby favoring sympatric speciation mechanisms. Allopatric mechanisms may be favored by the geographic mosaics patterns described above.

11.5. ANCESTRAL TRAITS AND PHYLOGENETIC CONSTRAINTS ON COEVOLUTION

Leptopilina parasitoid spp. studied for their virulence can be divided into two groups, *Heterotoma* and *Boulardi* (Allemand et al., 2002). *Heterotoma* group parasitoids have a high virulence toward the tested *Drosophila* host strains of the species *D. melanogaster*, *D. yakuba* and *D. simulans*, leading to zero to weak encapsulation rates whereas *Boulardi* group parasitoids are often encapsulated (Dupas, unpublished data). Such a difference is likely due to occurrence of an ancestral immune suppressive trait, efficient against multiple hosts, which appeared at the basis of the Heterotoma clade. Phylogenetic constraints might also be important in the host, preventing evolution of efficient immune resistance, as suggested above for *D. subobscura*. Constraints in one or the other partner can lead to long-term coevolutionary cold spots and their role in driving coevolutionary interactions strongly needs to be assessed.

11.6. CONCLUSION

We have discussed local coevolutionary dynamics between traits involved in the *Drosophila* immune response and in the parasitoid immune evasion in the context of the three main available classes of models: coevolutionary escalation, coevolutionary alternation and coevolutionary polymorphism. The latter model is not supported by available data: there is no described genotype specificity of interactions that could allow maintenance of polymorphism by negative frequency dependent selection. For most *Drosophila*–parasitoid interactions, there is a rather strong empirical evidence of coevolutionary escalation, a one-dimensional interaction where the investment of each partner depends on the selective pressures from the other partner. One of the possible issues of escalation would be host alternation since many *Drosophila* parasitoids are able to choose the most suitable host for their development. However, this model potentially applies to date only to the particular situation of *L. boulardi* parasitoids in tropical Africa where virulence polymorphism is encountered. In this situation, variation in ecological conditions, such as variation of the structure of the host community along the seasons, would narrow the adaptive potential of the parasitoid, limit its specialization and lead to locally different coevolutionary interactions.

The geographic mosaics of interactions, acting on local coevolutionary escalation, might thus explain much of the evolution of *Drosophila*–parasitoid interactions. However, in most situations, virulence traits are fixed and whether resistance will increase in turn or coevolutionary escalation is stopped remains to be determined. Fixation of virulence traits might be due to a lack of coevolutionary potential. At the phylogenetic level, some parasitoid clades may lack coevolutionary potential due to the presence of ancestral traits that would prevent occurrence of polymorphism (as the existence of nonspecific virulence traits in *L. heterotoma*). Finally, in clades and localities where coevolution occurs, the ability of the parasitoid to choose its host and the reproduction of the progeny on the host may favor the association between development and reproduction genes, thereby favoring speciation driven by coevolution.

A large set of data is still lacking to fully understand coevolution between hosts and parasitoids. Researches on G_H x G_P x E interactions and occurrence of genotype specificity, experimental speciation and estimation of gene flow between parasitoid populations will shed light on this evolutionary puzzle. No doubt that they will reinforce *Drosophila*–parasitoid interactions as a major model to understand the evolution of one of the most diverse and ecologically important animal interactions on Earth.

REFERENCES

Allemand, R., Lemaître, C., Frey, F., Bouletreau, M., Vavre, F., Nordlander, G., et al., 2002. Phylogeny of six African *Leptopilina* species (Hymenoptera: Cynipoidea, Figitidae), parasitoids of *Drosophila*, with description of three new species. Ann. Soc. Entomol. Fr. 38, 319–332.

Arnquvist, E., 2000. Sexual conflict promotes speciation in insects. Proc. Natl. Acad. Sci. USA 97, 10460–10464.

Carton, Y., Boulétreau, M., 1985. Encapsulation ability of *Drosophila melanogaster*: a genetic analysis. Dev. Comp. Immunol. 9, 211–219.

Carton, Y., Sokolowski, M.B., 1992. Interactions between searching strategies of *Drosophila* parasitoids and the polymorphic behavior of their hosts. J. Insect Behav. 5, 161–175.

Carton, Y., Capy, P., Nappi, A.J., 1989. Genetic variability of host parasite relationship traits: utilisation of isofemale lines in a *Drosophila* parasitic wasp. Genet. Select. Evol. 21, 437–447.

Carton, Y., Chibani, F., Marrakchi, M., 1987. Egg-laying strategy under natural conditions of *Leptopilina boulardi*, a hymenopteran parasitoid of *Drosophila* spp. Entomol. Exp. Appl. 43, 193–201.

Delpuech, J.M., 1993. Conserving genetic variability of a wild insect population under laboratory conditions. Entomol. Exp. Appl. 67, 233–239.

Dieringer, D., Nolte, V., Schlötterer, C., 2005. Population structure in African *Drosophila melanogaster* revealed by microsatellite analysis. Mol. Ecol. 14, 563–573.

Doury, G., Eslin, P., 2006. The fly *Drosophila subobscura*: a natural case of innate immunity deficiency. Dev. Comp. Immunol. 30, 977–983.

Dubuffet, A., Dupas, S., Frey, F., Drezen, J.M., Poirie, M., Carton, Y., 2007. Genetic interactions between the parasitoid wasp *Leptopilina boulardi* and its *Drosophila* hosts. Heredity 98, 21–27.

Dubuffet, A., Rodríguez Álvarez, C.I., Drezen, J.M., van Alphen, J.J.M., Poirie, M., 2006. Do parasitoid preferences for different host species match virulence? Physiol. Entomol. 31, 170–177.

Dupas, S., Boscaro, M., 1999. Geographic variation and evolution of immuno suppressive genes in a *Drosophila* parasitoid. Ecography 22, 284–291.

Dupas, S., Carton, Y., 1999. Two non-linked genes for specific virulence of *Leptopilina boulardi* against *Drosophila melanogaster* and *D. yakuba*. Evol. Ecol. 13, 211–220.

Dupas, S., Poirié, M., Carton, Y., 2003. The genetic dimension of the coevolution of virulence resistance in *Drosophila*–parasitoid wasp relationships. Heredity 90, 84–89.

Eslin, P., 1996. Variation in *Drosophila* concentration of haemocytes associated with different ability to encapsulate *Asobara tabida* larval parasitoid. J. Insect Physiol. 42, 549–555.

Fauvergue, X., Fleury, F., Lemaitre, C., Allemand, R., 1999. Parasitoid mating structures when hosts are patchily distributed: field and laboratory experiments with *Leptopilina boulardi* and *L. heterotoma*. Oïkos 86, 344–356.

Fellowes, M.D., Kraaijeveld, A.R., Godfray, H.C., 1998. Trade-off associated with selection for increased ability to resist parasitoid attack in *Drosophila melanogaster*. Proc. R. Soc. Lond. B 265, 1553–1558.

Fellowes, M.D.E., Travis, J.M.J., 2000. Linking the coevolutionary and population dynamics of host–parasitoid interactions. Popul. Ecol. 42, 195–203.

Fellowes, M.D.E., Kraaijeveld, A.R., Godfray, H.C.J., 1999. Association between feeding rate and parasitoid resistance in *Drosophila melanogaster*. Evolution 53, 1302–1305.

Fleury, F., Ris, N., Allemand, R., Fouillet, P., Carton, Y., Boulétreau, M., 2004. Ecological and genetic interactions in *Drosophila*-parasitoids communities: a case study with *D. melanogaster D. simulans* and their common *Leptopilina* parasitoids in south-eastern France. Genetica 120, 181–194.

Frank, S.A., 1993. Specificity versus detectable polymorphism in host–parasite genetics. Proc. R. Soc. Lond. B 254, 191–197.

Fytrou, A., Schoffield, P.G., Kraaijeveld, A.R., Hubbard, S.F., 2006. Wolbachia infection suppresses both host defence and parasitoid counterdefence. Proc. R. Soc. Lond. B 273, 791–796.

Green, D.M., Kraaijeveld, A.R., Godfray, H.C.J., 2000. Evolutionary interactions between *Drosophila melanogaster* and its parasitoid *Asobara tabida*. Heredity 85, 450–458.

Kawecki, T.J., 1998. Red queen meets santa rosalia: arms races and the evolution of host specialization in organisms with parasitic lifestyles. Am. Nat. 152, 635–651.

Kraaijeveld, A.R., 2001. Is there local adaptation in *Drosophila*–parasitoid interactions? Evol. Ecol. Res. 3, 107–116.

Kraaijeveld, A.R., Godfray, H.C.J., 1997. Trade-off between parasitoid resistance and larval competitive ability in *Drosophila melanogaster*. Nature 389, 278–280.

Kraaijeveld, A.R., Godfray, H.C.J., 1999. Geographic patterns in the evolution of resistance and virulence in *Drosophila* and its parasitoids. Am. Nat. 153, S61–S74.

Kraaijeveld, A.R., Hutcheson, K.A., Limentani, E.C., Godfray, H.C.J., 2001. Costs of counterdefenses to host resistance in a parasitoid of *Drosophila* larvae. Evolution 55, 1815–1821.

Kraaijeveld, A.R., Nowee, B., Najem, R.W., 1995. Adaptive variation in host selection behaviour in *Asobara tabida*, a parasitoid of *Drosophila* larvae. Funct. Ecol. 9, 113–118.

Kraaijeved, A.R., van Alphen, J.J.M., Godfray, H.C.J., 1998. The coevolution of host resistance and parasitoid virulence. Parasitology 116, S29–S45.

Lachaise, D., Cariou, M.L., David, J., Lemeunier, F., Tsacas, L., Ashburner, M., 1988. Historical biogeography of the *Drosophila melanogaster* species subgroup. Evol. Biol. 22, 159–225.

Lapchin, L., Guillemaud, T., 2005. Asymmetry in host and parasitoid diffuse coevolution: when the red queen has to keep a finger in more than one pie. Front. Zool. 2, 4.
Mousseau, T.A., Roff, D.A., 1987. Natural selection and the heritability of fitness components. Heredity 59, 181–197.
Rice, W.R., Linder, J.E., Friberg, U., Lew, T.A., Morrow, E.H., Stewart, A.D., 2005. Inter-locus antagonistic coevolution as an engine of speciation: assessment with hemiclonal analysis. Proc. Natl. Acad. Sci. USA 102, 6527–6534.
Sasaki, A., Godfray, H.C.J., 1999. A model for the coevolution of resistance and virulence in coupled host and parasitoid interactions. Proc. R. Soc. Lond. B 266, 455–463.
Smith, J.W., Benkman, C.W., 2007. A coevolutionary arm race causes ecological speciation in Crossbills. Am. Nat. 169, 455–465.
Sokolowski, M.B., 1980. Foraging strategies of *Drosophila melanogaster*: a chromosomal analysis. Behav. Genet. 10, 291–302.
Sorci, G., 1997. Genetics of host–parasite interactions. Trends Ecol. Evol. 12, 196–200.
Thompson, J.N., 2005. The Geographic Mosaic of Coevolution. The University of Chicago Press, Chicago.
Thompson, J.N., 2009. The coevolving web of life. Am. Nat. 173, 125–140.
Woolhouse, M.E., Webster, J.P., Domingo, E., Charlesworth, B., Levin, B.R., 2002. Biological and biomedical implications of the co-evolution of pathogens and their hosts. Nat. Genet. 32, 569–577.
van Alphen, J.J.M., 2002. Hotspots are rare in the evolution of parasitoid virulence and host resistance. Trends Ecol. Evol. 17, 499.

SECTION 4

SYMBIOTIC ORGANISMS OF *DROSOPHILA* PARASITOIDS

CHAPTER **12**

Drosophila–Parasitoid Communities as Model Systems for Host–*Wolbachia* Interactions

Fabrice Vavre,* Laurence Mouton,* and **Bart A. Pannebakker**[†]

Contents

* Université Lyon 1; CNRS, UMR5558, Laboratoire de Biométrie et Biologie Evolutive, F-69622, Villeurbanne, France
[†] Evolutionary Genetics, Centre for Ecological and Evolutionary Studies, University of Groningen, P.O. Box 14, NL-9750 AA Haren, The Netherlands

Advances in Parasitology, Volume 70
ISSN 0065-308X, DOI: 10.1016/S0065-308X(09)70012-0

Abstract

Wolbachia bacteria are cytoplasmic endosymbionts that infect a wide range of arthropod and nematode hosts. They are transmitted from mother to offspring via the eggs (vertical transmission) and enhance their transmission to the next generation by manipulating the reproductive system of their hosts. These manipulations occur in many forms, such as the induction of cytoplasmic incompatibility, feminization, male killing and parthenogenesis induction. *Wolbachia* is estimated to occur in up to 66% of all insect species, but the greatest diversity of reproductive manipulations is found in the order of the Hymenoptera. Studies of *Wolbachia* in *Drosophila–parasitoid* communities have allowed for important insights into different aspects of *Wolbachia* biology. The extensive knowledge available on *Drosophila* parasitoids provides a solid base on which to test new hypotheses on host–Wolbachia interactions. The large range of *Wolbachia* phenotypes present in *Drosophila* parasitoids, combined with the recent acquisition of the bacteria from their Drosophilid hosts, make them an ideal model system to study the evolution and dynamics of *Wolbachia* infections, both in the laboratory as in the field. In this chapter, we aim to review the current knowledge on the associations between *Wolbachia* and *Drosophila* parasitoids, and identify open questions and specify new research directions.

12.1. INTRODUCTION

Wolbachia bacteria are cytoplasmic endosymbionts (α-proteobacteria) that infect a wide range of arthropod and nematode hosts (Duron et al., 2008; O'Neill et al., 1997; Stouthamer et al., 1999; Werren et al., 2008). They are transmitted from mother to offspring via the eggs (vertical transmission). *Wolbachia* enhance their transmission to the next generation by manipulating the reproductive system of their hosts. This bacterial reproductive parasitism occurs in many forms, such as the induction of cytoplasmic incompatibility (Boyle et al., 1993), feminization (Rousset et al., 1992), male killing (Hurst et al., 1999) and parthenogenesis induction (Stouthamer et al., 1999). All these manipulations result in an increase in the number of infected females in the host population and maximize the transmission of the bacteria in the host population. These direct effects of *Wolbachia* on host reproduction can also have important indirect ecological and evolutionary consequences for the host, from structuring communities to mediating parasitoid–host interactions and life-history strategies.

Even though *Wolbachia* is estimated to occur in up to 66% of all insect species (Hilgenboecker et al., 2008), the greatest diversity of reproductive manipulations is found in the order of the Hymenoptera. Haplodiploid sex determination, where females develop from fertilized diploid eggs, and males from unfertilized haploid eggs, makes Hymenoptera especially prone to (bacterial) manipulation of their reproductive system. Among the Hymenoptera, *Drosophila* parasitoids are one of the best studied groups for host–symbiont interactions. Other Hymenopteran genera in which host–symbiont interactions are well described are *Nasonia* (Hymenoptera: Pteromalidae) (Bordenstein and Werren, 2007; Bordenstein et al., 2001; Breeuwer and Werren, 1990, 1995; Breeuwer et al., 1992; Tram et al., 2003, 2006) and *Trichogramma* (Hymenoptera: Trichogrammatidae) (Pintureau et al., 1999, 2002; Stouthamer and Kazmer, 1994; Stouthamer and Luck, 1993; Stouthamer et al., 1990). Here we present *Drosophila* parasitoids as a model system for studying host–Wolbachia interactions, with a focus on the *Leptopilina* and *Asobara* genera. Studies of *Wolbachia* in *Drosophila* parasitoid communities have allowed for important insights in different aspects of *Wolbachia* biology, such as the dynamics of *Wolbachia* in insect communities, the diversity of interactions between hosts and bacteria, the regulation of bacterial populations, and the evolutionary consequences of infection for the host. The knowledge available on *Drosophila* parasitoids, which is reflected in this issue and covers fields as diverse as developmental biology, immunology, physiology, ecology and evolution, forms a solid base on which to test new hypotheses on a wide array of host–Wolbachia interactions. In addition, the large range of *Wolbachia* phenotypes present in *Drosophila* parasitoids (Section 12.3), and the recent transfer of the bacteria from their Drosophilid hosts (Section 12.2) make them an ideal model system in which to study the evolution and dynamics of *Wolbachia* infections, both in the laboratory and in the field. In the next sections, we aim to (1) review the current knowledge on the associations between *Wolbachia* and *Drosophila* parasitoids, and (2) identify open questions and specify new research directions.

12.2. PATTERN OF INFECTION AND PHYLOGENETIC DIVERSITY OF *WOLBACHIA* IN *DROSOPHILA* PARASITOIDS

One of the most striking results obtained from phylogenetic studies of *Wolbachia* is the almost complete absence of congruence between the *Wolbachia* and the host phylogenies. This is interpreted as the possibility of *Wolbachia* to be horizontally transmitted from one species to another, which means *Wolbachia* has the ability to invade new host species regularly. Because of their intimacy, parasitoid–host interactions have immediately been suggested as a possible route for these transfers, a hypothesis that has been supported by some isolated cases (Werren et al., 1995).

While case studies are interesting to understand under which conditions horizontal transmission may occur, they provide only limited information on the general mechanisms that may favor or limit horizontal transmission. A more powerful way of tackling these questions is to study patterns of infection in many species simultaneously that share either ecological connections or phylogenetic ancestry. Both types of studies have been performed in *Drosophila* parasitoids and they shed light on important factors that determine the probability of horizontal transmission among species.

12.2.1. *Drosophila* parasitoids are highly susceptible to *Wolbachia* infection

While *Wolbachia* infection has been detected in about 20% of all insect species, 11 out of 16 *Drosophila* parasitoid species (around 69%) have been found infected (Table 12.1). This higher incidence of *Wolbachia* infection in *Drosophila* parasitoids may be caused by small sample sizes in most global surveys (often only one or two individuals per species), which undoubtedly underestimates the real *Wolbachia* incidence. Statistical inferences on these global surveys have recently estimated that *Wolbachia* incidence may be as high as 66% (Hilgenboecker et al., 2008), which is close to the incidence found in *Drosophila* parasitoids. However, in *Drosophila* parasitoids, infection is fixed or near fixation in most species, even though polymorphism has been detected in some of them, notably in *Leptopilina victoriae* and *Pachycrepoideus dubius* (Vavre et al., 2000, 2002). This means that either the incidence of *Wolbachia* is higher in *Drosophila* parasitoids than in other species, or that infection within species is more prevalent.

Another result that suggests a high susceptibility of *Drosophila* parasitoids to *Wolbachia* infection is that many species are infected by multiple *Wolbachia* strains. In many cases, these multiple strains are hosted within

TABLE 12.1 *Wolbachia* infection in *Drosophila* parasitoids

Genus	Species	Infection	Phenotype
Leptopilina	*heterotoma*	wLhet1, wLhet2, wLhet3	CI, CI, CI
	victoriae	wLvic	CI
	guineaensis	wLgui1, wLgui2	CI, unknown
	boulardi	–	
	freyae	–	
	orientalis	–	
	clavipes	wLcla	PI
	australis	wLaus	PI
Asobara	*tabida*	wAtab1, wAtab2, wAtab3	CI, CI, oogenesis
	rufescens	wAruf	CI suspected
	japonica	wAjap	PI
	persimilis	–	
	citri	–	
Trichopria	*Drosophilae*	wTdro	?
	nr. Drosophilae	wTsp1, wTsp2	CI
Pachycrepoideus	*dubius*	wPdub	No effect detected

Table shows species, *Wolbachia* strain, *Wolbachia* phenotype and infection type (CI-cytoplasmic incompatibility, PI-parthenogenesis induction, and oogenesis-obligatory for oogenesis.)

the same individuals, like in *Trichopria nr. drosophilae* that is infected by two *Wolbachia* strains, and *Leptopilina heterotoma* and *Asobara tabida* where three *Wolbachia* strains have been detected (Vavre et al., 1999). In *Leptopilina guinaensis*, however, two *Wolbachia* strains have been detected, but each of these occur in different populations (Vavre, unpublished data). Taken together, these results suggest that *Drosophila* parasitoids are quite susceptible to *Wolbachia* infection.

Finally, *Drosophila* parasitoids are also susceptible to other vertically transmitted bacteria. For example, *P. dubius* has been shown to host *Arsenophonus* (Gueguen and Duron, personal communication) and a *Rickettsia* has been recently detected in newly collected individuals in *A. tabida* (Zouache and Mavingui, personal communication), but their effects remain unknown to date. In addition, it has recently been shown that *Leptopilina boulardi* is host to a virus that can be either vertically or horizontally transmitted and that manipulates the strategy of host exploitation by increasing superparasitism by infected females (Varaldi et al., 2003; Chapter 13 by Varaldi et al.).

Taken together, these results show that symbionts, and especially *Wolbachia*, are frequent partners of *Drosophila* parasitoids and, because of their potential impact on the evolutionary trajectories of their hosts, they should not be ignored in these insects.

12.2.2. Phylogenetic analyses reveal frequent horizontal transmission

Because of their intimate links, hosts and parasitoids are especially prone to the occurrence of horizontal transmission. *Drosophila* and their parasitoids have been the first insect communities where a clear pattern of horizontal transmission has been detected at a community level (Vavre et al., 1999). In many cases, the same *Wolbachia* strain is found in *Drosophila* and their parasitoids (Fig. 12.1). For example, the strain wLhet1, infecting

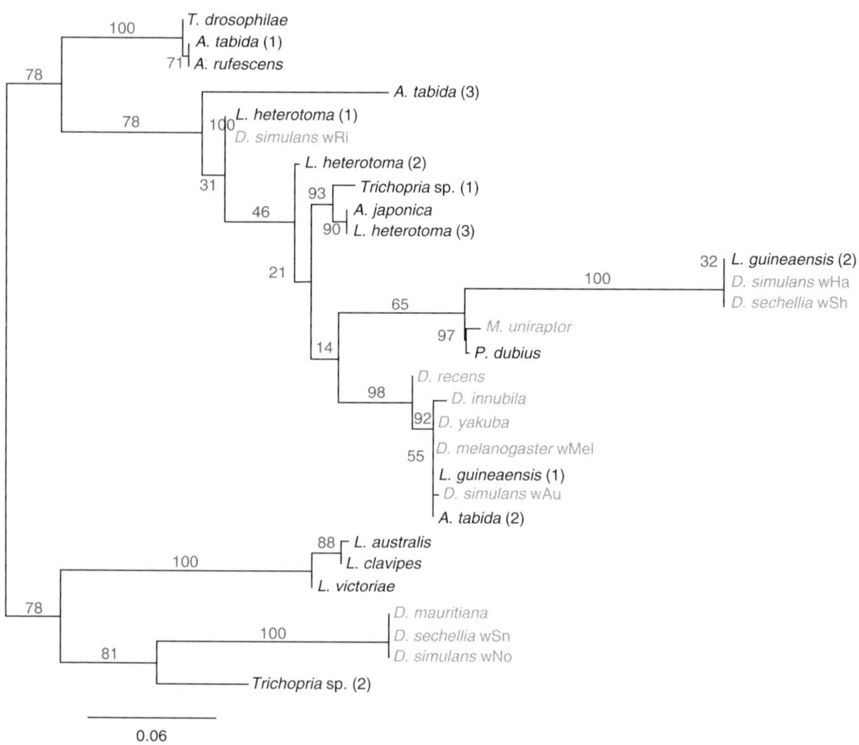

FIGURE 12.1 Phylogenetic tree of *Wolbachia* reconstructed by maximum likelihood using the wsp gene. Each *Wolbachia* strain is represented by the name of its host species and by a number or the strain name for multiply infected species. In grey are the *Wolbachia* in *Drosophila* hosts; in black, the *Wolbachia* in *Drosophila* parasitoids. *Note: Adapted from Vavre* et al. *(1999).*

L. heterotoma, has the same WSP sequence as the strain wRi infecting *Drosophila simulans* in continental areas. The same is true for the strain wAtab2 infecting *A. tabida* and the strain wMel infecting *Drosophila melanogaster*. The situation in *L. guinaensis* is also striking: populations from West Africa harbor a *Wolbachia* strain similar to a strain found in *D. melanogaster*, while populations from East Africa harbor a *Wolbachia* strain similar to a strain harbored by *D. simulans* in islands of the Indian Ocean. Parasitoids may also share the same infection, like for instance *A. tabida* and *Trichopria drosophilae*, *L. heterotoma* and *T. nr drosophilae* and *Pachycrepoideus dubius* and *Muscidifurax uniraptor*, a generalist parasitoid of Diptera (Vavre et al., 1999).

One important restriction of these data was that they were obtained using only the *wsp* gene, which has been shown to recombine frequently among *Wolbachia* strains (Baldo et al., 2005). Preliminary additional data confirm some of these strains have been recently horizontally transmitted. Using a multiple locus strain typing (MLST) approach (Baldo et al., 2006b), no variation was found for wLhet1 and wRi, and wAtab2 and wMel, confirming that these strains are very closely related. However, the close relationships observed between wAjap and WLhet3 on wsp was not sustained by MLST analysis, suggesting recombination events among these strains rather than horizontal transmission of the bacterium. Clearly, ongoing characterization of *Wolbachia* strains infecting *Drosophila* parasitoids will allow a more thorough analysis of strain exchanges and/or recombination among strains infecting interconnected species at a community level.

Overall, these results strongly suggest that interactions at the community level favor horizontal transmission of *Wolbachia* in *Drosophila–parasitoid* communities. Interestingly, Heath et al. (1999) experimentally transferred *Wolbachia* from *D. simulans* to *L. boulardi* with a success rate of 0.7%, showing that transfers do occur during parasitism. However, infection was lost in *L. boulardi* during subsequent generations, suggesting this species is somehow resistant to *Wolbachia* infection. Similar experiments using *L. heterotoma* and *D. melanogaster* could not detect any horizontal transfer (Vavre, unpublished data). This can be due either to insufficient sampling effort (only 250 wasp lines were tested) or to the fact that density of *Wolbachia* in *D. melanogaster* is lower than in *D. simulans* (Boyle et al., 1993), which would reduce the efficiency of transfer. Thus, while horizontal transmission is common at an evolutionary scale, it does not occur at high frequency at the individual level in these systems.

12.2.3. Pattern of infection in the genus *Leptopilina*: Coincidence or constraint

Additional species have recently been described in the *Leptopilina* genus (Allemand et al., 2002), allowing for a thorough analysis of the pattern of *Wolbachia* infection at the scale of the entire genus (Vavre and Henri,

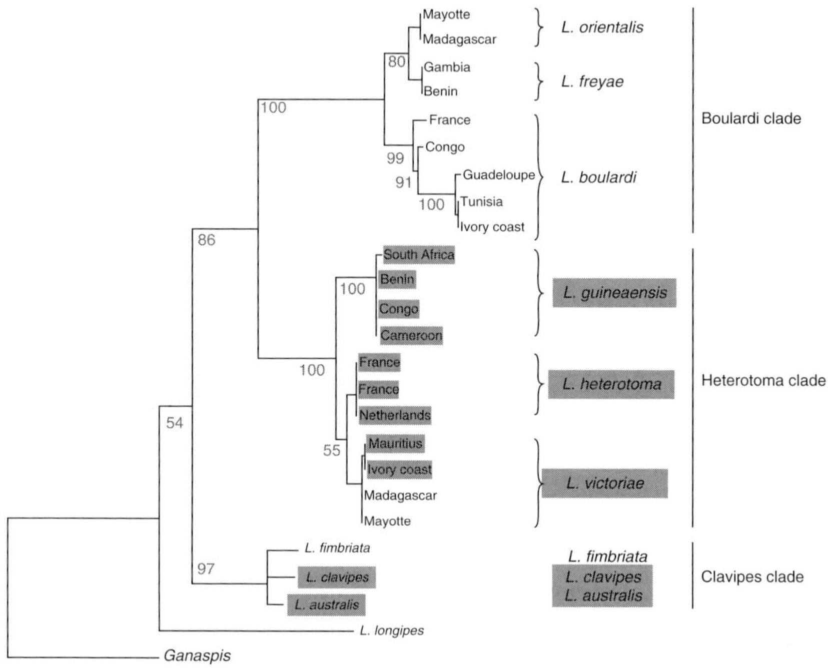

FIGURE 12.2 Phylogenetic tree of the *Leptopilina* genus reconstructed by maximum likelihood using ITS1. In dark grey, populations and species infected with CI-inducing *Wolbachia*; in light grey, species infected with PI-inducing *Wolbachia*. For each species origin of the individuals is indicated.

unpublished data). The *Leptopilina* genus is composed of three clades: Heterotoma, Boulardi and Clavipes, each composed of three species (Fig. 12.2). The three species in the Heterotoma clade are infected with *Wolbachia* strains inducing CI, and two are infected by more than one strain. In many cases, these strains are closely related to the strains that also infect *Drosophila* species. Within the Clavipes clade two of the three species are singly infected with closely related *Wolbachia* strains inducing parthenogenesis, which could suggest cospeciation. Finally, three species belong to the Boulardi clade, but none of them is infected. The probability that *Wolbachia* infection and phenotypes cluster as they do is only 0.007, which suggests that susceptibility to *Wolbachia* infection is determined at the phylogenetic level.

12.2.4. What determines the infection status of a species?

For horizontal transmission to occur, different factors must be fulfilled, which can be considered as a series of filters the infection must pass through (Vavre et al., 2003). First, the *encounter* filter must be passed, where an uninfected species must come into contact with an infected species, so that the transfer can occur. Second, the newly acquired *Wolbachia* must be able to pass a *compatibility* filter that corresponds to its ability to escape the novel hosts immune system and to multiply in this new host in order to get transmitted to the next generation. Third, the infection must pass an *invasion* filter and spread into the host population. This is an especially important step for CI-*Wolbachia* because infection must reach a certain threshold in order to be maintained in the population (Turelli, 1994).

The parasitoid way of life obviously opens up the encounter filter, and facilitates the transfer of *Wolbachia* from the *Drosophila* host to the wasp as each wasp develops in a potentially infected *Drosophila* host. The reverse, however, is not so easy. First, most *Drosophila* that reach the adult stage have not been parasitized. Second, those that have been parasitized but that were resistant to parasitism have mounted an immune response leading to the encapsulation of the wasp egg. This physical barrier probably limits the ability of *Wolbachia* to reach the *Drosophila* tissues. Hence, while parasitoid–host interaction certainly increases the chance that wasps are infected, parasitoids may not be major vectors of *Wolbachia* for their *Drosophila* hosts.

Given that all *Drosophila* parasitoids share the same way of life, all of them should be equally susceptible to *Wolbachia* primary infection. Why then do we observe important variations for infection among the *Leptopilina* clades? A first obvious possibility is that the *Drosophila* hosts differ among these various species and that these hosts differ for their infection status. It has been proposed that fungivorous *Drosophila* may be less infected than frugivorous *Drosophila* because of the presence of natural antibiotics produced by the substrate on which fungivorous *Drosophila* live (Haine et al., 2005; Jaenike et al., 2006). Interestingly, parasitoids developing in these hosts are mostly found in the Clavipes group where no obvious case of horizontal transmission could be detected. However, parasitoids form the Heterotoma and Boulardi clades share similar hosts, in particular *D. melanogaster* and *D. simulans*, but these two clades show very different susceptibility to *Wolbachia* infection. These differences can only be accounted for by differences in the compatibility or invasion filters. Experimental infections of *L. boulardi* by *Wolbachia* showed that infection was lost in a few generations in this species (Heath et al., 1999). This suggests that *Wolbachia* is able to infect *L. boulardi* but is not able to proliferate or be transmitted in this species, which could reflect a closure of the compatibility filter. The origin of this is unknown, and it would be difficult to investigate. The recent discovery of a virus in *L. boulardi* (Varaldi et al., 2003), which is not found in the sister species *L. heterotoma* that does harbor *Wolbachia*, opens up the possibility that resistance is mediated by a third party. Interestingly, two independent studies have recently shown that *Wolbachia* protects *Drosophila* against infection with RNA viruses (Hedges et al., 2008; Teixeira et al., 2008). Whether viruses can also protect from *Wolbachia* infection would thus be interesting to test. Another possibility is that the ancestor of the Boulardi group has been infected previously by *Wolbachia* and subsequently acquired resistance to *Wolbachia*. Finally, it is suspected that the effective population size of *L. boulardi* is higher than in *L. heterotoma*, which could also limit the spread of CI-*Wolbachia* in the former species, by closing the invasion filter. Unfortunately, ecological data are missing for the other species of these two clades.

Finally, other communities have been studied to detect patterns of horizontal transmission. While some have found such patterns (Dedeine et al., 2005a; Kittayapong et al., 2003), others have failed to detect them (Schilthuizen and Stouthamer, 1998; West et al., 1998). *Drosophila*–parasitoid communities and their variations in *Wolbachia* susceptibility may help to explain these variations among communities.

12.3. PHENOTYPIC DIVERSITY OF *WOLBACHIA* IN *DROSOPHILA* PARASITOIDS

12.3.1. Cytoplasmic incompatibility and its variability

Wolbachia-induced cytoplasmic incompatibility is a postzygotic isolation mechanism, which in its simplest form occurs in the crosses between infected males and uninfected females (unidirectional CI, Fig. 12.3). While the molecular mechanism of CI remains unknown, cytological observations indicate that in this cross, male chromosomes are improperly condensed precluding their normal participation to karyogamy (reviewed in Poinsot et al. 2003). The proposed model is that *Wolbachia* "imprint" chromosomes during spermatogenesis of infected males (the so-called modification (*mod*) function). These chromosomes can only get properly condensed when the *Wolbachia* present in the eggs rescue them (*resc* function). In diploids, CI results in the death of embryos. In haplodiploids, CI has first been described in the genus *Nasonia* (Breeuwer and Werren, 1990). Based on these results, it was thought that CI resulted in male-biased sex-ratio without reduction in offspring production (we will refer to this type of CI as the male development (MD) type of CI, Fig. 12.3). This could be explained by two factors. First, unfertilized eggs are obviously not exposed to CI and develop normally into males. Second, complete elimination of the paternal chromosome set from fertilized eggs restores haploidy, and hence allow their male development. Breeuwer (1997), however, showed that in the haplodiploid mites *Tetranychus urticae* and *T. turkestanica*, CI resulted in a male-biased sex-ratio, but that was

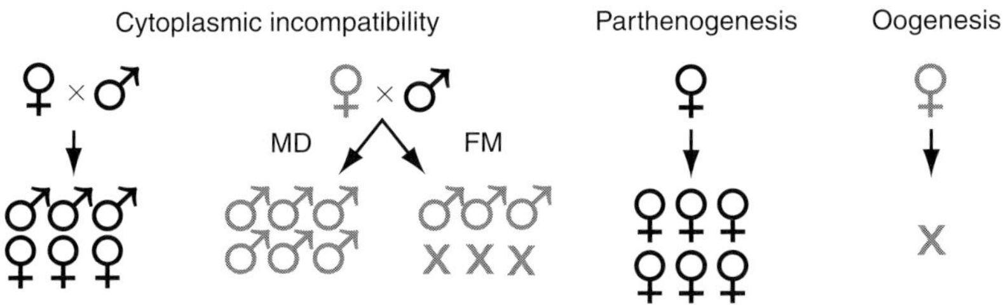

FIGURE 12.3 Schematic representation of the effects of *Wolbachia* found in *Drosophila* parasitoids. In black, infected individuals; in grey, uninfected individuals. MD, male development CI type; FM, female mortality CI type.

accompanied by a reduction in offspring production. In fact, male production in incompatible crosses was only due to the normal development of unfertilized eggs, while fertilized eggs died, as in diploids. We will refer to this CI type as the female mortality (FM) type (Fig. 12.3). The proposed mechanism was that male chromosomes were not entirely eliminated, leading to aneuploid, unviable eggs, a hypothesis recently confirmed by precise cytological analyses (Tram et al., 2006). Two hypotheses were proposed to explain this phenomenon: either this was due to the holokinetic structure of mite chromosomes, or to a reduced efficiency of the *mod* function in males.

Observations in *Drosophila* parasitoids, and especially the description of the CI phenotype in *L. heterotoma*, allowed for important advances in the description of CI in haplodiploids. First, in crosses between triply infected males and uninfected females, a FM phenotype was observed for the first time in Hymenoptera (Vavre et al., 2000). Second, crosses between males infected by a subset of these three *Wolbachia* strains showed that FM and MD phenotypes are only the extreme of a continuous gradient between FM and MD phenotypes (Mouton et al., 2005; Vavre et al., 2001), suggesting that the CI type is a quantitative rather than a qualitative trait. These two results reinforce the hypothesis that the variability was due to quantitative variations in the *mod* intensity, which is also corroborated by some indications that reduced *Wolbachia* density in males could lead to aneuploid unviable eggs in incompatible crosses in *Nasonia* (Breeuwer and Werren, 1993). Since these first observations, all new reported cases of bacterial-induced CI in haplodiploids (*Wolbachia* or *Cardinium)* have been shown to be of the FM phenotype (Hunter et al., 2003; Mochiah et al., 2002; Perlman et al., 2006), among which two in *Drosophila* parasitoids, in *A. tabida* (Dedeine et al., 2004) and in *Trichopria nr. drosophilae* (Vavre et al., 2002). Even in the *Nasonia* genus, a reanalysis of the CI phenotypes showed that the MD phenotype is only restricted to *N. vitripennis*, while a FM phenotype is expressed in *N. giraulti* and *N. longicornis* (Bordenstein et al., 2003).

Interestingly, studies in *L. heterotoma* and in the *Nasonia* genus highlight that both host and *Wolbachia* can affect the CI type. In *L. heterotoma*, CI type was measured using a single host genotype, but various compositions of *Wolbachia* strains. These experiments showed that the fraction of dying fertilized eggs (FM type) increased when the number of *Wolbachia* strains inducing CI is smaller (Fig. 12.4), thus showing that variation in the bacterial community alone is sufficient to affect the CI phenotype (Mouton et al., 2005; Vavre et al., 2001). On the contrary, in *Nasonia*, crosses between the different sister species showed that the host genotype plays a crucial role in the CI phenotype in this genus (Bordenstein et al., 2003). Understanding the CI diversity in haplodiploids and the evolutionary forces acting on this phenotype thus requires to take into account both partners.

From a mechanistic point of view, the results obtained on *L. heterotoma* are counter-intuitive. One hypothesis to explain the different phenotypes was that a reduced *mod* function could limit the ''imprinting'' of paternal chromosomes resulting in an incomplete destruction of the chromosomes and subsequent to aneuploid unviable embryos. Because bacterial density can be related to the intensity of CI (e.g., Boyle et al., 1993; Veneti et al., 2004), it has been proposed that a reduced bacterial density might result in a FM phenotype (Breeuwer and Werren, 1993). However, in *L. heterotoma* an increase in bacterial density, as observed when more *Wolbachia* strains are present, induces an increase in the number of dying embryos,

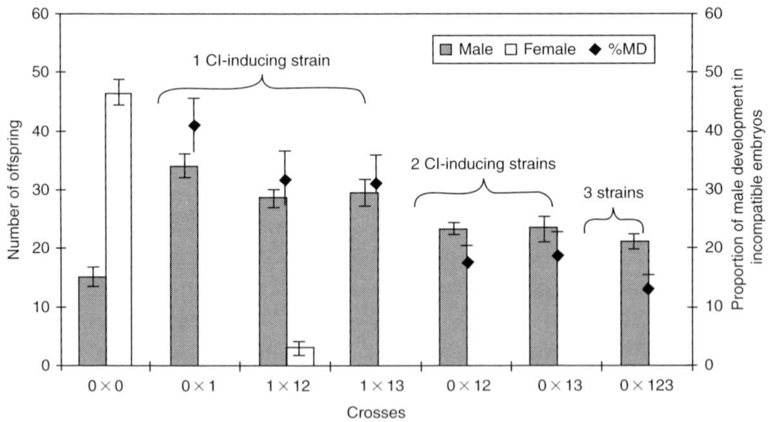

FIGURE 12.4 Male and female offspring production (left axis) and proportion of incompatible embryos developing into males (right axis) in crosses between *Leptopilina heterotoma* strains infected by different compositions of *Wolbachia* strains. Crosses are classified according to the number of strains that induce the CI phenotype.
Note: Adapted from Mouton et al. *(2005).*

thus in the FM phenotype (Mouton et al., 2005). Unfortunately, it is impossible to distinguish between the effect of bacterial diversity versus the effect of bacterial density in these experiments. However, the *Wolbachia* strains each exhibit a specific CI phenotype, and no direct interaction could be observed among them. This unexpected result makes *L. heterotoma* a suitable species to further investigate the mechanism of CI in haplodiploids, and cytological studies as those performed in *Nasonia* (Tram et al., 2006) would be interesting to conduct in this species.

The diversity of CI phenotypes in haplodiploids is not only important to understand the molecular basis of CI, but also has important consequences on the epidemiology and stability of *Wolbachia* infection, which could explain the higher incidence of the FM CI type. Indeed, spread and maintenance of *Wolbachia* in natural populations depend on the counter-selection exerted on uninfected females by incompatible crosses, whose frequency in turn depend on the frequency of infected males in the population. In haplodiploids, incompatible crosses result in the production of uninfected males, but more so under the MD CI type. As a consequence, CI in haplodiploids is less efficient than in diploid species, and among haplodiploids, the MD type is less efficient than the FM CI type (Egas et al., 2002; Vavre et al., 2000, 2003). This raises the possibility that at the start of a new *Wolbachia* infection, *Wolbachia* inducing the FM type has more chance to spread in the population than a *Wolbachia* inducing the MD type, that is, the invasion filter is less restrictive to the FM type. This selection for more invasive *Wolbachia* strains has been referred to as clade selection by Hurst and McVean (1996). In addition, it is more and more accepted that *Wolbachia* infections do not necessarily last for long periods in the same host. Theoretical models have shown that the MD type is more susceptible to infection loss than the FM type, and this could also explain why the MD type has not been as commonly found (Vautrin et al., 2007; Vavre et al., 2003).

12.3.2. Parthenogenesis induction and sexual degradation

Parthenogenesis-inducing *Wolbachia* (Fig. 12.3) are known in three *Drosophila* parasitoid species: *Leptopilina clavipes* (Schidlo et al., 2002), *Leptopilina australis* (Werren et al., 1995) and *Asobara japonica* (Kremer et al., 2009). Since PI-*Wolbachia* was only recently discovered in the latter species, and the infection is poorly described in *L. australis*, the focus in this section will be on *L. clavipes*, for which most information is available. *L. clavipes* is a parasitoid of *Drosophila* larvae that has a pan-European distribution (Nordlander, 1980). It occurs in woodlands where it parasitizes *Drosophila* larvae living in fungal fruit bodies (Driessen et al., 1990; Vet, 1983). In northern-western Europe *D. phalerata* is its main host, but larvae of *D. kuntzei*, *D. transversa* and *D. subobscura* are also parasitized

(Driessen et al., 1990). Southern European *L. clavipes* parasitize *D. melanogaster* larvae as well, a species that mainly breeds in fermenting fruits (Pannebakker et al., 2004c, 2008). Two modes of reproduction occur within Europe: all north-western European populations (Denmark, Sweden, Germany, The Netherlands, England and France) reproduce thelytokously (Nordlander, 1980; Pannebakker et al., 2004c; Vet, 1983). In contrast, populations south of the Pyrenees reproduce arrhenotokously (Pannebakker et al., 2004c). Thelytoky in *L. clavipes* is induced by infection with *Wolbachia* bacteria (Schidlo et al., 2002; Werren et al., 1995). The southern European arrhenotokous populations are uninfected by *Wolbachia*.

Thelytoky in *L. clavipes* is induced by diploidization of the haploid eggs through anaphase restitution during the first somatic mitosis (Pannebakker et al., 2004b). This mechanism is a form of gamete duplication that results in the generation of completely homozygous offspring (Suomalainen et al., 1987). Because gamete duplication potentially reduces infected populations to clones without genetic exchange, it can have large consequences for the population genetic structure of the wasps. However, *Wolbachia*-induced parthenogenesis does not necessarily result in a reduction of genetic variation, since the parthenogenetic populations are initially derived from uninfected populations. An analysis of the population genetic structure of uninfected arrhenotokous and infected thelytokous populations of *L. clavipes* in Europe did show similar levels of genetic variation in the uninfected and infected populations, but also a clear division between the two modes of reproduction (Pannebakker et al., 2004c). The infected wasps show two distinct haplotypes that are present at different collection sites and that sometimes co-occur at the same locality. The coexistence of multiple clones in the same habitat is likely to be ephemeral, as one clone will eventually replace the others through competitive exclusion or clonal drift (Jaenike et al., 1980). The current coexistence of clonal haplotypes is a direct consequence of the infection history of European *L. clavipes*. Because both clonal haplotypes *L. clavipes* are infected by the same *Wolbachia* strain, multiple infection events by different bacterial strains can be excluded (Pannebakker et al., 2004c). Analysis of the mitochondrial DNA of infected wasps revealed the presence of multiple mitochondrial haplotypes, which suggests the initial *Wolbachia* infection was horizontally transmitted from infected to uninfected wasps (Kraaijeveld, personal communication). Horizontal transmission in the initial stages of the infection will effectively "freeze" the genetic variation available in the uninfected population, after which only the best adapted clones survive (Simon et al., 2003).

Besides its drastic effects on population structure, PI-*Wolbachia* also has profound effects for the individual hosts. In the short term, parthenogenetic reproduction can be considered to be more efficient than sexual reproduction as no resources need to be invested in males or costly mating behavior (Maynard Smith, 1978). The long-term effects, however, are not necessarily beneficial. Parthenogenetic reproduction reduces selection on traits involved in sexual reproduction. Genes coding for these traits are no longer maintained by selection, and mutations in these genes can accumulate freely or even be favored by selection either because they improve the performance of the infected females or in response to the nucleo-cytoplasmic conflict that favors male production (Huigens and Stouthamer, 2003; Pijls et al., 1996; Werren, 1998). *Wolbachia*-induced parthenogenesis offers a unique possibility to study these mutational processes, because high temperature or antibiotic treatment can cure PI-*Wolbachia*-infected females from their infection (Stouthamer et al., 1990). This results in the production of uninfected males and females, in which the decay of sexual functionality can then be studied. The presence of uninfected populations in *L. clavipes* allows for a full comparison of both male and female sexual function, as opposed to similar studies in species where infection has gone to fixation (De Barro and Hart, 2001; Gottlieb and Zchori-Fein, 2001; Weeks and Breeuwer, 2001; Zchori-Fein et al., 1992, 1995).

Antibiotic curing of females from thelytokous populations resulted in males that were able to complete courtship behavior with arrhenotokous females, resulting in successful copulation and sperm transfer (Pannebakker et al., 2004a, 2005). Mating by these "cured" males also resulted in a full inhibition of female mating receptivity (Reumer et al., 2007) as observed in *L. heterotoma* (van den Assem, 1969) and *L. boulardi* (Kopelman and Chabora, 1986). Interestingly, for all the thelytokous lines tested, the sex ratio (proportion of males) of the offspring resulting from these crosses was significantly higher than that from crosses between arrhenotokous individuals (in both intra- and interpopulation crosses). Because of haplodiploid sex determination, an increase in sex ratio implies a lower fertilization success for the thelytokous males. *Wolbachia*-infected females were willing to mate with arrhenotokous males but they did not use the received sperm (Pannebakker et al., 2005). Hence, restored males from thelytokous populations are sexually only partially functional, and females from thelytokous populations apparently lost their sexual functionality. *Wolbachia* has become an obligate partner for survival and reproduction in thelytokous *L. clavipes* populations, which makes the transition to thelytoky in *L. clavipes* irreversible. Rather than being infected by a facultative symbiont, *L. clavipes* has become dependent on its reproductive parasite (see Section 12.3.4).

A similar situation is observed in *A. japonica*. In this species, populations from the main islands of Japan are thelytokous and *Wolbachia*-infected, while populations from the subtropical islands are arrhenotokous and uninfected (Kremer et al., 2009; Mitsui et al., 2007). No other symbiont could be detected in this species. Antibiotic treatment allows the restoration of male production in thelytokous females, suggesting that *Wolbachia* is the causative agent of thelytoky (Kremer et al., 2009). Analysis of reproductive behaviors showed that as in *L. clavipes,* females originating from thelytokous populations are not able to reproduce sexually. However, in contrast to *L. clavipes,* this is due to the complete absence of courtship between males (both from arrhenotokous and thelytokous populations) and thelytokous females. As in *L. clavipes,* males originating from thelytokous populations are still able to reproduce sexually with females from arrhenotokous populations, even though a reduction of fertility might occur. This situation reinforces the idea that when a PI-*Wolbachia* spreads and persists in a population, sexual decay is quicker in females than in males, and is a good indication that selective forces act to promote sexual decay in females. It also shows that when a species gets infected with a PI-*Wolbachia,* the evolution of dependence is swift and frequent.

12.3.3. Oogenesis

The involvement of *Wolbachia* in host oogenesis was only recently discovered in *Asobara tabida*, where it is obligate for successful oogenesis (Fig. 12.3; Dedeine et al., 2001) and is thus far the only such case within the *Drosophila* parasitoids. *A. tabida* is infected with three *Wolbachia* strains of which only the wAtab3 strain is involved in host oogenesis, while the other two strains induce CI of the FM type (Dedeine et al., 2004). Removal of this strain reduces female oocyte production in European strains of *A. tabida* from 260 to 300 to zero oocytes, resulting in complete sterility. The pattern is different for North American *A. tabida* strains, where symbiotic females have a lower oocyte number (approximately 220 oocytes) than the European strains, but removal of *Wolbachia* still leaves females capable of producing about 80 oocytes. These oocytes, however, are smaller than the symbiotic oocytes, and the larvae that emerge from these eggs die before completing development (Dedeine et al., 2005b). Thus, although the underlying mechanisms are likely to be different, failure to produce viable offspring makes *Wolbachia* an obligate partner for reproduction in North American *A. tabida* strains like it is for European strains. Introgression experiments have shown that these two distinct ovarian phenotypes are under the sole control of the host genotype.

The mechanisms underlying *Wolbachia*-dependent oogenesis have been explored in further detail by cytological observations. A detailed description and discussion of these observations can be found in Vavre et al. (2009). Briefly, removal of *Wolbachia* results in the occurrence of extensive programmed cell death (PCD) in the ovarioles of females from both the American and European populations (Pannebakker et al., 2007; Pannebakker, unpublished data). PCD is a vital part of insect oogenesis, both as a structural developmental process, as well as checkpoint processes at early and mid-oogenesis that regulate oocyte production in response to intrinsic or environmental cues (Buszczak and Cooley, 2000; McCall, 2004; see Vavre et al. (2009) for a brief overview of the role of PCD in insect oogenesis). In *A. tabida*, removal of *Wolbachia* does not induce general apoptosis, but apoptosis is restricted to mid-oogenesis egg chambers, suggesting a specific interaction of *Wolbachia* with the mid-oogenesis PCD pathway (Vavre et al., 2009). Because the mid-oogenesis PCD pathway is also dependent on external signals, such as nutrient deprivation, it is not clear whether PCD in *A. tabida* is directly controlled by *Wolbachia*, or whether induction of PCD is the by-product of another manipulation of the wasp. The involvement of the *Wolbachia* outer surface protein (WSP) in the inhibition of apoptosis of human granulocytes (Bazzocchi et al., 2007) combined with the ability of *Rickettsia* bacteria – close relatives of *Wolbachia* – to manipulate PCD of their vertebrate hosts (Clifton et al., 1998), suggests that direct manipulation of PCD in *A. tabida* is possible (but see Braig et al., 2009).

The discovery of bacterial manipulation of PCD could provide some insight into the evolutionary scenario that resulted in *Wolbachia*-dependent oogenesis in *A. tabida*. Immune responses against intracellular bacteria often involve apoptosis of infected cells (Zychlinsky and Sansonetti, 1997) and, in turn, several intracellular bacteria, including close relatives of *Wolbachia* have evolved apoptosis-inhibiting mechanisms to enable and sustain their own growth environment (Batut et al., 2004; Gao and Abu Kwaik, 2000). Bacterial inhibition of apoptosis can severely reduce the functionality of infected host tissues where apoptosis plays a crucial role, such as in the ovaries. In this coevolutionary process, hosts are then selected to compensate for this bacterial manipulation, and adjust their own gene expression and physiology to the presence of the bacteria (Aanen and Hoekstra, 2007; Pannebakker et al., 2007). Both host and bacteria benefit from this status quo, but will suffer equally when *Wolbachia* is removed and PCD is deregulated. Several other hypothesis have been proposed to explain the evolution of *Wolbachia*-dependent oogenesis in *A. tabida*, and we would like to refer the interested reader to the several papers discussing these in detail, that is, Aanen and Hoekstra (2007), Braig et al. (2009), Dedeine et al. (2001), Pannebakker et al. (2007), Vavre et al. (2009). Irrespective of the exact evolutionary pathway, the outcome of the interaction between *A. tabida* and *Wolbachia* shows that the evolution of obligate mutual dependence between host and parasite can be swift and does not have to involve initial fitness advantages to either partner.

12.3.4. Recurrent evolution of dependence

As discussed in the previous sections, dependence of *Drosophila* parasitoids on *Wolbachia* has at least evolved several times independently: in *A. tabida* and *A. japonica*, and in *L. clavipes* and *L. australis*. In the latter two species, close phylogenetic relationships among *Wolbachia* and parasitoids may suggest a cospeciation event, and make it difficult to consider these two cases as an independent event. Interestingly, dependence in these cases is not associated with a new function to the host, but rather to a loss of autonomy of the host to accomplish a function that obviously pre-existed the infection by *Wolbachia* (i.e., oogenesis and reproduction). These results suggest that evolution of dependence can occur swiftly through different mechanisms that include the evolution of tolerance, but also possibly the resolution of the nucleo-cytoplasmic conflict associated with sex-ratio distortion by microorganisms. These remarkable cases contrast with the classic scenarios of evolution of insect–symbiont interactions. It is generally thought that host dependence has evolved secondarily to mutualism through specialization and coevolution between hosts and symbionts. On the contrary, we suggest that the evolution of dependence can precede the evolution of mutualism. The arguments for this scenario are the following: while *Wolbachia* bacteria are selected to provide an advantage to their hosts, very few cases of "true" mutualism, where *Wolbachia* provides an additional function to the host, have been reported, suggesting evolution of mutualism is not straightforward. One of the clearest cases is in filarial nematodes where *Wolbachia* is obligate and where both partners have cospeciated (Fenn and Blaxter, 2006; Foster et al., 2005). However, even there, the benefit *Wolbachia* provides is not as obvious as it is in long-term mutualisms in insects (Moran and Baumann, 2000). In addition, it remains possible that the evolution of tolerance is at the origin of the patterns observed in nematodes as well. This suggests the evolution towards mutualism is not straightforward. Another interesting pattern that emerges from insect–symbiont associations is that many of them are labile. Host–Wolbachia cospeciation is not the rule, and if horizontal transmission is able to explain the observed pattern, it is clear that host–Wolbachia associations do not persist long enough within a host for cospeciation to occur.

When combining these two observations, that is, the not so straightfoward evolution towards mutualism, and the observed labile associations, the question arises why so many insects depend on symbionts for their reproduction and development. One option is that mutualism has evolved from facultative associations. The alternative scenario that we propose is that dependence precedes the evolution of mutualism. By consolidating the fate of the two partners, dependence can stabilize host–symbiont associations, leaving more time for the evolution of mutualism.

12.4. STABILITY, REGULATION AND CONSEQUENCES OF MULTIPLE *WOLBACHIA* INFECTIONS

Multiple infections have been extensively studied in horizontally transmitted parasites and were shown to play a major role in the evolution of host–parasite interactions and virulence (e.g. Chao et al., 2000; Frank, 1996). The central idea is that hosts provide a limited space and amount of resources, resulting in competition among parasites. Depending on the type of competition (direct competition among parasites or competition by interference) both an increase or a decrease in virulence can be expected. In both cases, the consequence of multiple infections is a departure from the expected optimal level of virulence when infections occur individually.

This issue has not received much attention in vertically transmitted symbionts. One reason for this is that, until recently, multiple infections with maternally inherited symbionts were thought to be rare, notably due to the bottleneck during transmission that should result in the homogenization of the bacterial population within the host. This image has changed drastically in recent years, with more and more descriptions of multiply infected systems involving different strains of *Wolbachia*, but also different bacterial species (reviewed in Vautrin and Vavre, 2009). Numerous questions arise from these observations: how do these multiple infections invade populations, how are they maintained, how are they regulated, and how do they affect the host? Because *Drosophila* parasitoids are frequently infected by multiple *Wolbachia* strains, that is, both *L. heterotoma* and *A. tabida* each harbor three *Wolbachia* strains, they are excellent systems to study these questions.

12.4.1. Invasion and stability of multiple infections

In all the populations of *A. tabida* and *L. heterotoma* studied to date, which originated from France, Spain and the UK, triple infections have been found (Dedeine et al., 2005b; Haine et al., 2005; Vavre et al., 1999). In addition, a recent survey of field-collected individuals in France showed that all *L. heterotoma* individuals are triply infected, regardless of the collection site or the time in the season they were collected (Vautrin, 2008). Multiple infections thus seem very stable in these associations.

In an experimental setup, parasitoid lines infected by different subsets of *Wolbachia* strains were established for these two species, which allowed the determination of the effect of the individual bacterial strains. In *L. heterotoma*, the three strains induce CI, and are mutually incompatible, even though it is still unknown whether wLhet2 and wLhet3 are able to rescue CI induced by wLhet1 because the attempts to remove wLhet1 without eliminating the two other strains failed (Mouton et al., 2005; Vavre et al., 2001). These results demonstrate that these strains are "real" strains that differ not only in molecular markers, but also in their phenotype. In addition, the fact that all these strains are mutually incompatible can explain the spread and stability of multiple infections. Only females infected by all three bacterial strains are able to mate with all males in the population, which allows multiple infections to spread in the population (Frank, 1998; Vautrin et al., 2007). In *A. tabida*, wAtab1 and wAtab2 induce CI and are mutually incompatible (Dedeine et al., 2004). wAtab3 is required for oogenesis and is unable to rescue the CI induced by the two other strains (see also Section 12.3.3). Because uninfected females are sterile, the ability of wAtab3 to induce CI cannot be tested. This shows that also in *A. tabida*, each strain has a particular phenotype (different types of CI, oogenesis) and again this can account for the success and stability of these multiple infections. In this case, however, it is not known what mechanism drove wAtab3 in the population. Obviously, *A. tabida* was able to produce eggs before infection with wAtab3. Either this strain was or still is inducing CI, allowing it to spread in the population, and the involvement in oogenesis and dependence evolved only at a later time. Alternatively, wAtab3 was able to confer an advantage to *A. tabida* by increasing its fecundity, thereby allowing its spread and subsequent evolution of dependence (see also Section 12.3.3).

12.4.2. Regulation of multiple infections and phenotypic consequences

When studying host–microbial associations, one must keep in mind that the interaction involves a single host but a population of symbionts. The size of this bacterial population plays a key role on fundamental parameters of the association. Intuitively, sustaining a higher number of symbionts will increase the cost of infection to the host, which indirectly can have consequences for the bacteria. On the other hand, it might also increase the efficiency of transmission and increase the level of CI. Because of this tradeoff between transmission and infection cost, an optimal density is expected. The situation is more complex when multiple symbionts share the same host since, in addition to this tradeoff, competition among symbionts might change the optimal size of the total bacterial population, but also for individual strains.

Using lines with a controlled genetic background and infected with different subsets of *Wolbachia* strains, it was shown in *L. heterotoma* and *A. tabida* that the total bacterial density increases with the diversity of the bacterial community, while in the mean time the specific density of each CI-inducing strain remained constant, regardless of the composition of the bacterial community (Mouton et al., 2003, 2004). For wAtab3, a slight increase in its density was observed in lines harboring other strains, suggesting a positive effect of other strains on wAtab3. These results indicate that there is no competition among CI-inducing *Wolbachia* strains within a host, and that some cooperation might exist between wAtab3 and the two other strains in *A. tabida*. Differential localization of the three strains is unlikely since the relative proportion of the different strains is constant in different parts of the body (Mouton et al., 2003, 2004). Interestingly, strain-specific regulation has also been found in other insects infected by different CI-inducing *Wolbachia* strains, suggesting a general phenomenon (Ikeda et al., 2003; Rousset et al., 1999; but see Kondo et al., 2005). This result is surprising given that various studies showed variation in the specific densities of each symbiont when hosts are infected with symbionts belonging to different species (e.g., Goto et al., 2006; Oliver et al., 2006).

Analysis of the phenotypic consequences of density variations revealed a positive correlation between density and infection cost in *A. tabida* (Mouton et al., 2004). A similar trend was also observed in *L. heterotoma* (Mouton, 2003). Similar results have also been obtained in *Drosophila* species (McGraw et al., 2002), suggesting a general phenomenon. The pattern suggested by this correlation is that sustaining a higher number of *Wolbachia* strains increases the infection cost, which in turn should be selected against. However, the infection costs are generally low, and their expression under field conditions is probably limited. In addition, loosing one of the *Wolbachia* strains may result in the exposure of females to CI and these females will be strongly selected against in populations where *Wolbachia* reaches high prevalence, such as in *L. heterotoma* and *A. tabida*. Therefore, specific regulation of *Wolbachia* at the strain level could be seen as a way to limit stochastic loss of some *Wolbachia* strains and hence exposure to CI. In addition, by limiting competition among strains this might also limit the evolution of increased levels of virulence.

How specific regulation is achieved remains a complex issue. Bacterial density is influenced by the bacterial genotype, the host genotype, the environment, and the interactions among these factors (Mouton et al., 2006, 2007). In addition, interactions among symbionts might also take place. This seems to be the case for wAtab3, whose density increases when other strains are present. Interactions are also suspected in *L. heterotoma* where infected lines not harboring wLhet1 have never been obtained despite numerous attempts, suggesting that this strain is obligate for the maintenance of the two other strains. A recent theoretical study demonstrated that such positive or dependence relationships among symbionts can evolve in these systems (Vautrin et al., 2008). Indeed, vertical transmission not only locks one symbiont within a host, but also consolidates the different symbiotic genotypes that are cotransmitted from one generation to the other. This situation creates extreme partner fidelity among symbionts, which is one of the conditions for cooperation or dependence to evolve (Sachs et al., 2004).

12.4.3. Evolution of *Wolbachia* genomes in multiply infected hosts

Recombination in *Wolbachia* genomes have now been found repeatedly (Baldo et al., 2006a). Multiple infections in a single host create favorable conditions for genetic exchanges among *Wolbachia* strains. The results obtained in *L. heterotoma* and *A. tabida* so far did not prove any exchanges between strains, although it should be noted that only few studies have tackled this question. First, for all cases observed so far, there is complete linkage between the *wsp* sequence of a strain and its induced phenotype. However, the most striking result has been obtained on the WO bacteriophage that infects *Wolbachia*. This phage has been shown to be frequently horizontally transmitted between *Wolbachia* strains (Gavotte et al., 2007). However, within *L. heterotoma*, each *Wolbachia* strain harbors a single and specific phage (Gavotte et al., 2004), showing complete linkage between two specific markers (WSP and WO) that are known to be frequently involved in recombination in other systems (Baldo et al., 2006a; Gavotte et al., 2007). Thus, while recombination between strains does occur, its frequency at the individual host level might be rather limited.

In addition, genomes of intracellular bacteria are known to undergo reductive evolution where functions related to the free-living state, or that are provided by the host, are rapidly eliminated from the genome (Wernegreen, 2002). Interestingly, in the case of multiple infections some other functions might also be dispensable because they are provided by coinherited symbionts. This could create dependence among symbionts such has been recently observed in the aphid *Cinara cedri* (Gosalbes et al., 2008). We have no information whether genome erosion is more pronounced in multiply infected hosts, but it would be interesting to test this hypothesis.

12.5. THE ROLE OF *WOLBACHIA* IN THE INTERACTION BETWEEN PARASITOIDS AND HOSTS

Besides playing an important role in the biology of the parasitoid, *Wolbachia* can potentially mediate the *Drosophila–parasitoid* interaction. By manipulating the reproduction of its hosts (both *Drosophila* and parasitoids), *Wolbachia* has the potential to alter the parasitoid–*Drosophila* dynamics. The consequences of *Wolbachia* infection in this dynamics will depend on its fitness effects for both parasitoid and *Drosophila*. In general, vertically transmitted parasites can only be maintained in host populations if they do not reduce the fitness of their host (Anderson and May, 1982; Ebert and Herre, 1996). By contrast, *Wolbachia* and other reproductive parasites, can spread through their hosts population, even if they induce a physiological cost to their hosts (O'Neill et al., 1997; Turelli, 1994). Across its range of infection, the effect of *Wolbachia* ranges from an increase to decrease in host fitness (e.g., Min and Benzer, 1997; Teixeira et al., 2008). Because of the intimate interaction between parasitoid and its *Drosophila* host, *Wolbachia* mediated alteration in fitness, can potentially impact both of them. Below, we discuss these fitness effects on *Drosophila* parasitoids and put them in the context of parasitoid–host dynamics.

12.5.1. Effects on host physiology and their consequences on the *Drosophila*–parasitoid interaction

The cost of infection with *Wolbachia* can be expressed in many different traits. Across taxa, *Wolbachia* has been found to negatively impact a wide range of traits, such as longevity in *D. melanogaster* (Min and Benzer, 1997), competitive ability in *Trichogramma kaykai* (Huigens et al., 2004), fecundity in *Tetranychus urticae* mites (Perrot-Minnot et al., 2002) or sperm competitive ability in *D. simulans* (de Crespigny and Wedell, 2006). However, the effects are not always strongly expressed (Harcombe and Hoffmann, 2004) and, at least in *Drosophila*, appear to depend on the genotype of the host (Fry et al., 2004).

Within the *Drosophila* parasitoids, the cost of *Wolbachia* infection has been studied most extensively for *L. heterotoma*. Fleury et al. (2000) were the first to investigate the *Wolbachia*-related physiological and behavioral costs in a *Drosophila* parasitoid. They found a negative impact of infection on female fecundity and adult survival, in addition to a strong reduction in locomotor activity in both sexes. Because locomotor activity is a good proxy for the overall physiological state of individuals, the observed reduction suggests a heavy cost of infection (Fleury et al., 2000). Detailed analysis of the *Wolbachia* density further revealed a positive correlation between bacterial density and infection cost, measured as tibia length, fresh weight and longevity (Mouton, 2003). In addition, *Wolbachia* was found to have negative effects on virulence-related traits in *L. heterotoma*. Elimination of *Wolbachia* from the parasitoid resulted in lower encapsulation rates by *D. simulans* (Fytrou et al., 2006). In *Leptopilina*, suppression of the hosts immune response involves the injection of virus-like particles (VLPs) in the host upon oviposition (see also Chapter 5 in this issue). VLPs render the hosts lamellocytes unable to encapsulate the parasitoid egg (Labrosse et al., 2003). Fytrou et al. (2006) proposed that the observed higher encapsulation rates of *Wolbachia*-infected parasitoids could be due to the bacteria influencing the VLP production in the parasitoid. While the exact mechanism remains to be determined, their suggestion was corroborated with the recent discovery of *Wolbachia*-induced resistance to viral infections in *D. melanogaster* (Hedges et al., 2008; Teixeira et al., 2008), even though the exact nature of VLP is still controversial.

While *Wolbachia* results in a higher encapsulation rate of *L. heterotoma* eggs, the infection is also harmful to *D. simulans* that are less resistant to parasitoid attacks when infected and this likely affects parasitoid–host dynamics in this system (Fytrou et al., 2006). In species where *Wolbachia* is fixed or nearly fixed (e.g., *L. heterotoma, A. tabida, D. simulans*) the impact on the *Drosophila–parasitoid* community will be constant over all populations. However, in species where infection is polymorphic such as in *D. melanogaster* or *P. dubius*, complex interactions between *Wolbachia* dynamics and host–parasitoids relationship might be expected. In other words, in species where reproductive manipulation is strong, the impact of the infection cost on *Wolbachia* dynamics will be limited to the invasive phase of the symbionts. On the opposite, in species where reproductive manipulation is milder and does not allow fixation of the symbiont, infection cost may play an important role on the prevalence of the infection, and even on the maintenance of the symbiont. In addition, if infection frequency varies among populations, it might locally modify the coevolutionary dynamics of hosts and parasitoids.

The relation between bacterial density and infection costs was studied in more detail in *A. tabida* (see also Section 12.4.2). Bacterial density and diversity were found to be negatively correlated with dry weight, adult survival and locomotor activity (Mouton et al., 2004). The observed link between bacterial density and infection costs can partly be explained by the higher energy requirements of having more symbionts (Thompson, 1988). However, in multiply infected species such as *A. tabida*, different *Wolbachia* strains can induce different fitness costs on their hosts, resulting in complex interactions between the cost of infection and strain-specific and/or total bacterial density.

Infection with *Wolbachia* does not always result in a cost to its host. For instance, in *D. melanogaster*, *Wolbachia* infection enhances the survival and fecundity of some fly strains (Fry et al., 2004) and in other strains the infection can induce resistance to viral infections (Hedges et al., 2008; Teixeira et al., 2008). For the *Drosophila* parasitoids, not many fitness benefits have been described. To our knowledge, the only known case of *Wolbachia*-induced fitness benefits is found in *L. clavipes*. Here, PI-*Wolbachia* infected females show an increase in longevity compared to naturally uninfected females (Reumer et al., 2007). However, the strains used in this study originated from two different localities (northern and southern Europe). The observed differences could also be the result of adaptations to local ecological conditions, rather than be induced by differences in *Wolbachia* infection.

12.5.2. Indirect effects of *Wolbachia* infection

An indirect short-term benefit potentially provided by infection with PI-*Wolbachia* could be a reduction in the cost of sexual reproduction. Because males are absent, parthenogenetically reproducing organisms do not pay the twofold cost of sex (Maynard Smith, 1978) and are able to sustain a higher population growth. Because PI-*Wolbachia* infected populations consist only of females, they are potentially more effective in attacking *Drosophila* host populations than uninfected parasitoids. This can completely modify the structure of the local community through modifications of the coevolutionary arms race between hosts and parasitoids, but also of competitive interactions among parasitoids. On the long term, however, *Wolbachia*-induced parthenogenesis is not necessarily beneficial to the parasitoids. Because of gamete duplication, the mechanism involved in PI-*Wolbachia*, all genetic variation is effectively frozen in the infected females and recombination among genotypes is impossible (see Section 12.3.2). This lack of genetic variation and recombination could potentially be a handicap in the ongoing arms race between parasitoids and their hosts.

A reduction in genetic diversity of the parasitoid is not only expected for PI-*Wolbachia*. Also for the other phenotypes of *Wolbachia*, presence of the reproductive parasite can drive rapid changes in allele frequencies in the host population (Charlat et al., 2007; Hornett et al., 2006, 2008). *Wolbachia* dynamics can result in a selective sweep in the host genome, potentially fixating deleterious alleles, some of which might be involved in parasitoid–host interactions. However, the exact long-term impact of *Wolbachia* on parasitoid–host interactions is hard to predict and requires additional studies.

12.6. CONCLUSION

Drosophila parasitoids provide insight into a variety of questions related to host–*Wolbachia* interactions. These include, among others, patterns of horizontal transmission, the diversity of *Wolbachia* phenotypes and their consequences on host evolution, and the regulation of bacterial populations in complex systems. The facility to combine field and laboratory experiments makes *Drosophila* parasitoids perfect models to tackle questions on these host–symbiont associations ranging from molecules to community levels. However, the parasitoid way of life also has some constraints. For example, cytological studies on embryos are quite difficult to realize and time-consuming, especially in solitary parasitoids. These studies would, however, provide necessary answers on the mechanisms maintaining the variation in CI phenotypes in haplodiploids. In addition, artificial *Wolbachia* transfers are also very difficult since the most powerful techniques for such artificial transfers is the injection of *Wolbachia* directly in the embryos (Boyle et al., 1993), which is impossible in endoparasitoids such as *Drosophila* parasitoids. Injection in nymphae or adults may, however, be possible as has been demonstrated in *Drosophila* (Frydman et al., 2006) and it is worth to explore this technique for the parasitoids as well. Clearly, studies of the genetic bases of *Wolbachia* phenotypic diversity would gain enormously from the possibility to create new parasitoid–*Wolbachia* associations by artificial transfers. For example, transfer of wAtab3 in different species would allow the determination of the respective roles of the host and bacterial genotypes in the dependence for oogenesis. Similarly, artificial transfers of CI and PI-*Wolbachia* in different species would be extremely interesting.

One other caveat is the current paucity of genomic data available for *Drosophila* parasitoids, limiting detailed functional studies. However, the recent development of new sequencing technologies is expected to rapidly fill that gap. Such information would allow for the study of host–parasitoid interactions at a molecular level, but would also help studying parasitoid–*Wolbachia* interactions. The current sequencing of the *Nasonia* genome will already revolutionize parasitoid research, but more such efforts are required on other parasitoid systems. Such attempt has been started for *A. tabida*, where more than 30,000 ESTs have been sequenced (Vavre, unpublished data) for *Wolbachia* infected and uninfected individuals to study the basis of *Wolbachia* dependence for oogenesis. In return, this extreme phenotype can be used as a mutant to study the genetic pathways underlying parasitoid oogenesis. Similarly, more data are required on the genomics of *Wolbachia* strains infecting *Drosophila* parasitoids. Very few data are available apart from sequences obtained for phylogenetic studies and WO phage infection. A noticeable exception is the full characterization of the genes encoding the type IV secretion system in wAtab3 (Rances et al., 2008) that could also pave the way to determine key effectors mediating the host–symbiont interaction. Access to *Wolbachia* genomes is also becoming easier, with currently two fully annotated *Wolbachia* genomes available and 14 genomes currently undergoing sequencing or full annotation (Werren et al., 2008). In addition, several *Wolbachia* genomes have been discovered as "by-product" from genome sequencing projects of their *Drosophila* hosts (Salzberg et al., 2005), further emphasizing the case for the sequencing of *Drosophila* parasitoids. Having genome information on *Wolbachia* infecting *Drosophila* parasitoids may help to study functional aspects of the interactions with the host but also among *Wolbachia* strains. It might also shed light on the consequences of multiple infections in the genome dynamics of symbionts.

Finally, an important effort is needed at the community level. While patterns of horizontal transmission have been well established at the phylogenetic level, explanation of the variation of *Wolbachia* infection at a wider scale is still required. In addition, the impact of *Wolbachia* infection at the community level has not received much attention. A promising question is certainly how the acquisition of thelytoky impacts the composition of communities. Field surveys and/or population cages experiments could provide some insights on this question.

ACKNOWLEDGMENTS

We thank Bregje Werthem, Hélène Henri and Genevieve Prevost for valuable comments on the manuscript. FV, LM and BAP were supported by the Centre National de la Recherche Scientifique, the Agence Nationale de la Recherche (ANR-06-BLAN-0316), and the Netherlands Organization for Scientific Research NWO (Veni-grant no. 863.08.008).

REFERENCES

Aanen, D.K., Hoekstra, R.F., 2007. The evolution of obligate mutualism: if you can't beat 'em, join 'em. Trends Ecol. Evol. 22, 506–509.

Allemand, R., Lemaitre, C., Frey, F., Bouletreau, M., Vavre, F., Nordlander, G., et al., 2002. Phylogeny of six African *Leptopilina* species (Hymenoptera: Cynipoidea, Figitidae), parasitoids of *Drosophila*, with description of three new species. Ann. Soc. Entomol. Fr. 38, 319–332.

Anderson, R., May, R., 1982. Coevolution of hosts and parasites. Parasitology 85, 411–426.

Baldo, L., Bordenstein, S., Wernegreen, J., Werren, J., 2006a. Widespread recombination throughout *Wolbachia* genomes. Mol. Biol. Evol. 23, 437–449.

Baldo, L., Dunning Hotopp, J.C., Jolley, K.A., Bordenstein, S.R., Biber, S.A., Choudhury, R.R., et al., 2006b. Multilocus Sequence Typing System for the Endosymbiont *Wolbachia* pipientis. Appl. Environ. Microbiol. 72, 7098–7110.

Baldo, L., Lo, N., Werren, J.H., 2005. Mosaic nature of the *Wolbachia* surface protein. J. Bacteriol. 187, 5406–5418.

Batut, J., Andersson, S.G.E., O'Callaghan, D., 2004. The evolution of chronic infection strategies in the alpha-proteobacteria. Nat. Rev. Microbiol. 2, 933–945.

Bazzocchi, C., Comazzi, S., Santoni, R., Bandi, C., Genchi, C., Mortarino, M., 2007. *Wolbachia* surface protein (WSP) inhibits apoptosis in human neutrophils. Parasite Immunol. 29, 73–79.

Bordenstein, S.R., Werren, J.H., 2007. Bidirectional incompatibility among divergent *Wolbachia* and incompatibility level differences among closely related *Wolbachia* in *Nasonia*. Heredity 99, 278–287.

Bordenstein, S.R., O'Hara, F.P., Werren, J.H., 2001. *Wolbachia*-induced incompatibility precedes other hybrid incompatibilities in *Nasonia*. Nature 409, 707–710.

Bordenstein, S.R., Uy, J.J., Werren, J.H., 2003. Host genotype determines cytoplasmic incompatibility type in the haplodiploid genus *Nasonia*. Genetics 164, 223–233.

Boyle, L., Oneill, S.L., Robertson, H.M., Karr, T.L., 1993. Interspecific and intraspecific horizontal transfer of *Wolbachia* in *Drosophila*. Science 260, 1796–1799.

Braig, H.R., Turner, B.D., Perotti, M.A., 2009. Symbiotic *Rickettsia*. In: Bourtzis, Miller, (Eds.), Insect Symbiosis 3. CRC Press, Boca Raton, Florida, pp. 221–250.

Breeuwer, J.A.J., 1997. *Wolbachia* and cytoplasmic incompatibility in the spider mites *Tetranychus urticae* and *T. turkestani*. Heredity 79, 41–47.

Breeuwer, J.A.J., Werren, J.H., 1990. Microorganisms associated with chromosome destruction and reproductive isolation between two insect species. Nature 346, 558–560.

Breeuwer, J.A.J., Werren, J.H., 1993. Cytoplasmic incompatibility and bacterial density in *Nasonia* vitripennis. Genetics 135, 565–574.

Breeuwer, J.A.J., Werren, J.H., 1995. Hybrid breakdown between two haplodiploid species: the role of nuclear and cytoplasmatic genes. Evolution 49, 705–717.

Breeuwer, J.A.J., Stouthamer, R., Barns, S.M., Pelletier, D.A., Weisburg, W.G., Werren, J.H., 1992. Phylogeny of cytoplasmic incompatibility micro-organisms in the parasitoid wasp genus *Nasonia* (Hymenoptera: Pteromalidae) based on 16S ribosomal DNA sequences. Insect Mol. Biol. 1, 25–36.

Buszczak, M., Cooley, L., 2000. Eggs to die for: cell death during *Drosophila* oogenesis. Cell Death Differ. 7, 1071–1074.

Chao, L., Hanley, K., Burch, C., Dahlberg, C., Turner, P., 2000. Kin selection and parasite evolution: Higher and lower virulence with hard and soft selection. Q. Rev. Biol. 75, 261–275.

Charlat, S., Hornett, E., Fullard, J., Davies, N., Roderick, G., Wedell, N., et al., 2007. Extraordinary flux in sex ratio. Science 317, 214.

Clifton, D., Goss, R., Sahni, S., van Antwerp, D., Baggs, R., Marder, V., et al., 1998. NF-kappa B-dependent inhibition of apoptosis is essential for host cell survival during *Rickettsia rickettsii* infection. Proc. Natl. Acad. Sci. USA 95, 4646–4651.

De Barro, P.J., Hart, P.J., 2001. Antibiotic curing of parthenogenesis in *Eretmocerus mundus* (Australian parthenogenic form). Entomol. Exp. Appl. 99, 225–230.

de Crespigny, F., Wedell, N., 2006. *Wolbachia* infection reduces sperm competitive ability in an insect. Proc. R. Soc. B 273, 1455–1458.

Dedeine, F., Ahrens, M., Calcaterra, L., Shoemaker, D., 2005a. Social parasitism in fire ants (*Solenopsis* spp.): a potential mechanism for interspecies transfer of *Wolbachia*. Mol. Ecol. 14, 1543–1548.

Dedeine, F., Bouletreau, M., Vavre, F., 2005b. *Wolbachia* requirement for oogenesis: occurrence within the genus *Asobara* (Hymenoptera, Braconidae) and evidence for intraspecific variation in *A. tabida*. Heredity 95, 394–400.

Dedeine, F., Vavre, F., Fleury, F., Loppin, B., Hochberg, M.E., Bouletreau, M., 2001. Removing symbiotic *Wolbachia* bacteria specifically inhibits oogenesis in a parasitic wasp. Proc. Natl. Acad. Sci. USA 98, 6247–6252.

Dedeine, F., Vavre, F., Shoemaker, D.D., Bouletreau, M., 2004. Intra-individual coexistence of a *Wolbachia* strain required for host oogenesis with two strains inducing cytoplasmic incompatibility in the wasp *Asobara tabida*. Evolution 58, 2167–2174.

Driessen, G., Hemerik, L., van Alphen, J.J.M., 1990. *Drosophila* species, breeding in the stinkhorn (*Phallus impudicus* Pers.) and their larval parasitoids. Neth. J. Zool. 40, 409–427.

Duron, O., Bouchon, D., Boutin, S., Bellamy, L., Zhou, L., Engelstadter, J., et al., 2008. The diversity of reproductive parasites among arthropods: *Wolbachia* do not walk alone. BMC Biol. 6, 27.

Ebert, D., Herre, E., 1996. The evolution of parasitic diseases. Parasitol. Today 12, 96–101.

Egas, M., Vala, F., Breeuwer, J.A.J., 2002. On the evolution of cytoplasmic incompatibility in haplodiploid species. Evolution 56, 1101–1109.

Fenn, K., Blaxter, M., 2006. *Wolbachia* genomes: revealing the biology of parasitism and mutualism. Trends Parasitol. 22, 60–65.

Fleury, F., Vavre, F., Ris, N., Fouillet, P., Bouletreau, M., 2000. Physiological cost induced by the maternally-transmitted endosymbiont *Wolbachia* in the *Drosophila* parasitoid *Leptopilina heterotoma*. Parasitology 121, 493–500.

Foster, J., Ganatra, M., Kamal, I., Ware, J., Makarova, K., Ivanova, N., et al., 2005. The *Wolbachia* genome of *Brugia malayi*: endosymbiont evolution within a human pathogenic nematode. PLoS Biol. 3, 599–614.

Frank, S.A., 1996. Models of parasite virulence. Q. Rev. Biol. 71, 37–78.

Frank, S.A., 1998. Dynamics of cytoplasmic incompatibility with multiple *Wolbachia* infections. J. Theor. Biol. 192, 213–218.

Fry, A., Palmer, M., Rand, D., 2004. Variable fitness effects of *Wolbachia* infection in *Drosophila melanogaster*. Heredity 93, 379–389.

Frydman, H., Li, J., Robson, D., Wieschaus, E., 2006. Somatic stem cell niche tropism in *Wolbachia*. Nature 441, 509–512.

Fytrou, A., Schofield, P., Kraaijeveld, A., Hubbard, S., 2006. *Wolbachia* infection suppresses both host defence and parasitoid counter-defence. Proc. R. Soc. B 273, 791–796.

Gao, L.Y., Abu Kwaik, Y., 2000. The modulation of host cell apoptosis by intracellular bacterial pathogens. Trends Microbiol. 8, 306–313.

Gavotte, L., Henri, H., Stouthamer, R., Charif, D., Charlat, S., Bouletreau, M., et al., 2007. A survey of the bacteriophage WO in the endosymbiotic bacteria *Wolbachia*. Mol. Biol. Evol. 24, 427–435.

Gavotte, L., Vavre, F., Henri, H., Ravallec, M., Stouthamer, R., Boulétreau, M., 2004. Diversity, distribution and specificity of WO phage infection in *Wolbachia* of four insect species. Insect Mol. Biol. 13, 147–153.

Gosalbes, M., Lamelas, A., Moya, A., Latorre, A., 2008. The striking case of tryptophan provision in the cedar aphid *Cinara cedri*. J. Bacteriol. 190, 6026–6029.

Goto, S., Anbutsu, H., Fukatsu, T., 2006. Asymmetrical interactions between *Wolbachia* and *Spiroplasma* endosymbionts coexisting in the same insect host. Appl. Environ. Microbiol. 72, 4805–4810.

Gottlieb, Y., Zchori-Fein, E., 2001. Irreversible thelytokous reproduction in *Muscidifurax uniraptor*. Entomol. Exp. Appl. 100, 271–278.

Haine, E., Pickup, N., Cook, J., 2005. Horizontal transmission of *Wolbachia* in a *Drosophila* community. Ecol. Entomol. 30, 464–472.

Harcombe, W., Hoffmann, A., 2004. *Wolbachia* effects in *Drosophila melanogaster*: in search of fitness benefits. J. Invert. Pathol. 87, 45–50.

Heath, B.D., Butcher, R.D.J., Whitfield, W.G.F., Hubbard, S.F., 1999. Horizontal transfer of *Wolbachia* between phylogenetically distant insect species by a naturally occurring mechanism. Curr. Biol. 9, 313–316.

Hedges, L.M., Brownlie, J.C., O'Neill, S.L., Johnson, K.N., 2008. *Wolbachia* and virus protection in insects. Science 322, 702.

Hilgenboecker, K., Hammerstein, P., Schlattmann, P., Telschow, A., Werren, J.H., 2008. How many species are infected with *Wolbachia*? A statistical analysis of current data. FEMS Microbiol. Lett. 281, 215–220.

Hornett, E., Duplouy, A., Davies, N., Roderick, G., Wedell, N., Hurst, G., et al., 2008. You can't keep a good parasite down: Evolution of a male-killer suppressor uncovers cytoplasmic incompatibility. Evolution 62, 1258–1263.

Hornett, E.A., Charlat, S., Duplouy, A.M.R., Davies, N., Roderick, G.K., Wedell, N., et al., 2006. Evolution of male-killer suppression in a natural population. PLoS Biol. 4, 1643–1648.

Huigens, M., Hohmann, C., Luck, R., Gort, G., Stouthamer, R., 2004. Reduced competitive ability due to *Wolbachia* infection in the parasitoid wasp *Trichogramma kaykai*. Entomol. Exp. Appl. 110, 115–123.

Huigens, M.E., Stouthamer, R., 2003. Parthenogenesis associated with *Wolbachia*. In: Bourtzis, Miller, (Eds.), Insect Symbiosis. CRC Press, Boca Raton, Florida, pp. 247–266.

Hunter, M.S., Perlman, S.J., Kelly, S.E., 2003. A bacterial symbiont in the Bacteroidetes induces cytoplasmic incompatibility in the parasitoid wasp *Encarsia pergandiella*. Proc. R. Soc. Lond. B 270, 2185–2190.

Hurst, G.D.D., Jiggins, F.M., von der Schulenburg, J.H.G., Bertrand, D., West, S.A., Goriacheva, I. I., et al., 1999. Male-killing *Wolbachia* in two species of insect. Proc. R. Soc. Lond. B 266, 735–740.

Hurst, L., McVean, G., 1996. Clade selection, reversible evolution and the persistence of selfish elements: the evolutionary dynamics of cytoplasmic incompatibility. Proc. R. Soc. Lond. B 263, 97–104.

Ikeda, T., Ishikawa, H., Sasaki, T., 2003. Infection density of *Wolbachia* and level of cytoplasmic incompatibility in the Mediterranean flour moth, *Ephestia kuehniella*. J. Invert. Pathol. 84, 1–5.

Jaenike, J., Elizondo, S., Delahanty, N., 2006. Mushrooms as potential sources of *Wolbachia*-curing antibiotics. Evol. Ecol. Res. 8, 1291–1299.

Jaenike, J., Parker, E.D., Selander, R.K., 1980. Clonal niche structure in the parthenogenetic earthworm *Octolasion tyrtaeum*. Am. Nat. 166, 196–205.

Kittayapong, P., Jamnongluk, W., Thipaksorn, A., Milne, J., Sindhusake, C., 2003. *Wolbachia* infection complexity among insects in the tropical rice-field community. Mol. Ecol. 12, 1049–1060.

Kondo, N., Shimada, M., Fukatsu, T., 2005. Infection density of *Wolbachia* endosymbiont affected by co-infection and host genotype. Biol. Lett. 1, 488–491.

Kopelman, A., Chabora, P., 1986. Aspects of the reproductive-biology of *Leptopilina boulardi* (Hymenoptera, Eucoilidae). Ann. Entomol. Soc. Am. 79, 808–813.

Kremer, N., Charif, D., Henri, H., Bataille, M., Prevost, G., Kraaijeveld, K., et al., 2009. A new case of *Wolbachia* dependence in the genus *Asobara*: evidence for parthenogenesis induction in *Asobara japonica*. Heredity (in press). http://dx.doi.org/10.1038/hdy.2009.63.

Labrosse, C., Carton, Y., Dubuffet, A., Drezen, J.M., Poirie, M., 2003. Active suppression of *D. melanogaster* immune response by long gland products of the parasitic wasp *Leptopilina boulardi*. J. Insect Physiol. 49, 513–522.

Maynard Smith, J., 1978. The Evolution of Sex. Cambridge University Press, Cambridge.

McCall, K., 2004. Eggs over easy: cell death in the *Drosophila* ovary. Dev. Biol. 274, 3–14.

McGraw, E.A., Merritt, D.J., Droller, J.N., O'Neill, S.L., 2002. *Wolbachia* density and virulence attenuation after transfer into a novel host. Proc. Natl. Acad. Sci. USA 99, 2918–2923.

Min, K.T., Benzer, S., 1997. *Wolbachia*, normally a symbiont of *Drosophila*, can be virulent, causing degeneration and early death. Proc. Natl. Acad. Sci. USA 94, 10792–10796.

Mitsui, H., Van Achterberg, K., Nordlander, G., Kimura, M., 2007. Geographical distributions and host associations of larval parasitoids of frugivorous Drosophilidae in Japan. J. Nat. Hist. 41, 1731–1738.

Mochiah, M., Ngi-Song, A., Overholt, W., Stouthamer, R., 2002. *Wolbachia* infection in *Cotesia sesamiae* (Hymenoptera: Braconidae), causes cytoplasmic incompatibility: implications for biological control. Biol. Control 25, 74–80.

Moran, N.A., Baumann, P., 2000. Bacterial endosymbionts in animals. Curr. Opin. Microbiol. 3, 270–275.

Mouton, L., 2003. Diversité et Densité Bactériennes dans les Symbioses à Infections Multiple: Régulation et Effets sur l'Hôte. Unversité de Lyon 1, Lyon, France Cas des Associations *Wolbachia*-Insectes [PhD thesis].

Mouton, L., Dedeine, F., Henri, H., Bouletreau, M., Profizi, N., Vavre, F., 2004. Virulence multiple infections and regulation of symbiotic population in the *Wolbachia-Asobara tabida* symbiosis. Genetics 168, 181–189.

Mouton, L., Henri, H., Bouletreau, M., Vavre, F., 2003. Strain-specific regulation of intracellular *Wolbachia* density in multiply infected insects. Mol. Ecol. 12, 3459–3465.

Mouton, L., Henri, H., Bouletreau, M., Vavre, F., 2005. Multiple infections and diversity of cytoplasmic incompatibility in a haplodiploid species. Heredity 94, 187–192.

Mouton, L., Henri, H., Bouletreau, M., Vavre, F., 2006. Effect of temperature on *Wolbachia* density and impact on cytoplasmic incompatibility. Parasitology 132, 49–56.

Mouton, L., Henri, H., Charif, D., Bouletrea, M., Vavre, F., 2007. Interaction between host genotype and environmental conditions affects bacterial density in *Wolbachia* symbiosis. Biol. Lett. 3, 210–213.

Nordlander, G., 1980. Revision of the genus *Leptopilina* Forster 1869, with notes on the status of some other genera (Hymenoptera, Cynipoidea: Eucoilidae). Entomol. Scand. 11, 428–453.

Oliver, K., Moran, N., Hunter, M., 2006. Costs and benefits of a superinfection of facultative symbionts in aphids. Proc. R. Soc. B 273, 1273–1280.

O'Neill, S.L., Hoffmann, A.A., Werren, J.H., 1997. Influential Passengers: Inherited Microorganisms and Arthropod Reproduction. Oxford University Press, Oxford.

Pannebakker, B.A., Garrido, N.R., Zwaan, B.J., van Alphen, J.M., 2008. Geographic variation in host-selection behaviour in the *Drosophila* parasitoid *Leptopilina clavipes*. Entomol. Exp. Appl. 127, 48–54.

Pannebakker, B.A., Beukeboom, L.W., van Alphen, J.J.M., Brakefield, P.M., Zwaan, B.J., 2004a. The genetic basis of male fertility in relation to haplodiploid reproduction in *Leptopilina clavipes* (Hymenoptera: Figitidae). Genetics 168, 341–349.

Pannebakker, B.A., Loppin, B., Elemans, C.P.H., Humblot, L., Vavre, F., 2007. Parasitic inhibition of cell death facilitates symbiosis. Proc. Natl Acad. Sci. U.S.A. 104, 213–215.

Pannebakker, B.A., Pijnacker, L.P., Zwaan, B.J., Beukeboom, L.W., 2004b. Cytology of *Wolbachia*-induced parthenogenesis in *Leptopilina clavipes* (Hymenoptera: Figitidae). Genome 47, 299–303.

Pannebakker, B.A., Schidlo, N.S., Boskamp, G.J.F., Dekker, L., Van Dooren, T.J.M., et al., 2005. Sexual functionality of *Leptopilina clavipes* (Hymenoptera : Figitidae), after reversing *Wolbachia*-induced parthenogenesis. J. Evol. Biol. 18, 1019–1028.

Pannebakker, B.A., Zwaan, B.J., Beukeboom, L.W., van Alphen, J.J.M., 2004c. Genetic diversity and *Wolbachia* infection of the *Drosophila* parasitoid *Leptopilina clavipes* in western Europe. Mol. Ecol. 13, 1119–1128.

Perlman, S.J., Kelly, S.E., Zchori-Fein, E., Hunter, M.S., 2006. Cytoplasmic incompatibility and multiple symbiont infection in the ash whitefly parasitoid, *Encarsia inaron*. Biol. Control 39, 474–480.

Perrot-Minnot, M., Cheval, B., Migeon, A., Navajas, M., 2002. Contrasting effects of *Wolbachia* on cytoplasmic incompatibility and fecundity in the haplodiploid mite *Tetranychus urticae*. J. Evol. Biol. 15, 808–817.

Pijls, J., van Steenbergen, H.J., van Alphen, J.J.M., 1996. Asexuality cured: the relations and differences between sexual and asexual *Apoanagyrus diversicornis*. Heredity 76, 506–513.

Pintureau, B., Chapelle, L., Delobel, B., 1999. Effects of repeated thermic and antibiotic treatments on a *Trichogramma* (Hym., Trichogrammatidae) symbiont. J. Appl. Entomol. Z. Ang. Entomol. 123, 473–483.

Pintureau, B., Lassabliere, F., Daumal, J., Grenier, S., 2002. Does a cyclic natural thermal cure occur in *Wolbachia*-infected *Trichogramma* species? Ecol. Entomol. 27, 366–372.

Poinsot, D., Charlat, S., Mercot, H., 2003. On the mechanism of *Wolbachia*-induced cytoplasmic incompatibility: confronting the models with the facts. Bioessays 25, 259–265.

Rances, E., Voronin, D., Tran-Van, V., Mavingui, P., 2008. Genetic and functional characterization of the type IV secretion system in *Wolbachia*. J. Bacteriol. 190, 5020–5030.

Reumer, B., Kraaijeveld, K., van Alphen, J., 2007. Selection in the absence of males does not affect male–female conflict in the parasitoid wasp *Leptopilina clavipes* (Hymenoptera: Figitidae). J. Insect Physiol. 53, 994–999.

Rousset, F., Bouchon, D., Pintureau, B., Juchault, P., Solignac, M., 1992. *Wolbachia* endosymbionts responsible for various alterations of sexuality in arthropods. Proc. R. Soc. Lond. B 250, 91–98.
Rousset, F., Braig, H., O'Neill, S., 1999. A stable triple *Wolbachia* infection in *Drosophila* with nearly additive incompatibility effects. Heredity 82, 620–627.
Sachs, J.L., Mueller, U.G., Wilcox, T.P., Bull, J.J., 2004. The evolution of cooperation. Q. Rev. Biol. 79, 135–160.
Salzberg, S., Hotopp, J., Delcher, A., Pop, M., Smith, D., Eisen, M., et al., 2005. Serendipitous discovery of *Wolbachia* genomes in multiple *Drosophila* species. Genome Biol. 6, R23.
Schidlo, N.S., Pannebakker, B.A., Zwaan, B.J., Beukeboom, L.W., Van Alphen, J.J.M., 2002. Curing thelytoky in the *Drosophila* parasitoid *Leptopilina clavipes* (Hymenoptera: Figitidae). Proc. Exp. Appl. Entomol. Neth. Entomol. Soc. 13, 93–96.
Schilthuizen, M., Stouthamer, R., 1998. Distribution of *Wolbachia* among the guild associated with the parthenogenetic gall wasp *Diplolepis rosae*. Heredity 81, 270–274.
Simon, J.C., Delmotte, F., Rispe, C., Crease, T., 2003. Phylogenetic relationships between parthenogens and their sexual relatives: the possible routes to parthenogenesis in animals. Biol. J. Linn. Soc. 79, 151–163.
Stouthamer, R., Kazmer, D.J., 1994. Cytogenetics of microbe-associated parthenogenesis and its consequences for gene flow in *Trichogramma* wasps. Heredity 73, 317–327.
Stouthamer, R., Luck, R.F., 1993. Influence of microbe-associated parthenogenesis on the fecundity of *Trichogramma deion* and *T. pretiosum*. Entomol. Exp. Appl. 67, 183–192.
Stouthamer, R., Breeuwer, J.A.J., Hurst, G.D.D., 1999. *Wolbachia pipientis*: Microbial manipulator of arthropod reproduction. Ann. Rev. Microbiol. 53, 71–102.
Stouthamer, R., Luck, R.F., Hamilton, W.D., 1990. Antibiotics cause parthenogenetic Trichogramma (Hymenoptera/Trichogrammatidae) to revert to sex. Proc. Natl. Acad. Sci. USA 87, 2424–2427.
Suomalainen, E., Saura, A., Lokki, J., 1987. Cytology and Evolution in Parthenogenesis. CRC Press, Boca Raton, Florida.
Teixeira, L., Ferreira, Á., Ashburner, M., 2008. The bacterial symbiont *Wolbachia* induces resistance to RNA viral infections in *Drosophila melanogaster*. PLoS Biol. 6, e2.
Thompson, J.N., 1988. Variation in interspecific interactions. Ann. Rev. Ecol. Syst. 19, 65–87.
Tram, U., Ferree, P.A., Sullivan, W., 2003. Identification of *Wolbachia*–host interacting factors through cytological analysis. Microbes Infect. 5, 999–1011.
Tram, U., Fredrick, K., Werren, J.H., Sullivan, W., 2006. Paternal chromosome segregation during the first mitotic division determines *Wolbachia*-induced cytoplasmic incompatibility phenotype. J. Cell Sci. 119, 3655–3663.
Turelli, M., 1994. Evolution of incompatibility-inducing microbes and their hosts. Evolution 48, 1500–1513.
Varaldi, J., Fouillet, P., Ravallec, M., Lopez-Ferber, M., Bouletreau, M., Fleury, F., 2003. Infectious behavior in a parasitoid. Science 302, 1930.
Vautrin, E., 2008. Les Infections Multiples dans les Symbioses Insectes / Bactéries: Évolution Intra-Hôte et Conséquences Populationnelles [PhD thesis]. Université de Lyon 1, Lyon, France.
Vautrin, E., Vavre, F., 2009. Interactions between vertically transmitted symbionts: cooperation or conflict? Trends Microbiol. 17, 95–99.
Vautrin, E., Charles, S., Genieys, S., Vavre, F., 2007. Evolution and invasion dynamics of multiple infections with *Wolbachia* investigated using matrix based models. J. Theor. Biol. 245, 197–209.
Vautrin, E., Genieys, S., Charles, S., Vavre, F., 2008. Do vertically transmitted symbionts co-existing in a single host compete or cooperate? A modelling approach. J. Evol. Biol. 21, 145–161.

Vavre, F., Dedeine, F., Quillon, M., Fouillet, P., Fleury, F., Bouletreau, M., 2001. Within-species diversity of *Wolbachia*-induced cytoplasmic incompatibility in haplodiploid insects. Evolution 55, 1710–1714.

Vavre, F., Fleury, F., Lepetit, D., Fouillet, P., Bouletreau, M., 1999. Phylogenetic evidence for horizontal transmission of *Wolbachia* in host–parasitoid associations. Mol. Biol. Evol. 16, 1711–1723.

Vavre, F., Fleury, F., Varaldi, J., Fouillet, P., Bouletreau, M., 2000. Evidence for female mortality in *Wolbachia*-mediated cytoplasmic incompatibility in haplodiploid insects: Epidemiologic and evolutionary consequences. Evolution 54, 191–200.

Vavre, F., Fleury, F., Varaldi, J., Fouillet, P., Bouletreau, M., 2002. Infection polymorphism and cytoplasmic incompatibility in Hymenoptera–*Wolbachia* associations. Heredity 88, 361–365.

Vavre, F., Fouillet, P., Fleury, F., 2003. Between- and within-host species selection on cytoplasmic incompatibility-inducing *Wolbachia* in haplodiploids. Evolution 57, 421–427.

Vavre, F., Kremer, N., Pannebakker, B.A., Loppin, B., Mavingui, P., 2009. Is symbiosis evolution influenced by the pleiotropic role of programmed cell death in immunity and development? In: Miller, Bourtzis, (Eds.), Insect Symbiosis 3. CRC Press, Boca Raton, Florida, pp. 57–75.

Veneti, Z., Clark, M.E., Karr, T.L., Savakis, C., Bourtzis, K., 2004. Heads or tails: host–parasite interactions in the *Drosophila–Wolbachia* system. Appl. Environ. Microbiol. 70, 5366–5372.

Vet, L.E.M., 1983. Host-habitat location through olfactory cues by *Leptopilina clavipes* (Hartig) (Hym: Eucoilidae), a parasitoid of fungivorous *Drosophila*: the influence of conditioning. Neth. J. Zool. 33, 225–248.

Weeks, A.R., Breeuwer, J.A.J., 2001. *Wolbachia*-induced parthenogenesis in a genus of phytophagous mites. Proc. R. Soc. Lond. B 268, 2245–2251.

Wernegreen, J.J., 2002. Genome evolution in bacterial endosymbionts of insects. Nat. Rev. Genet. 3, 850–861.

Werren, J.H., 1998. *Wolbachia* and speciation. In: Howard, (Ed.), Endless Forms: Species and Speciation. Oxford University Press, New York, pp. 245–260.

Werren, J., Baldo, L., Clark, M., 2008. *Wolbachia*: master manipulators of invertebrate biology. Nat. Rev. Microbiol. 6, 741–751.

Werren, J.H., Zhang, W., Guo, L.R., 1995. Evolution and phylogeny of *Wolbachia*: reproductive parasites of arthropods. Proc. R. Soc. Lond. B 261, 55–63.

West, S.A., Cook, J., Werren, J.H., Godfray, H., 1998. *Wolbachia* in two insect host–parasitoid communities. Mol. Ecol. 7, 1457–1465.

Zchori-Fein, E., Faktor, O., Zeidan, M., Gottlieb, Y., Czosnek, H., Rosen, D., 1995. Parthenogenesis-inducing microorganisms in *Aphytis* (Hymenoptera, Aphelinidae). Insect Mol. Biol. 4, 173–178.

Zchori-Fein, E., Roush, R.T., Hunter, M.S., 1992. Male production induced by antibiotic treatment in *Encarsia formosa* (Hymenoptera: Aphelinidae), an asexual species. Experientia 48, 102–105.

Zychlinsky, A., Sansonetti, P., 1997. Apoptosis in bacterial pathogenesis. J. Clin. Invest. 100, 493–496.

van den Assem, J., 1969. Reproductive behaviour of *Pseudocoila bochei* (Hymenoptera: Cynipidae). Neth. J. Zool. 19, 641–648.

CHAPTER 13

A Virus-Shaping Reproductive Strategy in a *Drosophila* Parasitoid

Julien Varaldi,* **Sabine Patot,*** **Maxime Nardin,***
and **Sylvain Gandon**†

Contents

Abstract

Insect parasitoids are often infected with heritable viruses. Some of them, such as polydnaviruses, have evolved toward an obligatory relationship with the parasitoid because they are necessary to protect the parasitoid egg from the host immune reaction.

* Laboratoire de Biométrie et Biologie Evolutive, Université Lyon 1; CNRS; UMR 5558, 43 boulevard du 11 novembre 1918, F-69622 Villeurbanne, France
† Centre d'Ecologie Fonctionnelle et Evolutive (CEFE) – UMR 5175, 1919 route de Mende, F-34293 Montpellier cedex 5, France

Advances in Parasitology, Volume 70
ISSN 0065-308X, DOI: 10.1016/S0065-308X(09)70013-2

However, recent and past discoveries have revealed the presence of facultative inherited viruses in parasitoids for which no clear phenotypic effect was observed. In this chapter, we present how such an inherited virus was recently discovered in the *Drosophila* parasitoid, *Leptopilina boulardi*. We show that this virus is responsible for an increase in the superparasitism tendency of the infected females. This alteration is beneficial for the virus, since superparasitism conditions permit the horizontal transmission of the virus. We review theoretical developments suggesting that this leads to a conflict of interest between the parasitoid and the virus. The direct and indirect influence of the virus on several other fitness traits has also been studied both empirically and theoretically, in particular the egg load. Finally, because the frequency of horizontal transmission is a crucial parameter for the evolution of the superparasitism manipulation, we present an attempt to select the virus for high or low manipulation intensity.

13.1. INTRODUCTION

During the last decades, it has been discovered that host–parasitoid interactions are often directly or indirectly influenced by symbiotic organisms, such as bacteria and viruses. For instance, the symbiotic bacteria *Hamiltonella* infecting aphids confers resistance against parasitoid attack (Oliver et al., 2003, 2007). Although such phenomenon has not been documented to date in *Drosophila* spp., it surely indicates that symbionts have to be taken into account when studying *Drosophila*–parasitoid interactions. This idea finds further support in the recent literature, since symbiotic *Wolbachia* infecting *Drosophila melanogaster* have been found to confer resistance against viral pathogens. It is worth mentioning that this result has been obtained independently by two research groups (Hedges et al., 2008; Teixeira et al., 2008). Parasitoids have also evolved intimate associations with symbiotic bacteria (reviewed in Chapter 12 for *Drosophila*–parasitoids) deeply affecting their reproductive behavior. However, one of the most outstanding mutualistic relationships in parasitoids involves viral particles. Indeed, seven monophyletic subfamilies of Braconidae (the microgastroid complex), and two subfamilies of Ichneumonidae are associated with polydnaviruses (PDV), which replicates in females' reproductive organs without any detrimental effects to the wasp (Glatz et al., 2004). PDVs are injected into the parasitoid host during oviposition and alter host physiology thus allowing parasitoid larvae to circumvent the host immune reaction. It is likely that PDV symbiosis have arisen three times independently (giving rise to Bracovirus, Espagne et al., 2004; Ichnovirus and to a new genus recently proposed, Lapointe et al., 2007), afterward leading to longstanding coevolution between the ancestral viruses and the parasitoids.

Nowadays, all wasp species of these groups have obligate associations with PDVs. PDVs have completely lost their infectious capacity and are only vertically transmitted as an autosomal locus because of their integration within the wasp genome. The origin of PDVs have been debated since they were discovered. Recently, the ancestral bracovirus has been identified as a nudivirus, based on the expression of a large set of nudivirus related genes in the braconid wasp ovaries (Bezier et al., 2009). The ancestral state of the other PDVs is still to be determined. Although PDVs have not been found in *Drosophila* parasitoids, some proteins showing viral-like structure are also injected into the host haemolymph by *Leptopilina* spp. (Dupas et al., 1996; Rizki and Rizki, 1990). Although they do not contain deoxyribonucleic acid (DNA; as opposed to PDVs), these virus-like particles (VLPs) also circumvent the host immune reaction and may have a viral evolutionary origin. To understand the origin and mechanisms of virus or VLP incorporation into the wasps' genomes, it may be useful to study nowadays infectious viruses that are able to infect parasitoids. In *Leptopilina boulardi*, we have found that some females are infected by an inherited virus that manipulates the behavior of the wasp (Varaldi et al., 2003, 2006b). This virus, called LbFV for *L. boulardi* filamentous virus, forces the infected females to accept to lay their eggs in already parasitized hosts (a behavior called superparasitism). This behavioral manipulation benefits to the virus spread since superparasitism allows its horizontal transmission (transmission between unrelated parasitoid lineages). The peculiar transmission mode of this virus allows it to maintain and reach high frequencies in natural populations. The present chapter reviews the different features of this parasitoid/virus association.

13.2. MAIN EFFECT AND TRANSMISSION OF LbFV

As mentioned in previous chapters, all *Drosophila* parasitoids are solitary parasitoids, meaning that one *Drosophila* larva allows the development of a single parasitoid, whatever the number of parasitoid eggs. Females are usually able to recognize parasitized from unparasitized hosts (host discrimination) and normally avoid laying eggs in already parasitized host. If a female oviposits in a parasitized host, a behavior called superparasitism, parasitoid larval competition ends up in the death of all but one larva. Usually the second larva is most likely to be out-competed and its survival depends on the interval between the first and second ovipositions (van Alphen and Visser, 1990). If a parasitoid female accepts several times the same host (a behavior called self-superparasitism), she will waste some eggs since brothers and sisters will compete for the possession of the host until all but one die. Superparasitism is thus expected to be strongly counter selected in most ecological conditions. One remarkable feature of *L. boulardi* was that in some populations, females showed a huge tendency to superparasitize, while in others most females laid only one egg per host. In the related *L. heterotoma*, however, few superparasitism was observed (Varaldi et al., 2005b). In *L. boulardi*, we were thus able to derive stable ''nonsuperparasitizing'' lines (NS) and ''superparasitizing'' lines (S). From these lines, we studied the genetic determinism. Surprisingly, the variations in the superparasitism phenotype were strictly maternally inherited: whatever the nuclear genotype, females adopted the phenotype of their mother. Furthermore, when both S and NS lines laid their eggs inside the same host, in the case where NS lines won the within-host competition, the emerging (female) offspring did adopt the ''superparasitizing'' phenotype, despite the NS phenotype of its line of origin (Varaldi et al., 2003)! All is happening as if some unknown infectious element was causing the ''superparasitizing'' phenotype and was passed from S-infected lines to NS-uninfected lines during the short time they coexisted inside the *Drosophila* larva. The newly acquired S phenotype was stably transmitted over generations (Varaldi et al., 2006b). The infectious nature of the S-inducing element was further confirmed by injecting solutions derived from S individuals into *Drosophila* larvae parasitized by NS females. Solutions of S females proved its ability to induce the S phenotype on the emerging parasitoid females (originating from an NS line), whereas NS control injections did not induce any behavioral change (Varaldi et al., 2006b). The hypothesis that the causative agent was a bacterium was tested and clearly ruled out using antibiotic treatments (Varaldi et al., 2006b). The nature of the infectious element was finally determined by electron microscopy investigations inside the ovaries of *L. boulardi* females. It was evident that in S lines, a virus was replicating in cells bordering the lumen of the oviduct, contrary to NS females (Varaldi et al., 2003, 2006b). Based on its morphology, the superparasitism-inducing virus was called LbFV (for *Leptopilina boulardi* Filamentous Virus). The virus LbFV is thus vertically transmitted through the female line, and also horizontally in conditions of superparasitism.

To date, the precise means of transmissions are not known, but our working hypothesis is that the virus is injected in addition to the egg into the host during oviposition and that it infects the emerging parasitoid during its larval life (during which the parasitoid consumes the infected host hemocoel). If the infected parasitoid develops alone, then vertical transmission occurs (with a very efficient rate, near 100% under laboratory conditions), while if superparasitism occurs, horizontal transmission may occur. We suspect that the efficiency of the horizontal transmission depends critically on the delay between successive ovipositions: if an S female superparasitizes soon after an NS female has laid her egg, then the efficiency will be high, while if this delay is important the efficiency drops (Varaldi et al., 2006c). Accordingly, when we inject extracts of

S ovaries inside *Drosophila* larvae previously parasitized by NS females, the efficiency of the contamination is high if the delay is low (<24 h: 44% ($n = 9$)), and drops to zero when we increased the delay (24 – 48 h, 0% ($n = 17$); 48–72 h: 0% ($n = 21$) temperature: 26 °C).

LbFV has been discovered using electron microscopy and thus we lacked any genomic data. This precludes from identifying its phylogenetic position and from developing molecular tools, such as markers. Since LbFV could be either a DNA or ribonucleic acid (RNA) virus, we focused our attention on the identification of viral messenger RNA (mRNA; because both viral types should produce mRNAs). We performed a suppressive subtractive hybridization (SSH) between two lines sharing the same genotypic background but differing in their superparasitism behavior. This work permitted to identify an 809 base pairs (bp) mRNA that was S specific. From this mRNA sequence, we derived a simple polymerase chain reaction (PCR) test that showed amplification on all 14 independent S lines whereas no amplification was observed for all 11 independent NS lines, starting with DNA extracts as templates (Patot et al., 2009). This perfect correlation between superparasitism phenotype and PCR-amplification validates the viral origin of this sequence. Furthermore, it shows that LbFV has at least an intermediate DNA step during its replication cycle or that, more likely, LbFV has a DNA genome. This is consistent with the electron microscopy investigations showing apparent viral replication within the nuclei of the cells. This work (Patot et al., 2009) also indicates that the virus reaches very high prevalence in natural populations (around 70% in both sampled populations in the South of France), despite the fact that the penetrance of the extended-phenotype was incomplete (only 80% of the infected females expressed signs of behavioral modification).

13.3. ADAPTIVE SIGNIFICANCE OF SUPERPARASITISM ALTERATION: A MODELIZATION APPROACH

The vertical transmission of the virus implies that the virus and the parasitoid share some fitness components (they both benefit from female fecundity). It thus remains unclear whether this induced superparasitism behavior is actually adaptive for the virus (Gandon, 2005; Varaldi et al., 2003). To demonstrate the adaptive nature of the alteration of the parasitoid behavior one must show that a virus increasing superparasitism can invade a virus population that does not alter the behavior of its parasitoid host. In other words, one must demonstrate that the evolutionarily stable strategy (ESS) of superparasitism for the virus is higher than the ESS superparasitism for the parasitoid (in the absence of the virus). To address this question, we developed a model that allows to analyze both the dynamics and the evolution of a population of parasitoids (a proovigenic and solitary species) parasitizing a population of hosts (Gandon et al., 2006). This model includes the potential benefit of superparasitism (the possibility that parasitoid larvae developing in an already parasitized host win the within-host competition) and both the classical costs of superparasitism (the costs of time and the cost of eggs). We first used this model in the absence of any virus, to predict the fate of a mutant parasitoid with superparasitism strategy s^* appearing in a parasitoid population dominated by a resident with strategy s (where s indicates the rate of acceptance of parasitized hosts). As expected, the model predicts that the ESS of superparasitism is zero when the probability to win the within-host competition (c) is low but increases with an increase in c. This further confirmed previous models showing the potential adaptive value of superparasitism under conditions of host scarcity (van Alphen and Visser, 1990).

We extended the model to include a virus, based on LbFV biology. When females are infected, it is assumed that the parasitoid behavior is strictly under the control of the virus. In other words, the rate of acceptance of parasitized hosts of an infected female is no more s (the superparasitism strategy when the female is uninfected), but instead σ which is a feature of the virus. The virus is vertically transmitted with a rate of tv (<1), and will gain extra routes of transmission via the potential horizontal transmission that may occur between a larva infected with the virus and an uninfected larva (with probability τ_h). To allow direct competition between viral strains, it is assumed that a viral strain can replace another one when they compete inside the same *Drosophila* larva with a probability ε. However, no multi-infections at the adult stage are allowed. The model can be used to derive an expression of the fitness of a mutant virus with a strategy σ^* appearing in a population dominated by a resident virus with strategy σ, at the epidemiologic equilibrium set by the resident virus and the strategy s adopted by the host. Note that here, only the virus is allowed to evolve, not the parasitoid (s is fixed). In a first part, we fixed $\varepsilon = 0$, that is, a viral strain is not able to replace a resident viral strain in competition within *Drosophila* larvae. The results indicate that the ESS superparasitism is always higher for the virus than that observed for the parasitoid (allowed to evolve to its optimal strategy in the absence of the virus) demonstrating the adaptive value of the behavioral modification from the virus point of view. The virus is always selected to increase the natural superparasitism tendency of the parasitoid. The presence of the virus thus induces an evolutionary conflict of interest between the parasitoid and the virus on superparasitism behavior (Fig. 13.1A). The intensity of the evolutionary conflict is even increased if both the virus and the parasitoid are allowed to coevolve: after coevolution, uninfected females (that are produced even in infected populations, due to imperfect vertical

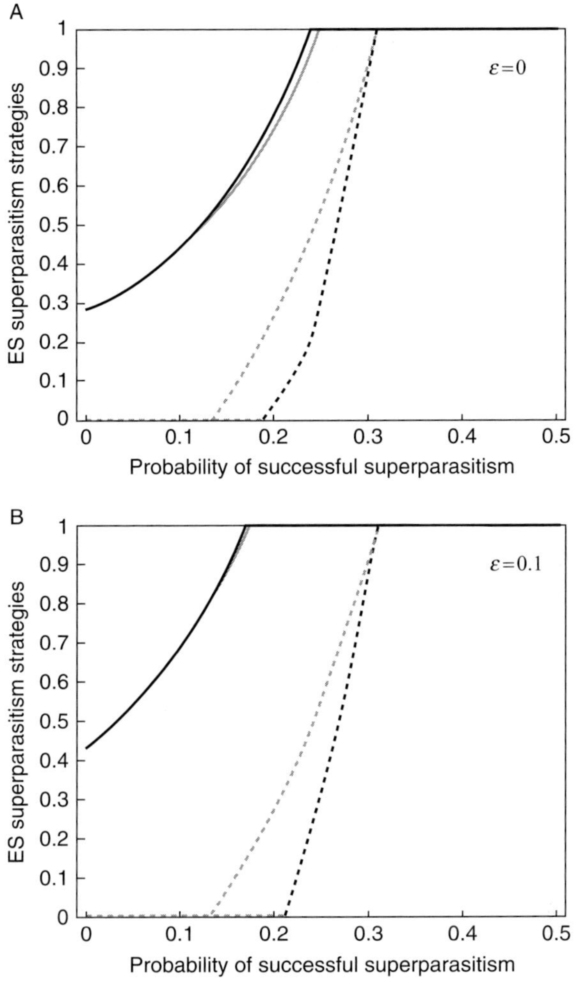

FIGURE 13.1 Evolutionarily stable superparasitism strategies of the virus (solid lines) and the parasitoid (dotted lines) versus the probability of successful superparasitism. In (A) $\varepsilon = 0$, in (B) $\varepsilon = 0.1$. The gray lines indicate a situation where the parasitoid does not coevolve with the virus. The black lines indicate the coevolutionary stable strategies of the virus and the parasitoid. Parameter values: $d = 0.2$; $e = 0.2$; $m = 0.1$; $t1 = 0$; $t2 = 0.1$; $a = 0.01$; $x_{tot} = 100$; tv $= 0.95$; th $= 0.75$; $e_{max} = 15$. (See Gandon et al., 2006 for details on parameter values).

transmission) should less superparasitize than uninfected females that did not coevolve with the virus (Fig. 13.1A). This shows that the presence of the virus in a population should indirectly modify the ESS of a trait for uninfected females. When we allowed direct competition between viral strains within *Drosophila* larvae ($\varepsilon > 0$), we found that the virus is even selected for much higher superparasitism strategies, thus strongly increasing the conflict of interest between the parasitoid and the virus (Fig. 13.1B). Coevolution between the virus and the parasitoid further increased the conflict of interest as has been found with $\varepsilon = 0$ (Fig. 13.1B). These results clearly show that increasing the superparasitism strategy of the parasitoid is an adaptive strategy from the virus point of view (whatever ε). To say it differently, there is a conflict of interest between the virus and the parasitoid on superparasitism behavior. However, the intensity of the conflict of interest depends critically on the ability of mutant virus strains to replace resident strains inside *Drosophila* larvae (ε) and also on coevolutionary processes.

13.4. EFFECT OF LbFV ON OTHER PHENOTYPIC TRAITS

It may be argued that *L. boulardi* females infected with LbFV adopt an aberrant behavior without any adaptive significance (neither for the host nor for the virus), because the virus disrupts indifferently several cognitive and possibly physiological properties (Poulin, 1995). However, it has been found that LbFV infection has no effect on parasitoid survival of females (but a negative impact on male survival), and only a slight negative impact on size (tibia length is reduced by 2%), and developmental speed (increased by 3% for both sexes). Nevertheless, the overall locomotor activity of infected females is reduced by 45% while no effect was detected on males. Interestingly, we found that egg load was even increased for infected females (+11%) compared to uninfected females (Varaldi et al., 2005a). Overall, the effect of LbFV on various traits is relatively moderate (except for locomotor activity) or even positive (egg load; Table 13.1). This surprising beneficial effect on egg load will be discussed in detail in the next section.

The influence of LbFV on several behaviors (apart from superparasitism) has also been investigated (Varaldi et al., 2006a). The behavioral components studied included sexual communication, circadian rhythms, ability of females to detect odors of hosts and trajectometric parameters of foraging females. None of these behavioral repertoires seemed to be perturbed by LbFV infection, demonstrating a specific action of LbFV on superparasitism behavior (Table 13.1).

How does the virus manage to have such a specific action? *L. boulardi* females need to pierce the skin of the host larvae with their ovipositor to detect chemical cues associated with a previous infestation. In effect, the ovipositor of parasitoids harbors chemoreceptors that are probably (all or some of them) involved in host discrimination. Their distribution and putative function has been investigated in great details on the related species *L. heterotoma* (van Lenteren et al., 2007). This species also needs to

TABLE 13.1 Effect of LbFV on several general traits and behavioral traits

		LbFV effect (%)	Ref
Physiology	Survival	0	1
	Size	–2	1
	Development speed	+3	1
	Egg load	+11	1
	Sex ratio	0	1
	Locomotor activity	–45	1
Behavior	Superparasitism	+++	3
	Circadian rhythm	0	2
	Perception of host odors by females	0	2
	Female searching paths	0	2
	Female interspecific discrimination	0	4
	Male detection of pheromones	0	2

Notes: 1: Varaldi et al. (2005a); 2: Varaldi et al. (2006a); 3: Varaldi et al. (2003); 4: this study.

pierce the skin of the host to detect the presence of a previous infestation. The authors found seven chemoreceptors at the tip of the ovipositor that come into contact with the *Drosophila* haemolymph during host probing. One single chemoreceptor was found on the unpaired valve, and three on each paired valve. Each chemoreceptor is innervated with six neurons. One tempting hypothesis would be that LbFV injures these neurons involved in the transmission of the nervous flux, either through cell lysis or through manipulation of gene expression. However, based on the work done in *L. heterotoma* (van Lenteren et al., 2007), it is unlikely that the gustatory receptor situated on the unpaired valve is the target of LbFV action since electrophysiologic investigations suggest that it is not involved in host discrimination. The perception of previous infestations is thus probably assured by some or all of the remaining six chemoreceptors present on the paired valve, and LbFV may interfere with some of them.

In addition to discriminating between parasitized and unparasitized hosts, female parasitoids usually make selective host choices when several potential related host species are available in the environment. The value of these different host species may differ in terms of parasitoid fitness and we expect that female parasitoids discriminate among them by preferentially laying their offspring in the most profitable host species. We would also predict that the virus should not interfere with this decision since both the virus and the parasitoid has interest in developing in a good host. However, the sensory capacities of the ovipositor are also probably involved in this decision process. In order to test

(1) whether *L. boulardi* females discriminate between good and bad host species, and (2) whether LbFV interferes with this ovipositor-based decision, we conducted a choice experiment in which we proposed a mix of *D. melanogaster* and *D. subobscura* to *L. boulardi* females. Both *Drosophila* spp. can be found in the same microhabitat, although *D. subobscura* is less frugivorous than *D. melanogaster*. While *D. melanogaster* offers a very good host for the development of *L. boulardi*, *D. subobscura* is reputed to be an unfavorable host (Carton et al., 1986). Indeed, based on the protocol described in Varaldi et al. (2005a), we estimated the preimaginal survival (probability of an egg to reach adulthood) of *L. boulardi* (strain Antibes) as 0.74 ± 0.09 (mean $\pm$ standard error, $n = 10$) on *D. melanogaster* and only 0.14 ± 0.09 ($n = 12$) on *D. subobscura* (at 25 °C). To test whether *L. boulardi* discriminates between *Drosophila* spp. and whether LbFV interferes with this decision, we did the following experiment. Isolated *L. boulardi* females (either infected or not, but sharing the same nuclear background as in Varaldi et al., 2005a) were provided with a mix of larvae that hatched from 75 *D. melanogaster* and 75 *D. subobscura* eggs in standard rearing tubes (at 21 °C). Because *D. subobscura* eggs needs more time to hatch than D. *melanogaster* and *D. subobscura* larvae grow slower than *D. melanogaster*, we used *D. subobscura* eggs collected 24 h before *D. melanogaster eggs*. Consequently, at the time that we added the parasitoid female within the tube, *D. melanogaster* were 24 h old (time since eggs were deposited within tubes), whereas *D. subobscura* were 48 h old. In these conditions, the size of larvae of both species is comparable (Varaldi et al., 2005b). Females were allowed to parasitize the larvae for 24 h. Starting from the moment at which the females were added to the vials, they were transferred at 24 °C ($\pm$ 1 °C) until the end of the experiment (this temperature was chosen because it was suitable for *D. melanogaster*, low enough for *D. subobscura* and was high enough to prevent the diapause of *L. boulardi*). Sixteen replicates of each test modality were simultaneously conducted, in addition to 12 controls kept without parasitoids that were manipulated exactly in the same way as test tubes. For each of the 44 tubes, we scored the number and identity of the *Drosophila* reaching adulthood, and the number of emerging *L. boulardi* in test tubes.

The choice of each female was indirectly measured by first calculating the parasitoid-induced mortality on each *Drosophila* spp. which is a measure of the attack rate (a). Indeed, neither *D. subobscura* nor this strain of *D. melanogaster* are able to get rid of parasitoids by mounting an efficient immune reaction, thus the parasitoid-induced mortality is a good estimation of the proportion of *Drosophila* spp. that have been attacked and parasitized (in accordance with this hypothesis, we found no capsule on all emerging adult *Drosophila*). Attack rates against each *Drosophila* species were then defined for each parasitoid female as:

$$a_{\text{mel}}i = (\text{mean number of } D.\ melanogaster \text{ in controls} - \text{number of } D.\ melanogaster \text{ in test tube } i)/\text{mean number of } D.\ melanogaster \text{ in controls}$$

$$a_{\text{sub}}i = (\text{mean number of } D.\ subobscura \text{ in controls} - \text{number of } D.\ subobscura \text{ in test tube } i)/\text{mean number of } D.\ subobscura \text{ in controls}$$

Based on this, we derived a choice index calculated for each female. There was a slight difference in the survival of *D. melanogaster* and *D. subobscura*, since a mean of 58.72 *D. melanogaster* emerged from the controls (without parasitoid) versus 45.45 *D. subobscura* (out of 75 eggs initially deposited). We made the assumption that the mortality occurred before *Drosophila* eggs were exposed to the wasps (considering that the mortality occurred after the exposition to the wasp gave very similar results). Thus in each test tube, we estimated that the parasitoid female was provided approximately 58.72 + 45.45 = 104.17 *Drosophila* larvae, including 56% (58.72/104.17) *D. melanogaster*. For each female i, we calculated the whole rate of attack of both *Drosophila* spp. as:

$$a_{\text{global}}i = (a_{\text{mel}}i \times 58.72 + a_{\text{sub}}i \times 45.45)/104.18$$

To quantify the choice of each female, we derived an index, using an analogy with the calculation of the linkage disequilibrium in population genetics: on the one hand, we know the proportion of both *Drosophila* spp. in tubes (56% *D. melanogaster* and 44% *D. subobscura*) and, on the other hand, we know for each female *i* the whole attack rate ($a_{\text{global}}i$). Under the hypothesis h0 that wasp attacks are randomly distributed among *Drosophila* spp., then we expect for a female *i*:

$$\begin{aligned}
\text{Proportion of } D.\ mel \text{ attacked} &= 0.56 \times a_{\text{global}}i \\
\text{Proportion of } D.\ mel \text{ nonattacked} &= 0.56 \times (1 - a_{\text{global}}i) \\
\text{Proportion of } D.\ sub \text{ attacked} &= (1 - 0.56) \times a_{\text{global}}i \\
\text{Proportion of } D.\ sub \text{ nonattacked} &= (1 - 0.56) \times (1 - a_{\text{global}}i).
\end{aligned}$$

We can then calculate a deviation from this null model by subtracting for instance the proportion of *D. melanogaster* effectively attacked by the wasp in tube *i* with the expected proportion of attacks on *D. melanogaster* under h0:

$$c = a_{\text{mel}}i - 0.56 \times a_{\text{global}}i.$$

If this choice index (c) is positive then the female preferentially attacked *D. melanogaster*, whereas if this is negative, the female preferentially attacked *D. subobscura*. Because the range of variations for this index may vary between females (because their whole attack rates vary), we scaled it to range between –1 and +1 for all females (as is done for the calculation of D′ in population genetics) by dividing the choice index by its minimal value (if negative) or maximal value (if positive). Minimal and maximal values can be obtained this way:

$$c_{\min} = \min\,(0.56, 1 - a_{\text{global}}i) - 0.56 \times (1 - a_{\text{global}}i)$$
$$c_{\max} = \min\,(0.56, a_{\text{global}}i) - 0.56 \times (a_{\text{global}}i).$$

And finally, the scaled choice index can be obtained this way:

$$c^* = c/c_{\min} \quad \text{if } c < 0$$
$$= c/c_{\max} \quad \text{if } c > 0.$$

This scaled choice index (c^*) varies between –1 when the female concentrated to the best her attacks on *D. subobscura*, and +1 when the female concentrated to the best her attacks on *D. melanogaster*, and equals zero when the female do not show any preference.

The survival from egg to adulthood was also estimated for the offspring of each parasitoid female (each test vial):

$$\text{Offspring survival} = \text{number of parasitoid reaching adulthood in tube } i/$$
$$\text{number of } Drosophila\ (mel + sub) \text{ killed due to parasitism in tube } i$$

where number of *Drosophila* (*mel* + *sub*) killed due to parasitism in tube i = (mean number of *D. melanogaster* in controls – number of *D. melanogaster* in tube *i*) + (mean number of *D. subobscura* in controls – number of *D. subobscura* in tube *i*).

The number of *Drosophila* emerging in each vial is plotted in Figure 13.2A. *D. melanogaster* had a higher preimaginal survival than *D. subobscura* and the parasitoids induced a significant mortality on both *Drosophila* spp. indicating that both species were attacked by the parasitoids. The results indicate that *D. melanogaster* suffered a higher parasitoid-induced mortality than *D. subobscura*, suggesting a choice in the direction of the former. This trend was confirmed by the calculation of the choice index, which was significantly above 0 for both infection status (Student t test respectively 6.52 and 6.13 for uninfected and infected wasps, degrees of freedom (df) = 14 and 15, both $P < 0.00001$, see Fig. 13.2B). Importantly, the choice indexes obtained for infected or uninfected wasps were very similar ($t = 0.62$, 29, $P = 0.27$). First, the results show that *L. boulardi* is able to discriminate between both *Drosophila* spp. This can be due to the perception at distance of larval kairomones (odors produced by the larva) differences or to contact differences. Since both species were mixed within the tubes, *Drosophila* odors should also mix and it is unlikely that *L. boulardi* was able to use volatile components to

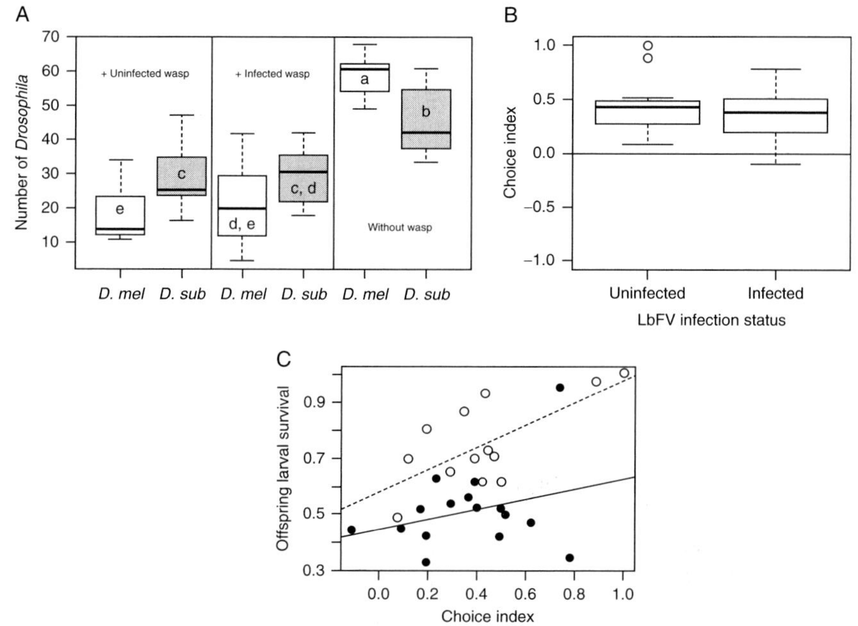

FIGURE 13.2 Host choice in *L. boulardi* and effect of LbFV. (A) Number of *Drosophila* emerging from vials containing initially 75 eggs of *D. melanogaster* (*D. mel*) and 75 eggs of *D. subobscura* (*D. sub*). Boxes with different letters (a–e) are significantly different at 0.05. (B) Choice index: negative values indicate preference for *D. subobscura*, whereas positive values indicate preference for *D. melanogaster*. (C) Open circles: uninfected, closed circles: infected. Relation between host choice and offspring survival.

discriminate in these conditions. Instead, the females probably used information obtained with their ovipositor, either by probing the medium close to the larvae or directly the larvae. There was a clear choice for *D. melanogaster* which is the most profitable host, suggesting an adaptive value for this trait. This conclusion was further supported by the global positive correlation between the choice index and the offspring larval survival ($F(1,27) = 6.91$, $P = 0.014$, Fig. 13.2C). The correlation was, however, only significant for uninfected wasps ($F(1,11) = 9.68$, $P < 0.01$ for uninfected and $F(1,14) = 1.35$, $P = 0.26$ for infected wasps) but the tendency was the same for both infection status (Fig. 13.2C). The more females chose *D. melanogaster*, the higher was their offspring survival. This confirmed previous results showing adaptive host choice obtained on *L. boulardi* or related species (Dubuffet et al., 2006; Pannebakker et al., 2008). Importantly, LbFV did not alter this adaptive host selection decision. This suggests that LbFV specifically impairs perception skills involved in superparasitism avoidance (possibly chemoreceptors) without impairing receptors involved in the discrimination among different host species, which is quite remarkable since both perception skills are probably due to chemoreceptors innervating the ovipositor.

13.5. ADAPTIVE SIGNIFICANCE OF THE PHENOTYPIC ALTERATION INDUCED (EXCEPT SUPERPARASITISM)

In Section 13.3, we presented a theoretical approach that shows that the viral-induced modification of superparasitism behavior is an adaptive trait for the virus. This conclusion is further supported by the fact that no other behavioral component is modified by the virus (Table 13.1), underlying the specificity of the behavioral modification. A conflict of interest arises between the parasitoid and the virus since they are selected for divergent superparasitism strategies. Consequently, both partners are in conflict of interest from an evolutionary point of view. What about other traits? "Physiology"-related traits appear to be relatively poorly affected by the virus except for locomotor activity which is reduced by 45%. This may result from an energetic cost induced by the replication of the virus, which may reduce the energy available for the insect movement. One surprising result concerns the egg load. How can the observed increase of egg load in LbFV-infected females be explained? Is it an adaptation of the parasitoid, in response to virus infection or an adaptation (another way to manipulate the reproductive behavior of the parasitoid) of the virus to increase its own transmission?

To address this question, we modified the model used to study the evolution and the manipulation of superparasitism (Gandon et al., in press). In this model, each parasitoid female is born with a fixed number of eggs and lacks the ability to mature additional oocytes later on (i.e., strictly proovigenic parasitoid). The initial egg load may be modified by the presence of the virus (either caused by a manipulation induced by the virus or by a plastic response of the host) and *Ez* and *E* refer to the egg load at emergence of infected and uninfected females, respectively. The evolution of the egg load of proovigenic parasitoid species, like any other life history trait, can be viewed as a resource allocation problem. Producing more eggs will divert resources from other important life history traits. In our model, we consider various tradeoffs between egg load and the probability of emergence, and adult survival. The ESS resource allocation strategy is the one that balances the benefits and the costs of producing more eggs.

This model can be used to study the evolution of egg load in the absence of a virus manipulating the behavior of the females. In this simple scenario, we recovered the main result of Rosenheim (1996) that the evolutionarily stable egg load increases with the rate of oviposition thus limiting the risk of egg limitation (i.e., the probability to exhaust its total number of eggs before dying). We can also use this model to consider the situation where a virus manipulating the superparasitism behavior is present in the population (and has reached an endemic equilibrium). In this case, the parasitoid population becomes heterogeneous. Some individuals are uninfected and have a low probability of superparasitism, while other individuals are infected by the virus and have large probabilities of superparasitism. We use our model to analyze different scenarios depending on the ability of the parasitoid females to adopt plastic strategies with regard to viral infection.

First, we consider that the egg load of the females is only determined by the female but not by the virus. If egg load is allowed to be conditional on the infectious status (i.e., two different egg loads may be expressed, depending on whether or not the female is infected), we found that the ESS egg load is to increase egg load when the female is infected. This is due to the fact that infected females lay a higher number of eggs because they also lay eggs in already parasitized hosts (because females infected by the virus are assumed to always superparasitize). They thus have a higher chance of being egg limited (to run out of eggs before dying) than uninfected hosts, and this is why they evolve higher egg loads. Second, if the egg load is assumed to be a fixed strategy (independent on whether or not the female is infected) we found that the evolution of the parasitoid egg load is mainly driven by the selection acting on infected parasitoids because of the often large prevalence of the virus in the population (due to high rates of vertical and horizontal transmission). As a consequence, the unconditional ESS is close to the conditional ESS of infected females, and is thus increased by the presence of the virus in the population.

Then we also considered the scenario where the egg load of infected females is actually governed by the virus, not the parasitoid. When the virus is allowed to manipulate parasitoid egg load we find that it always increases the number of eggs above the ESS level in the absence of the virus. Thus, the fact that infected females of *L. boulardi* tend to have a higher egg load than uninfected females could be explained by two adaptive scenarios. Under the first scenario, *L. boulardi* females have evolved the ability to increase their egg load only when they are infected. Indeed, infected wasps have a higher rate of oviposition (and higher risk of egg limitation) than uninfected ones due to the manipulation of superparasitism. It is thus adaptive for infected females to produce more eggs to reduce the risk of egg limitation (increased by superparasitism). This situation thus corresponds to adaptive phenotypic plasticity of the parasitoid. Under the second scenario, this increase of egg load is induced by a manipulation of the virus. For the virus, higher egg load is also adaptive because it offers additional opportunities of vertical and horizontal transmission. This increase in egg load would thus correspond to another side of the manipulation of the parasitoid phenotype by the virus. The only way to distinguish between the two alternatives would require an examination of the mechanism responsible for the shift in egg load. For example, one could demonstrate that it is a conditional response if it was possible to see a change in egg load in exposed-but-not-infected females (see Minchella, 1985, for a similar experiment in snails and trematodes).

Interestingly, thus, in contrast with our analysis of the evolution of superparasitism, the analysis of this model does not allow us to determine if the higher egg loads are an evolutionary response of the host or a manipulation by the virus. This results from the fact that there is no real conflict over the evolution of this trait between the parasitoid and the virus. Given that the virus manipulates the superparasitism of infected females, both partners benefit from increasing the egg load above the level in the absence of the virus. Another consequence of this alignment of interests can be seen when the parasitoid and the virus are allowed to coevolve. The optimal egg load strategies of the virus and of the uninfected females tend to be closer after coevolution. Again, this contrasts with the adaptive dynamics of superparasitism (Gandon et al. 2006), where coevolution increases the difference between the virus and the parasitoid strategies (Fig. 13.1).

13.6. EVOLUTION IN RELATION TO THE FREQUENCY OF HORIZONTAL VERSUS VERTICAL TRANSMISSION

The mode of transmission of a pathogen has long been recognized as a critical feature to consider in order to understand and predict its evolution (Ewald, 1987). It is clear that for a parasite with strict vertical transmission, host and parasite fitness are strongly correlated and any parasite feature that decrease host fitness will be counter selected. Consequently, vertical transmission is usually associated with low virulence or even mutualism (see however Chapter 12 by Vavre et al. for the special case of reproductive parasites). However, when a parasite is horizontally transmitted, host and parasite fitness are no more correlated and selection may promote highly virulent parasites, if increased virulence favors transmission. In the LbFV/*L. boulardi* system, both transmission modes may occur. Furthermore, depending on ecological conditions such as the ratio of parasitoids–hosts, the opportunities of horizontal or vertical transmission may vary. Indeed, if this ratio is low (numerous hosts for few parasitoids), then there will be few superparasitism and low horizontal transmission opportunities, whereas if this ratio is high (numerous parasitoids for few hosts), opportunities for horizontal transfer may be high (we will see later that this simple view is partly caricatural). This raises the question of the consequence of such ecological changes on the evolution of LbFV, and in particular on the evolution of superparasitism behavior. Based on the model described in Section 13.3 and in Gandon et al. (2006), we studied the ESS of superparasitism of the manipulating virus, as a function of th, which measures the probability of horizontal transmission between infected and uninfected parasitoids sharing the same host (superparasitism). Note that in the model, we assumed that the outcome of the competition between the resident and the newly arrived parasitoid larva is determined very rapidly. Consequently, the model does not keep track of superparasitized hosts because in those hosts, soon after superparasitism, only a single larva remains alive. In the model, parasitized hosts thus regroup hosts that have been parasitized once or several times. We identified two situations, depending on the value of ε, which measures the probability that a viral strain A replaces a resident viral strain B during the short period where both strains compete within the same superparasitized host (superinfection). Because we do not have any indication to date on the value of ε in reality, we derived the ESS of superparasitism for $\varepsilon = 0$, which corresponds to the case where no superinfection can occur and $\varepsilon = 0.5$, which corresponds to the situation where a supernumerary virus strain can outcompete the resident virus in 50% of the cases.

First, in Figure 13.3, it can be noted that below a certain value of th (0.2), the virus cannot maintain in the wasp population. This is due to the fact that at each generation, infected females produce only 95% infected offspring due to the incomplete vertical transmission. We have clear indications that vertical transmission is very efficient but imperfect (Varaldi et al., 2006c). In the absence of horizontal transmission or any

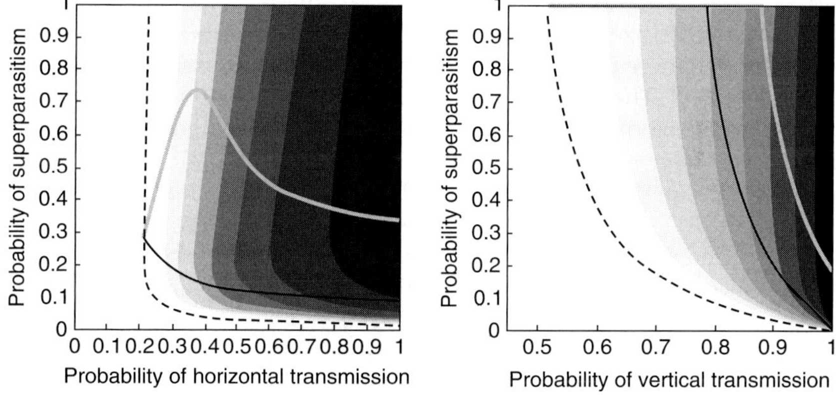

FIGURE 13.3 (A) Evolutionarily stable strategy (ESS) of superparasitism for the virus against the probability of horizontal transmission (th). (B) ESS of superparasitism for the virus against the probability of vertical transmission (tv). The shades of gray indicate the prevalence of the virus in the population, darker gray indicates more virus (10% difference in prevalence between each shade of gray). The dashed line indicates the threshold below which the virus goes extinct. Black line: $\varepsilon = 0$ (no superinfection). Gray line: $\varepsilon = 0.5$ (some superinfection may occur).

fitness advantage to being infected, because uninfected females will obviously produce 100% uninfected females, the frequency of infection in the whole population should decrease by a factor 0.95 from one generation to the next until disappearance. This verbal argument (but see Lipsitch et al., 1995 for a modelization of this simple problem) shows that without other compensating mechanism the virus cannot maintain in populations. The mechanism compensating for this incomplete vertical transmission is precisely the horizontal transmission, but in this situation (th < 0.2), it is not sufficient to compensate the incomplete vertical transmission and the virus is ousted from the population.

In the simplest situation where $\varepsilon = 0$ (no superinfection), increasing the probability of horizontal transfer (starting from 0.20) decreases the ESS of superparasitism for the virus (black line in Fig. 13.3A). This result may sound counterintuitive because it means that even if the probability of horizontal transfer is increased, the virus is selected for lower superparasitism, although superparasitism is precisely the mechanism necessary for horizontal transfer. The explanation lies in the fact that an increase in th has important epidemiologic consequences. Increasing the probability of horizontal transfer leads to a better diffusion of the virus between unrelated parasitoids and consequently to higher prevalence at the epidemiologic equilibrium (gray shading in Fig. 13.3A). Furthermore, an increase in the virus prevalence leads to an increase of the aggregation of wasp eggs inside *Drosophila* larvae and thus to a decrease in the proportion of parasitized hosts (with or without virus). Thus, increasing the probability of horizontal transmission has two main consequences. First, it decreases the number of parasitized hosts and thus limits the benefits of superparasitism. Second, it increases the prevalence of the virus among those hosts that are parasitized. This also selects against superparasitism when $\varepsilon = 0$ because no horizontal transmission can take place in this situation. Thus both these effects go in the same direction and explain why a small increase in th can lead to a decrease in the ESS superparasitism of the virus.

In contrast, if some superinfection is allowed (i.e., $\varepsilon > 0$) the pattern can be very different because horizontal transmission can take place even if the parasitoid already present in the host is infected by another strain of the virus. First, all parameter sets led to higher ESS values with $\varepsilon = 0.5$ than with $\varepsilon = 0$. This result makes sense since with $\varepsilon = 0.5$ an already parasitized host represents a potential wasp to colonize for a mutant virus even if it is already infected by a resident virus (contrary to the case where $\varepsilon = 0$). This leads to an increase in the payoff from superparasitism from the virus point of view. This result also confirms that increasing ε also increases the intensity of the conflict of interest between the parasitoid and the virus (see also fig. 5c and d in Gandon et al., 2006 and Fig. 13.1). With $\varepsilon = 0.5$, the ESS of superparasitism takes a humped shape, with an increase for low prevalence (or low probability of horizontal transmission) and a subsequent decrease for higher prevalences (high probability of horizontal transmission). The interpretation of this result also implies the correlative change in the viral prevalence. For low probability of horizontal transmission (but >0.2), the virus maintains at relatively low frequency (below 20%), and there is lots of opportunities for horizontal transfer. Conversely to the case where $\varepsilon = 0$, increasing the probability of horizontal transmission also increases the opportunities for horizontal transmission even at the epidemiologic equilibrium (where the prevalence reaches its equilibrium value) because one viral strain can replace another one within the host. This selects for higher superparasitism until a critical prevalence value is reached (about 20% with this parameter set) where the environment starts to saturate with the virus (which reduces the proportion of parasitized hosts due to egg aggregation), reducing drastically the opportunities for viral horizontal transfer (at the epidemiologic equilibrium) and also reducing the opportunities for vertical transmission during superparasitism. Consequently, this selects for reduced superparasitism.

When we fixed the probability of th, and varied the value of tv, the interpretation was much simpler (Fig. 13.3B). Here again, there was a minimal value for tv for the virus to maintain in the population (0.65). Above this threshold, the viral prevalence increased monotonously with an increase of tv. As expected, the ESS for the virus was high when tv is low and decreased afterward. This pattern was observed for both situations ($\varepsilon = 0$ and $\varepsilon = 0.5$). In Section 13.3, we have shown that the virus is selected for higher ESS values of superparasitism than the parasitoid (except in some peculiar combination of parameter sets, e.g., tv = 1 and $\varepsilon = 0$). In other words, the virus reduces the fitness of the parasitoid due to the wastage of eggs induced by the manipulation (classical cost of superparasitism). Consequently, increasing tv also increased the correlation of the fitness of both the virus and the parasitoid, thus reducing the conflict of interest. This selects for a reduction of the ESS of the virus. In the special case where tv = 1, the virus is selected to adopt the same strategy as the parasitoid (with this parameter set, the ESS for the parasitoid was to never superparasitize, not shown). However, this is true only with $\varepsilon = 0$, that is, when no superinfection is allowed. With superinfection ($\varepsilon = 0.5$), the conflict of interest between the virus and the parasitoid still holds.

The model shows three important features of the LbFV/L. *boulardi* system. The first is the importance of epidemiologic feedbacks. It was particularly visible when we varied the probability of horizontal transmission. Indeed the predictions were counterintuitive due to the indirect effect (i.e., epidemiologic effect) of an increase in th, through a decrease in the number of parasitized hosts, and an increase in the prevalence of the virus in the parasitoid population. Paradoxically, within a population with a high intensity of superparasitism and high viral prevalence, the frequency of horizontal transfer may be lower than within a population with fewer superparasitism but lower viral prevalence. However, this conclusion is deeply influenced by the superinfection parameter (ε). In this model, we were interested in epidemiologic equilibrium. However, the relative contribution of horizontal and vertical transmission in the course of the invasion process change substantially, with strong contribution of horizontal transmission at the beginning and a reduction with an increase in prevalence. This problem has been addressed in a general context in Lipsitch et al. (1996). Thus, highly manipulative strains are selected for at the beginning of the invasion process and less manipulative at the epidemiologic equilibrium. Another conclusion that can be drawn from the model, is that the value taken by the superinfection parameter (ε) is critical. In both Figures 13.3A and B, we found that increasing ε strongly increased the ESS of superparasitism and also modified the form of the relation between ESS and th. It is evident that the value of this parameter needs to be estimated in this system in order to predict correctly the ESS in natural populations.

13.7. EXPERIMENTAL EVOLUTION IN RELATION TO TRANSMISSION TYPE (HORIZONTAL OR VERTICAL)

The previous section showed how transmission type (horizontal or vertical) is a critical factor governing the evolution of the virus-induced superparasitism phenotype within natural populations. In this section, we describe an experiment in which we manipulated the transmission of the virus, either forcing it to spread vertically but not horizontally or forcing it to spread exclusively horizontally. Contrary to the model described above, this experiment did not include any epidemiologic feedback but only asked whether changing the transmission mode will select for alternative viral strategies. Our prediction was that forcing the virus to propagate exclusively by vertical means should select for lower superparasitism strategy, whereas forcing horizontal transmission should select for higher superparasitism strategy. In standard rearing conditions, three females are used to parasitize about 150 hosts in each vial. In these conditions, moderate superparasitism do occur (Varaldi et al., 2005a). Consequently, when the females are infected, it is likely that both vertical transmission (from mother to offspring) and horizontal transmission occur (horizontal transmission may occur if one viral strain is able to replace one other strain inside the *Drosophila* larva, e.g., $\varepsilon > 0$ in the previous model). However, in standard rearing tubes, if we use only a single female, then only vertical transmission will occur. Conversely, we can provide hosts already parasitized by uninfected females to (superparasitizing) infected females to maximize horizontal transfer. Under this condition, the offspring of uninfected females may become infected at the next generation. In this species, the reproduction is arrhenotokous parthenogenesis (males are haploid and obtained from unfertilized eggs whereas females are diploid and obtained from fertilized eggs). It is very simple to be sure that *all* transmission events are horizontal, by taking advantage of the fact that unfertilized females will only lay sons whereas fertilized females will lay sons and daughters. Consequently, by exposing hosts first to uninfected and fertilized females and subsequently to virgin infected females, we have the certainty that all infected female offspring is obtained through horizontal transfer. Based on this idea, we did the following experiment using an infected strain originating from Sienna, Italy (described in Varaldi et al., 2003; see Table 13.2). One hundred unparasitized *D. melanogaster* larvae were offered to three fertilized infected females in standard rearing tubes. At each generation, three emerging females were randomly selected and allowed to mate and used to maintain the line. Ten independent replicates were performed in parallel. At each generation, the superparasitism phenotype of two females emerging from each tube was tested according to a standard procedure (female isolated on 10 *D. melanogaster* larvae, see Varaldi et al., 2006b for details). This condition constitutes the control conditions where both vertical and horizontal transmission are likely to occur, because some superparasitism occurs (Table 13.2). A second modality forcing vertical transmission was performed, where a single female was

TABLE 13.2 Description of the experimental setup

	Hosts	Wasps	N lines	Tests per line
Control	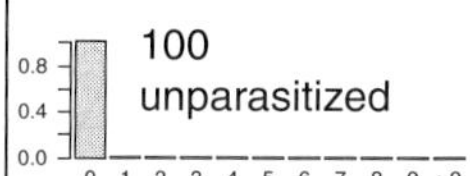 100 unparasitized	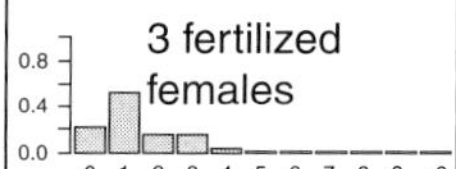 3 fertilized females	10	2
Vertical	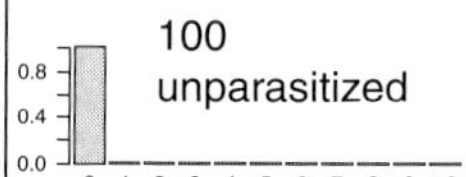 100 unparasitized	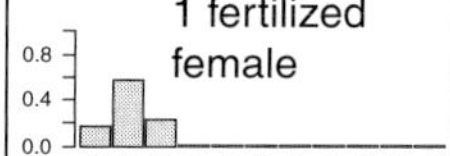 1 fertilized female	30	1
Horizontal	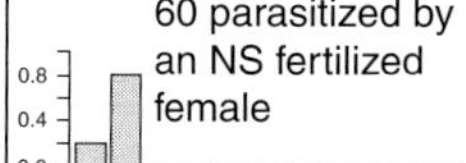 60 parasitized by an NS fertilized female	 10 virgin females	8	2

In each cell is indicated the frequency distribution of wasp eggs inside *Drosophila* larvae (left column: before infected wasp(s) were added, right column: after infected wasp(s) were removed).

provided with 100 unparasitized hosts. To allow a selective process to occur, we prepared 30 independent tubes, mixed all the emerging offspring (from all 30 tubes) at each generation and randomly selected 30 (fertilized) females to establish the next generation. The idea was that a virus that induces a low fitness cost and especially that induces few superparasitism will be selected since the infected wasp will not waste its eggs in (self-) superparasitism and will contribute more to the pool of emerging wasps. At each generation, the superparasitism phenotype of one female per line was tested. Finally, we provided 60 hosts already parasitized by an uninfected fertilized female to 10 virgin infected females. In this situation, harsh superparasitism occurs (Table 13.2), favoring horizontal transfer. To standardize the whole number of *Drosophila* larvae in the tubes for the three modalities (a total of 100), we added 40 unparasitized *Drosophila* larvae to the 60 (super)parasitized hosts. From the emerging wasp offspring, two females were used to test their superparasitism phenotype and 10 virgin females were used to continue the protocol (again they were provided with 60 hosts already parasitized by an uninfected fertilized female). Eight independent lines were performed (horizontal).

The controls (three fertilized females for 100 hosts) showed the expected superparasitism phenotype with a mean number of eggs per host between two and four (Fig. 13.4). Also, the horizontal transfers that were expected in the modality "horizontal" were evident, since the mean number of eggs per hosts was 1.54 and 4.63, compared with the phenotype

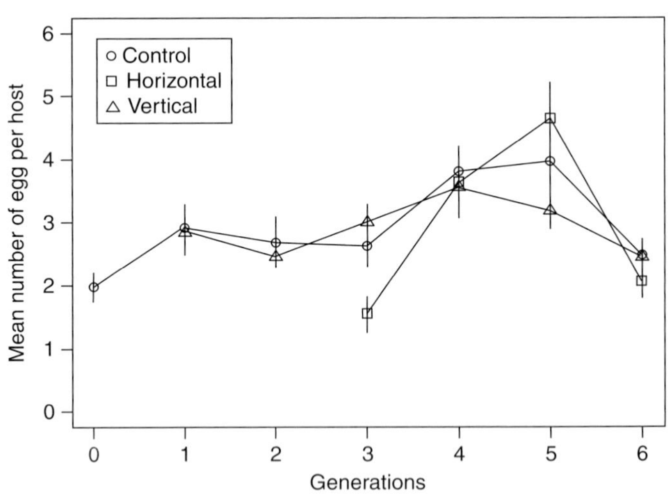

FIGURE 13.4 Experimental evolution of the intensity of the manipulation induced by LbFV.

of the uninfected line (mean = 1.07, $n = 5$). The superparasitism phenotype was tested in all three modalities (control, strict vertical and strict horizontal) starting from the third generation of selection (from generations three to six). In these data, there was evidence of between-generation variations ($F(3,199) = 9.81$, $P < 0.0001$), but no evidence of the type of transmission ($F(2,199) = 0.27$, $P = 0.75$) and no interaction between generation and the type of transmission ($F(6,199) = 2.12$, $P = 0.052$). As a conclusion, there was no evidence of evolution in this dataset. Several hypotheses can be formulated to explain this absence of response. The first is that the selective differential is not sufficient to observe a change in only six generations because of sampling errors. This explanation probably holds for the vertical transmission modality, because only 30 females out of around 180 (each of the 30 tubes produced around 60 females) were randomly selected at each generation to continue the experiment. Consequently, even if a female contributed more than the others to the whole emerging population (because its virus was more benevolent, induced less superparasitism), sampling errors may have cancelled the initial overrepresentation of this peculiar virus strain. However, this explanation is unlikely to hold for the horizontal transmission modality since at each generation, 10 wasps were randomly selected in a pool of only about 20 emerging females. This low number of emerging females (we recall that only 60 hosts were superparasitized in this modality) was due to the fact that when strong superparasitism occurs (as it is the case in this modality, see Table 13.2), both the host and the parasitoid incur a high risk of dying during the development (Varaldi et al., 2005a). This phenomenon constitutes a potential cost to the spread of highly manipulative strains and may explain part of the absence of response in the horizontal modality. Finally, one trivial hypothesis that may explain the absence of any selective response neither in the vertical nor the horizontal modality is that there was no sufficient genetic variability of the virus at the beginning of the experiment and that mutation alone did not generate enough polymorphism in the course of the experiment.

13.8. OTHER VIRUSES IN THE *DROSOPHILA*–PARASITOID COMMUNITY

Due to their very diverse genomic structure (DNA, RNA, single or double stranded) and to their high mutation rates, no simple systematic methods (such as PCR based) are available to detect the presence of viruses or even to detect all members of a given family. However, several viruses have been regularly discovered in several *Drosophila* spp., especially *D. melanogaster*. It is reasonable to think that most of these viruses have been discovered because *D. melanogaster* is a model system since the beginning of the twentieth century and has been extensively studied from all aspects of its biology (including immunity). L'Heritier and Teissier (1937) for instance discovered the σ virus because certain strains showed atypical (virus-induced) CO_2 sensitivity. Recently, molecular techniques have also provided additional means to reveal their presence. For instance, Asling et al. (1995) were interested in comparing the transcriptomes of *D. melanogaster* either "uninfected" or "challenged" with a pathogenic bacteria in order to identify immunity-related genes. They did found the induction of an antimicrobial peptide but they also detected an induced band presenting sequence similarity with viruses. They were in fact discovering a new single-stranded RNA (ssRNA) virus (picorna-like) apparently asymptomatic, called Nora virus (Habayeb et al., 2006). The virus was first detected in a huge quantity of fly stocks, but it was later found that a technical bias led to an overestimation of its prevalence (Habayeb et al., 2007). Table 13.3 presents a comprehensive list of the identified viruses infecting *Drosophila* spp. One striking pattern is that all of them are RNA viruses (or likely to be, when no genomic information are available), although other *Diptera* are infected by DNA viruses (Gratz, 2004). This surprising pattern remains unexplained. Understanding the biological reason for this (if any) may provide exciting insights on the enigmatic observation that the major virus genomic structures are clearly nonuniformly distributed among the main branches of hosts (Koonin et al., 2008). One other feature of *Drosophila* viruses that can be underlined is the diversity of transmission modes with strictly horizontally transmitted viruses (e.g., DCV), strictly vertically transmitted viruses (for instance σ virus) and viruses presenting both transmission modes (virus P and virus A). A remarkable feature of most of these viruses is that they have relatively mild pathological effects on their hosts. For instance, although DCV virus is highly pathogenic when artificially injected, Thomas (1974) and Gomariz-Zilber and Thomas-Orillard (1993) found that under natural infection routes (larval feeding on contaminated substrate), DCV reduces only slightly the survival of larvae and even induces an increase in the number of ovarioles and on adult longevity. However, Texeira (personal communication) found clear pathogenic effects of DCV even when larvae become infected by feeding. The reason for these somehow conflicting results remains unclear. The hereditary σ virus does not affect fertility, female longevity, but reduces egg viability (Fleuriet, 1981a) and overwintering survival probability (Fleuriet, 1981b). Sigma virus also induces CO_2 sensitivity, that is, *Drosophila* exposed for a while to CO_2 die instead of recovering from sleep. However, the ecological significance of this phenotype is probably negligible since CO_2 concentrations never reach such high concentration in the wild. It provides, however, a convenient way to identify infected flies, allowing population-level investigations (Bangham et al., 2008a,b; Carpenter et al., 2007).

A rough estimate of the overall viral prevalence in *D. melanogaster* has been given by Brun and Plus (1980). They found that among 49

TABLE 13.3 Viruses infecting *Drosophila* spp.

Virus	Host	Genome structure	Family	Genome sequence	Ref. genome	Transmission	Effects	Refs effects	Prevalences	Ref. prev
Sigma	*D. melanogaster*	ssRNA-	Rhabdoviridae	6477bp incomplete (ref genbank X91062)	1	Vertical through males and females gametes	No effect on fertility, female longevity, sexual selection and egg viability; reduced survival of eggs and overwintering survival (and CO_2 sensitivity)	1, 9	Up to 60%	17
DXV	*D. melanogaster*	dsRNA	Birnaviridae	6603bp (in 2 segments, ref genbank NC_004177, NC_004169)	2, 3	Horizontal (contact) apparently not vertical	Anoxia sensitivity reduction in survival (sometimes asymptomatic)	11	Never observed under natural conditions	
Virus C	*D. melanogaster* specific (16)	ssRNA	Dicistrioviridae	9264bp ref genbank NC_001834	4	Horizontal by feeding (adults or larvae)	Conflicting results. See text	12, 13, 14	6 populations infected out of 49	16
Virus P	*D. melanogaster*	ssRNA	Picornavirus-like superfamily	?	5	Horizontal by contact and ingestion and vertical by young females	Fitness reduction (survie, egg-laying)	15	?	
Virus A	*D. mel* but not only (16)	ssRNA	Picornavirus-like superfamily	4806bp NC_012958	19	Horizontal by contact and ingestion and vertical by young females	Low pathogeny	5	?	
Nora	*D. melanogaster*	ssRNA	Picornavirus-like superfamily	11908bp ref genbank NC_007919	6	Horizontal through feces	Slight reduction in survival and hatching	20	?	
Reovirus F	*D. mel* but not only (16)	dsRNA?	Reoviridae	?	7	Horizontal by contact, apparently not vertical	No signs	16	?	
Virus G	*D. mel* but not only (16)	RNA	?	?		Horizontal by contact, apparently not vertical	No signs	16	?	
DSV	*D. simulans*	dsRNA	Reoviridae	Around 8410bp (at least 8 segments)	8	Hereditary mainly maternal	Modification of cuticule (bristle) negative effects on fitness	18	?	
Iota virus	*D. immigrans*	RNA	Picornavirus-like superfamily	?		Transovarian	No signs. Induce CO_2 sensitivity in *D. melanogaster*	16	Up to 100%	16
RS virus	*D. ananassae* *D.montium*	?	?	?		?	?	16	?	

Notes: 1: Landès-Devauchelle et al. (1995); 2: Shwed et al. (2002); 3: Chung et al. (1996); 4: Johnson and Christian (1998); 5: Plus et al. (1976); 6: Habayeb et al. (2007); 7: Plus et al. (1981); 8: López-Ferber et al. (1989); 9: Fleuriet (1981a); 10: Fleuriet (1981b); 11: Teninges et al. (1979); 12: Thomas (1974); 13: Gomariz-Zilber (1993); 14: Jousset and Plus (1975); 15: David and Plus (1971); 16: Brun and Plus (1980); 17: Fleuriet and Periquet (1993); 18: Louis et al. (1988); 19: Ambrose et al. (in press); 20: Habayeb et al. (2009).

populations originating from Europe, Africa, North and South America, 19 populations were infected by at least one virus (39%). More detailed investigations have been done on the σ virus. The hereditary σ virus showed a frequency of up to 65% in some French populations (Fleuriet and Periquet, 1993), while a more recent study revealed that σ virus was present in five populations out of 12 originating from Greece, United Kingdom, Polynesia, United States of America, Kenya, Spain and Austria, with frequencies reaching 15% (Carpenter et al., 2007). These relatively high frequencies make them potential factors influencing the ecology and evolution of their hosts. It is interesting to note that in the aphid *Acirtosiphon pisum* several maternally transmitted bacterial secondary symbionts (facultative endosymbionts) reach high prevalence (but not fixation) in natural populations (Oliver et al., 2006). The ecological factors explaining their distribution has been elusive for a while. However, it has been shown that the secondary symbionts may increase the fitness of their aphid host in certain environments, because they confer resistance against heat stress, resistance to fungal pathogens, adaptation to host plant or protection against parasitoids (*Hamiltonella defensa*). However, they may be costly under alternative environments (Oliver et al., 2007; Russell and Moran, 2006), providing an explanation for their intermediate frequencies. In addition, secondary symbionts may benefit from natural horizontal transfer for instance during copulation (Moran et al., 2006), favoring the spread of infection and the occurrence of coinfection. There are evident similarities between both model systems (aphid secondary symbionts and *Drosophila* viruses) and we can ask whether some of these viruses have anything to do with the adaptation of *Drosophila* to their local environment, and especially to the presence of parasitoids. On this scale, it is interesting to note that the protective effect conferred by *Hamiltonella defensa* to its aphid host is probably caused by the presence of specific toxins encoded by its bacteriophage (Degnan and Moran, 2008).

It is clear that the parasitoids attacking *Drosophila* spp. have received much less attention than *Drosophila*. To our knowledge, the only virus described to date in *Drosophila* parasitoid is LbFV, apart from VLPs that may have a viral evolutionary origin. We argue that this apparent asymmetry between *Drosophila* and their parasitoids is probably a sampling bias, and we suspect that several other parasitoid viruses will be described in the near future. New molecular tools that are now available, especially high-throughput sequencing (Marioni et al., 2008; Vera et al., 2008) allowing for metagenomic analysis (Cox-Foster et al., 2007), will provide evidence of new infectious and/or heritable viruses in parasitoids. We can mention that another RNA virus have been fortuitously discovered in the Lepidoptera parasitoid wasp *Venturia canescens*, using transcriptomic analysis exactly the same way as was discovered the *Drosophila* Nora virus (Reineke and Asgari, 2005). Finally, it has been recently found that bacterial symbiont can confer protection against virus infection, suggesting possible interactions between virus and bacterial endosymbionts (Hedges et al., 2008; Teixeira et al., 2008). This result is particularly interesting since the phylogeny of *Leptopilina* spp. reveals that all *Leptopilina* spp. are infected by the endosymbiont *Wolbachia* (see Chapter 12), at the exception of *L. boulardi* where was found the manipulating virus (Allemand et al., 2002).

13.9. CONCLUSION

Viruses are ubiquitous. The *Drosophila*–parasitoids community is not an exception as several (RNA) viruses infecting *Drosophila* spp. have been identified. We discovered a new virus (probably a DNA virus) in the parasitoid wasp *Leptopilina boulardi* and we suspect that new viruses will be discovered in the near future, especially in parasitoids because the sampling effort in this group has been relatively low until now. Viruses may reach high prevalence in natural populations and are thus important players in the ecology and evolution of their hosts and on host–parasitoid interactions. Their possible ecological and evolutionary implications are illustrated by the LbFV/parasitoid interaction. Indeed, this virus specifically affects a critical foraging component of the wasp (superparasitism), allowing the virus to be horizontally transferred and to spread within wasp population. The behavior of most of the females of a population may then be deeply modified. This indirectly selects for different superparasitism strategies in uninfected females (Section 13.3), and also for higher investment in the egg load of infected females (Section 13.5). Because both the virus and the parasitoid share some fitness components due to vertical transmission, specific parasitoid virus combinations may be the target of selection, possibly leading to coadaptation and evolutionary innovation. In this respect, the discovery of LbFV may provide insights into the symbiogenesis at the origin of PDVs that protect parasitoids from the host immune response. Future investigations will target the molecular mechanisms allowing the virus to be maintained in wasp populations (superparasitism manipulation), the genetic response of their hosts and the ecological consequences on interspecific interactions.

REFERENCES

Allemand, R., Lemaitre, C., Frey, F., Boulétreau, M., Vavre, F., Nordlander, G., et al., 2002. Phylogeny of six African *Leptopilina* species (Hymenoptera: Cynipoidea, Figitidae), parasitoids of *Drosophila*, with description of three new species. Ann. Soc. Entomol. Fr. 38, 319–332.

Ambrose, R.L., Lander, G.C., Maaty, W.S., Bothner, B., Johnson, J.E., Johnson, K.N., in press. Drosophila A virus is an unusual RNA virus with a T = 3 icosahedral core and permuted RNA-dependent RNA-polymerase. J. Gen. Virol.

Asling, B., Dushay, M.S., Hultmark, D., 1995. Identification of early genes in the *Drosophila* immune response by PCR-based differential display: the Attacin A gene and the evolution of attacin-like proteins. Insect Biochem. Mol. Biol. 25, 511–518.

Bangham, J., Kim, K.W., Webster, C.L., Jiggins, F.M., 2008a. Genetic variation affecting host–parasite interactions: different genes affect different aspects of sigma virus replication and transmission in *Drosophila melanogaster*. Genetics 178, 2191–2199.

Bangham, J., Knott, S.A., Kim, K.W., Young, R.S., Jiggins, F.M., 2008b. Genetic variation affecting host–parasite interactions: major-effect quantitative trait loci affect the transmission of sigma virus in *Drosophila melanogaster*. Mol. Ecol. 17, 3800–3807.

Bezier, A., Annaheim, M., Herbinière, J., Wetterwald, C., Gyapay, G., Bernard-Samain, S., et al., 2009. Polydnaviruses of braconid wasps derive from an ancestral nudivirus. Science 323, 926–930.

Brun, P., Plus, N., 1980. The viruses of *Drosophila*. In: Ashburner, Wright, (Eds.), The Genetics and Biology of *Drosophila*. Academic Press, London, pp. 626–702.

Carpenter, J.A., Obbard, D.J., Maside, X., Jiggins, F.M., 2007. The recent spread of a vertically transmitted virus through populations of *Drosophila melanogaster*. Mol. Ecol. 16, 3947–3954.

Chung, H.K., Kordyban, S., Cameron, L., Dobos, P., 1996. Sequence analysis of the bicistronic *Drosophila* X virus genome segment A and its encoded polypeptides. Virology 225, 359–368.

Carton, Y., Boulétreau, M., Van Alphen, J.J. M., Van Lenteren, J.C., 1986. The *Drosophila* parasitic wasps. In : The Genetics and Biology of *Drosophila*. Ashburner, Carson, Thompson, (Eds.), Academic Press, Orlando, pp. 347–394.

Cox-Foster, D.L., Conlan, S., Holmes, E.C., Palacios, G., Evans, J.D., Moran, N.A., et al., 2007. A metagenomic survey of microbes in honey bee colony collapse disorder. Science 318, 283–287.

David, J., Plus, N., 1971. Le virus P de la drosophile: comparaison de la longevité et de la fecondité des mouches infectées par injection ou par contamination naturelle. Ann. Institut. Pasteur 120, 107–119.

Degnan, P.H., Moran, N.A., 2008. Diverse phage-encoded toxins in a protective insect endosymbiont. Appl. Environ. Microbiol. 74, 6782–6791.

Dubuffet, A., Alvarez, C.I., Drezen, J.M., van Alphen, J.J.M., Poirie, M., 2006. Do parasitoid preferences for different host species match virulence? Physiol. Entomol. 31, 170–177.

Dupas, S., Brehelin, M., Frey, F., Carton, Y., 1996. Immune suppressive virus-like particles in a *Drosophila* parasitoid: significance of their intraspecific morphological variations. Parasitology 113, 207–212.

Espagne, E., Dupuy, C., Huguet, E., Cattolico, L., Provost, B., Martins, N., et al., 2004. Genome sequence of a polydnavirus: insights into symbiotic virus evolution. Science 306, 286–289.

Ewald, P.W., 1987. Transmission modes and evolution of the parasitism-mutualism continuum. Ann. N. Y. Acad. Sci. 503, 295–306.

Fleuriet, A., 1981a. Comparison of various physiological traits in flies (*Drosophila melanogaster*) of wild origin, infected or uninfected by the hereditary rhabdovirus sigma. Arch. Virol. 69, 261–272.

Fleuriet, A., 1981b. Effect of overwintering on the frequency of flies infected by the rhabdovirus sigma in experimental populations of *Drosophila melanogaster*. Arch. Virol. 69, 253–260.

Fleuriet, A., Periquet, G., 1993. Evolution of the *Drosophila melanogaster*-sigma virus system in natural populations from Languedoc (southern France). Arch. Virol. 129, 131–143.

Gandon, S., 2005. Parasitic manipulation: a theoretical framework may help. Behav. Process. 68, 247–248.

Gandon, S., Rivero, A., Varaldi, J., 2006. Superparasitism evolution: adaptation or manipulation? Am. Nat. 167, E1–E22.

Gandon, S., Varaldi, J., Fleury, F., Rivero, A., in press. Evolution and manipulation of parasitoid egg load. Evolution.

Glatz, R.V., Asgari, S., Schmidt, O., 2004. Evolution of polydnaviruses as insect immune suppressors. Trends Microbiol. 12, 545–554.

Gomariz-Zilber, E., Thomas-Orillard, M., 1993. *Drosophila* C virus and *Drosophila* hosts: a good association in various environments. J. Evol. Biol. 6, 677–689.

Gratz, N.G., 2004. Critical review of the vector status of *Aedes albopictus*. Med. Vet. Entomol. 18, 215–227.

Habayeb, M.S., Cantera, R., Casanova, G., Ekström, J., Albright, S., Hultmark, D., 2009. The *Drosophila* Nora virus is an enteric virus, transmitted via feces. J. Invertebr. Pathol. 101, 29–33.

Habayeb, M.S., Ekengren, S.K., Hultmark, D., 2006. Nora virus, a persistent virus in *Drosophila*, defines a new picorna-like virus family. J. Gen. Virol. 87, 3045–3051.

Habayeb, M.S., Ekengren, S.K., Hultmark, D., 2007. Nora virus, a persistent virus in *Drosophila*, defines a new picorna-like virus family. J. Gen. Virol. 88, 3493.

Hedges, L.M., Brownlie, J.C., O'Neill, S.L., Johnson, K.N., 2008. *Wolbachia* and virus protection in insects. Science 322, 702.

Johnson, K.N., Christian, P.D., 1998. The novel genome organization of the insect picorna-like virus *Drosophila* C virus suggests this virus belongs to a previously undescribed virus family. J. Gen. Virol. 79, 191–203.

Jousset, F.X., Plus, N., 1975. Study of the vertical transmission and horizontal transmission of "*Drosophila melanogaster*" and "*Drosophila* immigrans" picornavirus. Ann. Microbiol. (Paris) 126, 231–249 [In French].

Koonin, E., Wolf, Y., Nagasaki, K., Dolja, V., 2008. The Big Bang of picorna-like virus evolution antedates the radiation of eukaryotic supergroups. Nat. Rev. Microbiol. 6, 925–939.

L'Heritier, P.H., Teissier, G., 1937. Une anomalie physiologique héréditaire chez la *Drosophile*. C. R. Acad. Sci. Paris 206, 1683–1685.

Landès-Devauchelle, C., Bras, F., Dezélée, S., Teninges, D., 1995. Gene 2 of the sigma rhabdovirus genome encodes the P protein, and gene 3 encodes a protein related to the reverse transcriptase of retroelements. Virology 213, 300–312.

Lapointe, R., Tanaka, K., Barney, W.E., Whitfield, J.B., Banks, J.C., Béliveau, C., et al., 2007. Genomic and morphological features of a banchine polydnavirus: comparison with bracoviruses and ichnoviruses. J. Virol. 81, 6491–6501.

Lipsitch, M., Nowak, M.A., Ebert, D., May, R.M., 1995. The population dynamics of vertically and horizontally transmitted parasites. Proc. Biol. Sci. B 260, 321–327.

Lipsitch, M., Siller, S., Nowak, M.A., 1996. The evolution of virulence in pathogens with vertical and horizontal transmission. Evolution 50, 1729–1741.

López-Ferber, M., Veyrunes, J.C., Croizier, L., 1989. *Drosophila* S virus is a member of the Reoviridae family. J. Virol. 63, 1007–1009.

Louis, C., Lopez-Ferber, M., Comendador, M., Plus, N., Kuhl, G., Baker, S., 1988. *Drosophila* S virus, a hereditary reolike virus, probable agent of the morphological S character in *Drosophila simulans*. J. Virol. 62, 1266–1270.

Marioni, J.C., Mason, C.E., Mane, S.M., Stephens, M., Gilad, Y., 2008. RNA-seq: an assessment of technical reproducibility and comparison with gene expression arrays. Genome Res. 18, 1509–1517.

Minchella, D.J., 1985. Host life history variation in response to parasitism. Parasitology 90, 205–216.

Moran, N.A., Dunbar, H.E., 2006. Sexual acquisition of beneficial symbionts in aphids. Proc. Natl Acad. Sci. USA 103, 12803–12806.
Oliver, K., Campos, J., Moran, N., Hunter, M., 2007. Population dynamics of defensive symbionts in aphids. Proc. R. Soc. B 275, 293–299.
Oliver, K.M., Moran, N.A., Hunter, M.S., 2006. Costs and benefits of a superinfection of facultative symbionts in aphids. Proc. Biol. Sci. 273, 1273–1280.
Oliver, K.M., Russell, J.A., Moran, N.A., Hunter, M.S., 2003. Facultative bacterial symbionts in aphids confer resistance to parasitic wasps. Proc. Natl. Acad. Sci. USA 100, 1803–1807.
Pannebakker, B.A., Garrido, N.R.T., Zwaan, B.J., van Alphen, J.J.M., 2008. Geographic variation in host-selection behaviour in the *Drosophila* parasitoid *Leptopilina clavipes*. Entomol. Exp. Appl. 127, 48–54.
Patot, S., Lepetit, D., Charif, D., Varaldi, J., Fleury, F., 2009. Molecular detection, penetrance and transmission of an inherited virus responsible for a behavioral manipulation of an insect parasitoid. Appl. Environ. Microbiol. 75, 703–710.
Plus, N., Croizier, G., Veyrunes, J.C., David, J., 1976. A comparison of buoyant density and polypeptides of *Drosophila* P, C and A viruses. Intervirology 7, 346–350.
Plus, N., Gissman, L., Veyrunes, J.C., Pfister, H., Gateff, E., 1981. Reoviruses of *Drosophila* and *Ceratitis* populations and of *Drosophila* cell lines: a new genus of the Reoviridae family. Ann. Virol. (Inst. Pasteur) 132E, 261–270.
Poulin, R., 1995. ''Adaptive'' changes in the behaviour of parasitized animals: a critical review. Int. J. Parasitol. 25, 1371–1383.
Reineke, A., Asgari, S., 2005. Presence of a novel small RNA-containing virus in a laboratory culture of the endoparasitic wasp Venturia canescens (Hymenoptera: Ichneumonidae). J. Insect Physiol. 51, 127–135.
Rizki, R.M., Rizki, T.M., 1990. Parasitoid virus-like particles destroy *Drosophila* cellular immunity. Proc. Natl. Acad. Sci. USA 87, 8388–8392.
Rosenheim, J.A., 1996. An evolutionary argument for egg limitation. Evolution 50, 2089–2094.
Russell, J.A., Moran, N.A., 2006. Costs and benefits of symbiont infection in aphids: variation among symbionts and across temperatures. Proc. Biol. Sci. 273, 603–610.
Shwed, P.S., Dobos, P., Cameron, L.A., Vakharia, V.N., Duncan, R., 2002. Birnavirus VP1 proteins form a distinct subgroup of RNA-dependent RNA polymerases lacking a GDD motif. Virology 296, 241–250.
Teixeira, L., Ferreira, A., Ashburner, M., 2008. The bacterial symbiont *Wolbachia* induces resistance to RNA viral infections in *Drosophila melanogaster*. PLoS Biol. 23, E2.
Teninges, D., Ohanessian, A., Richard-Molard, C., Contamine, D., 1979. Isolation and biological properties of *Drosophila* X virus. J. Gen. Virol. 42, 241–254.
Thomas, M., 1974. Contribution à l'Etude du Déterminisme du Nombre d'Ovarioles et de Quelques Autres Caractères Quantitatifs Chez *Drosophila melanogaster* Meigen [Thesis]. University Paris VI.
van Alphen, J.J., Visser, M.E., 1990. Superparasitism as an adaptive strategy for insect parasitoids. Annu. Rev. Entomol. 35, 59–79.
van Lenteren, J.C., Ruschioni, S., Romani, R., van Loon, J.J., Qiu, Y.T., Smid, H.M., et al., 2007. Structure and electrophysiological responses of gustatory organs on the ovipositor of the parasitoid *Leptopilina heterotoma*. Arthr. Struct. Dev. 36, 271–276.
Varaldi, J., Boulétreau, M., Fleury, F., 2005a. Cost induced by viral particles manipulating superparasitism behaviour in the parasitoid *Leptopilina boulardi*. Parasitology 131, 161–168.
Varaldi, J., Fouillet, P., Boulétreau, M., Fleury, F., 2005b. Superparasitism acceptance and patch-leaving mechanisms in parasitoids: a comparison between two sympatric wasps. Anim. Behav. 69, 1227–1234.
Varaldi, J., Fouillet, P., Ravallec, M., López-Ferber, M., Boulétreau, M., Fleury, F., 2003. Infectious behavior in a parasitoid. Science 302, 1930.

Varaldi, J., Gandon, S., Rivero, A., Patot, S., Fleury, F., 2006c. A newly discovered virus manipulates superparasitism behavior in a parasitoid wasp. In: Bourtzis, Miller, (Eds.), Insect symbiosis, Vol. 2. CRC Press, Boca Raton, FL, pp. 119–139.
Varaldi, J., Petit, S., Boulétreau, M., Fleury, F., 2006a. The virus infecting the parasitoid *Leptopilina boulardi* exerts a specific action on superparasitism behaviour. Parasitology 132, 747–756.
Varaldi, J., Ravallec, M., Labrosse, C., Lopez-Ferber, M., Boulétreau, M., Fleury, F., 2006b. Artifical transfer and morphological description of virus particles associated with superparasitism behaviour in a parasitoid wasp. J. Insect Physiol. 52, 1202–1212.
Vera, J.C., Wheat, C.W., Fescemyer, H.W., Frilander, M.J., Crawford, D.L., Hanski, I., et al., 2008. Rapid transcriptome characterization for a nonmodel organism using 454 pyrosequencing. Mol. Ecol. 17, 1636–1647.

INDEX

E

CONTENTS OF VOLUMES IN THIS SERIES

Volume 45

Volume 46

Volume 47

Volume 48

Volume 55

Volume 56

Volume 57

Volume 58

Volume 59

Volume 60

Volume 61

Volume 62

Volume 63

Volume 64

Volume 65

Volume 66

Volume 67

Volume 68

Volume 69

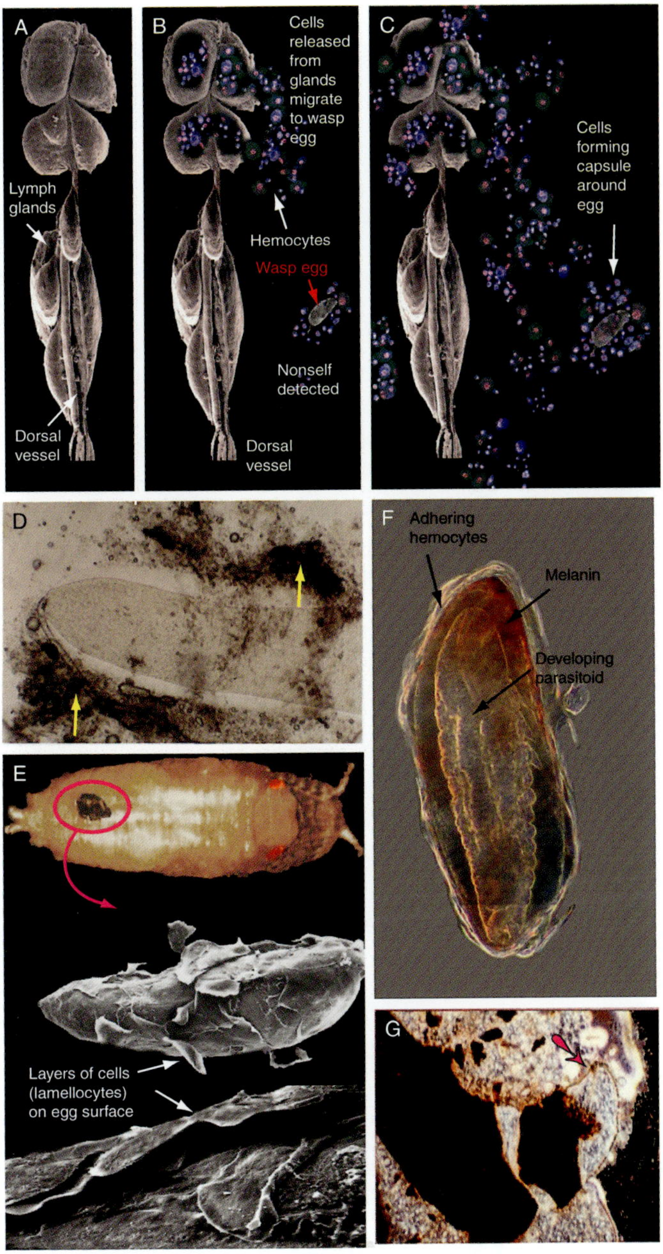

Chapter 4, Figure 4.2 (See page 104 of this volume.)

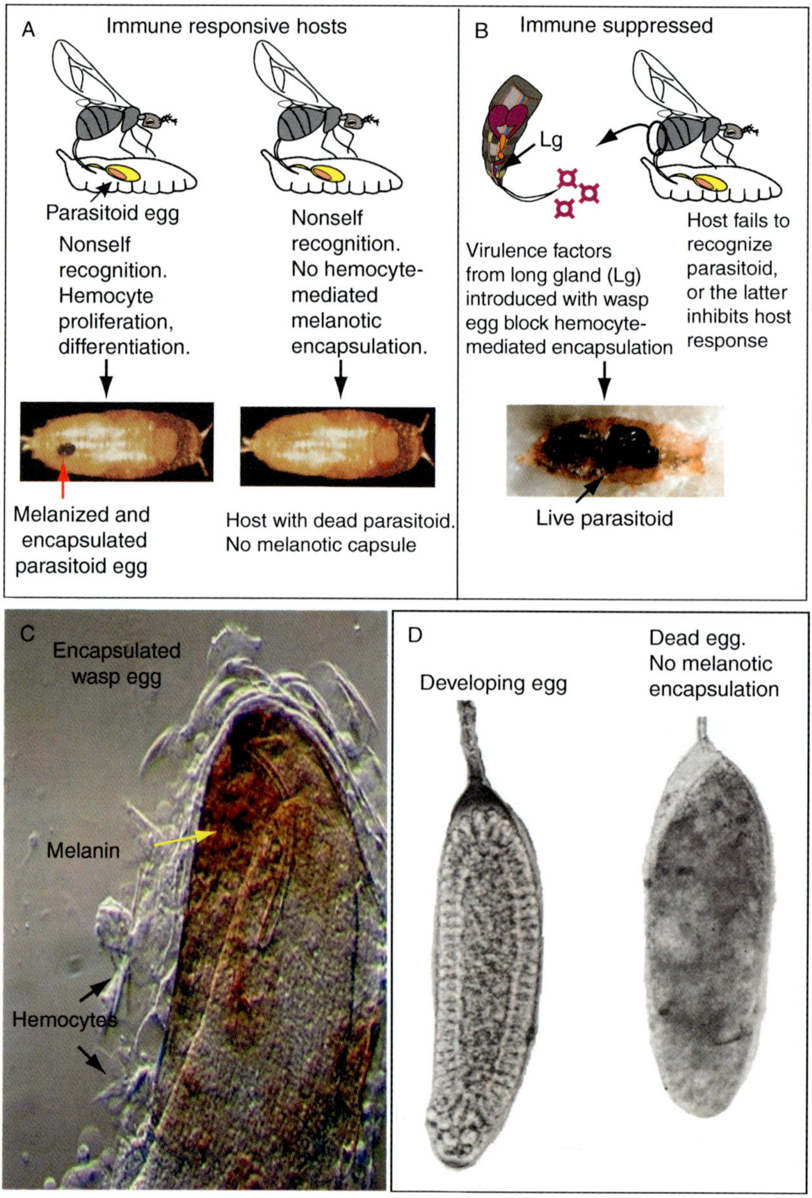

Chapter 4, Figure 4.4 **(See page 109 of this volume.)**

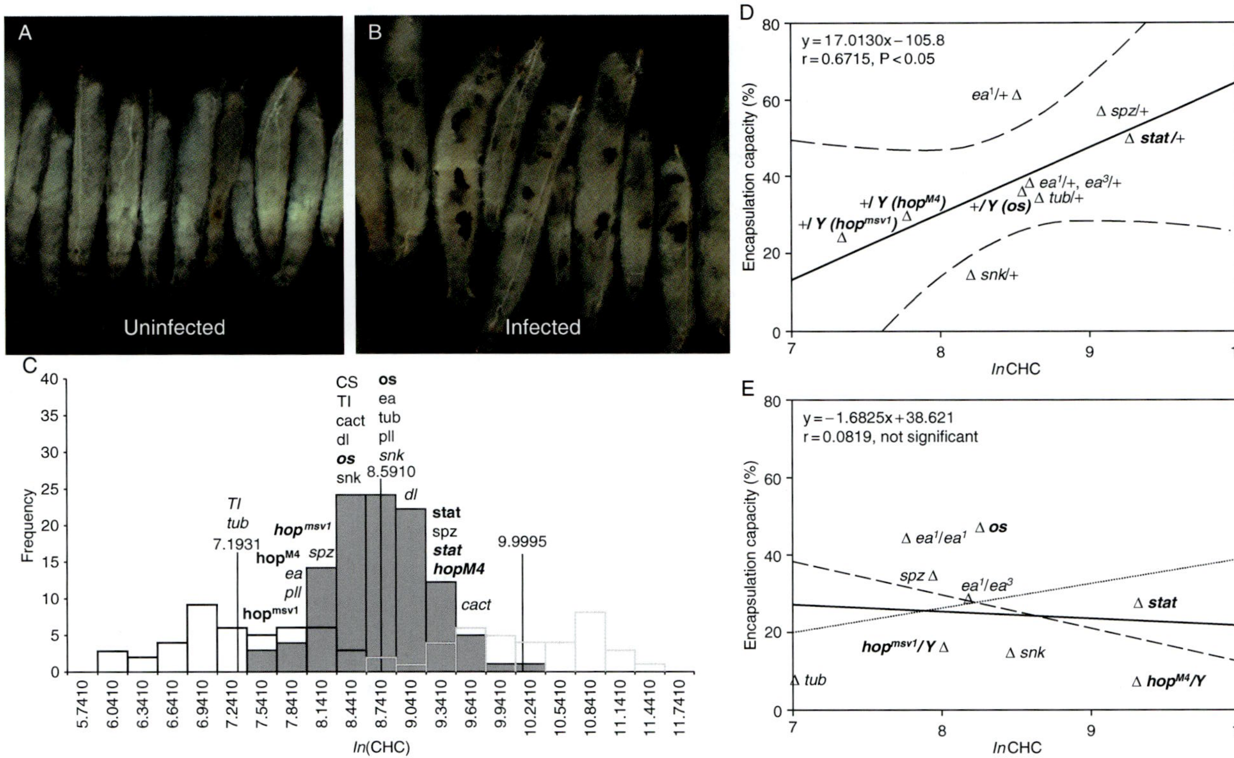

Chapter 5, Figure 5.2 (See page 129 of this volume.)

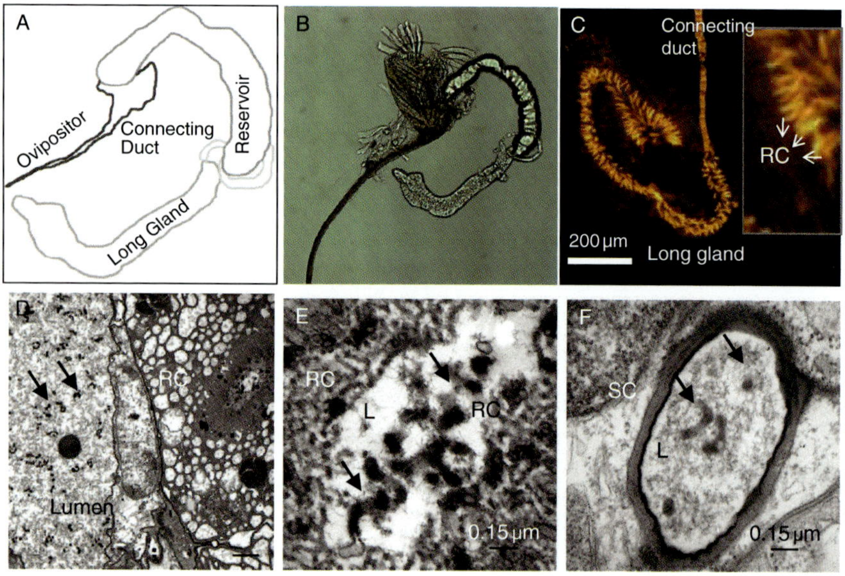

Chapter 5, Figure 5.5 (See page 135 of this volume.)

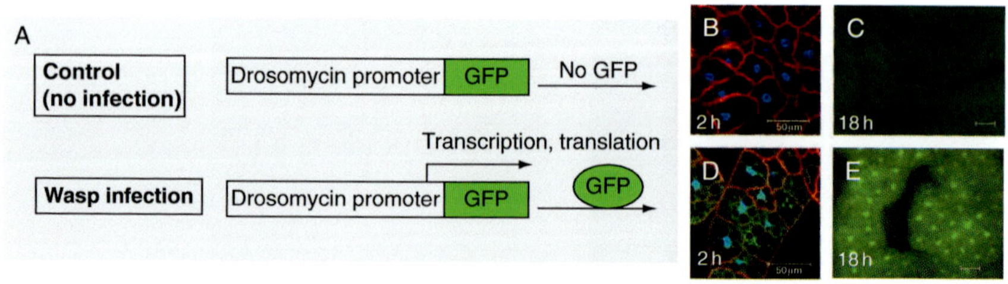

Chapter 5, Figure 5.7 (See page 141 of this volume.)

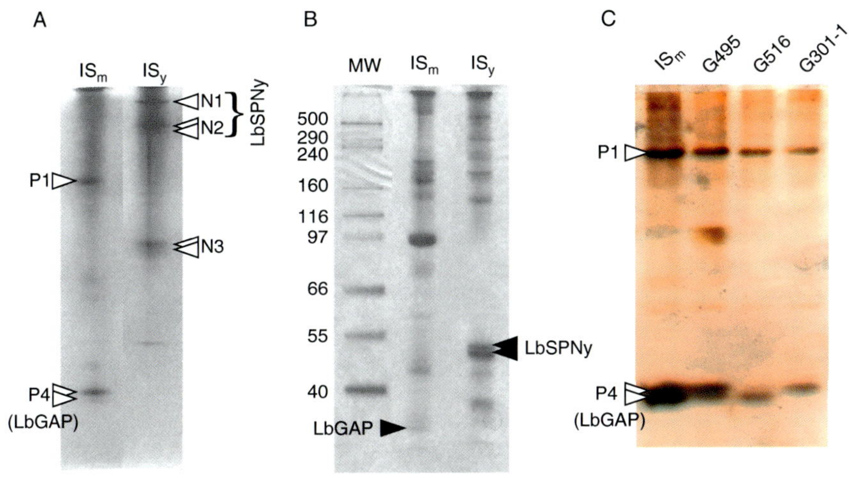

Chapter 6, Figure 6.7 (See page 171 of this volume.)

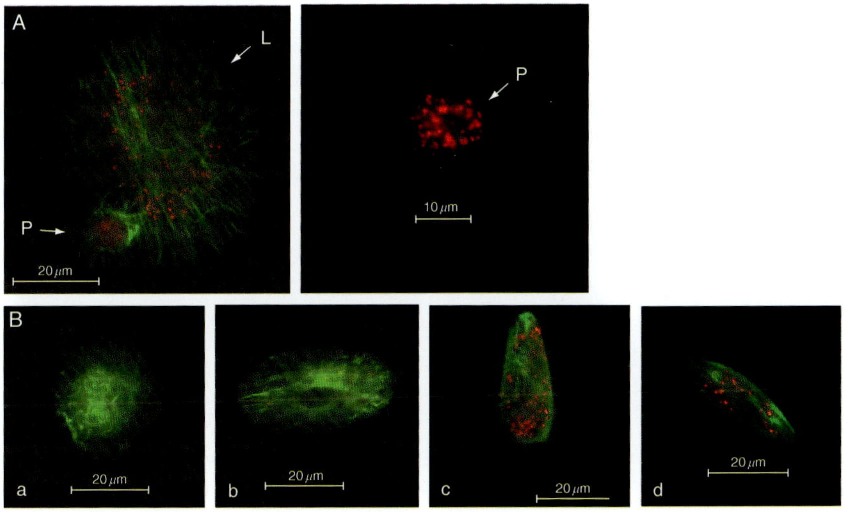

Chapter 6, Figure 6.8 (See page 172 of this volume.)

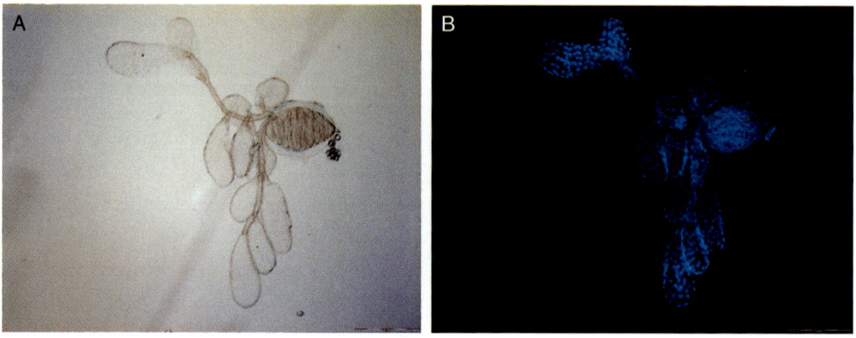

Chapter 8, Figure 8.3 (See page 221 of this volume.)

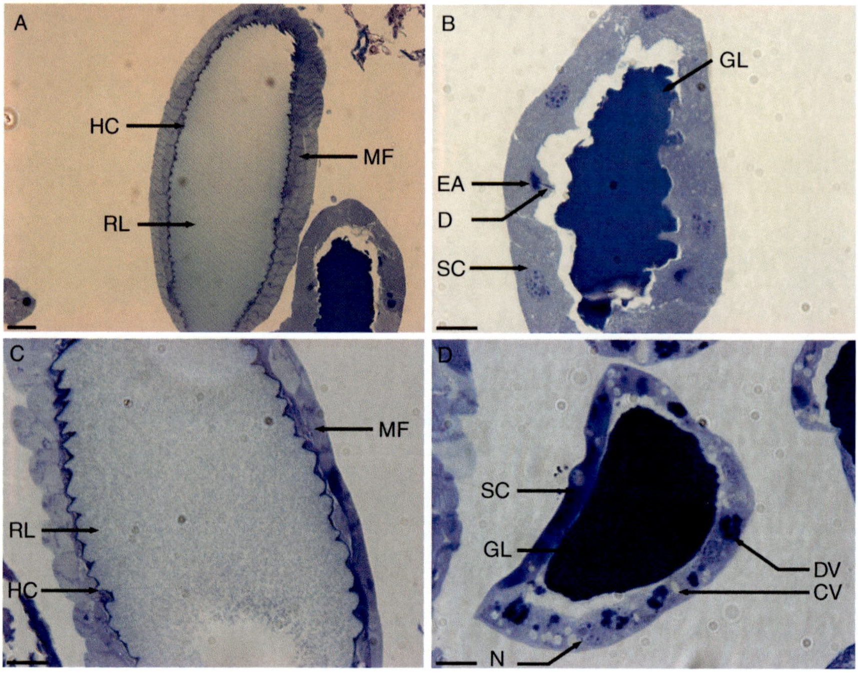

Chapter 8, Figure 8.4 (See page 222 of this volume.)

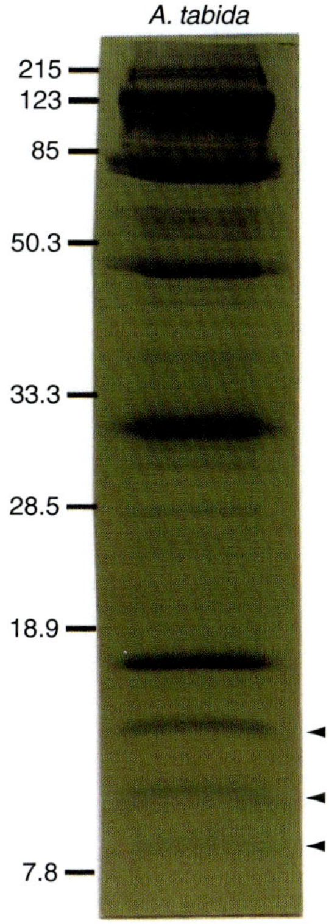

Chapter 8, Figure 8.6 (See page 226 of this volume.)

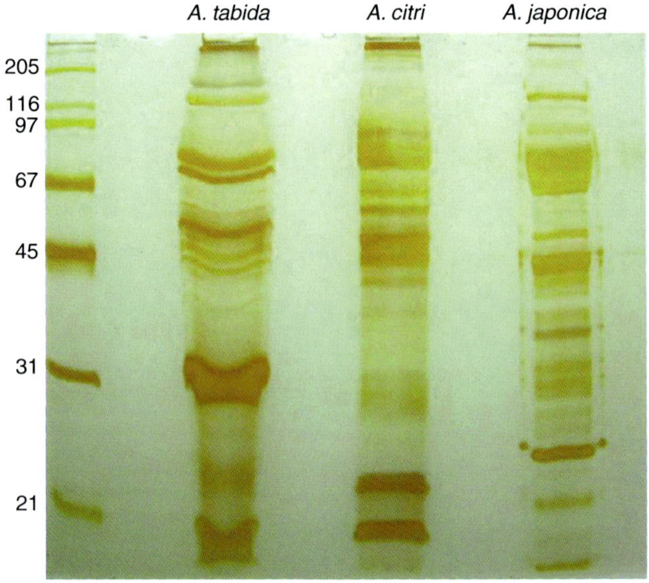

Chapter 8, Figure 8.8 (See page 229 of this volume.)